GETTING PUBLISHED IN THE LIFE SCIENCES

GETTING PUBLISHED IN THE LIFE SCIENCES

RICHARD J. GLADON
Iowa State University

WILLIAM R. GRAVES
Iowa State University

and

J. MICHAEL KELLY
Virginia Polytechnic Institute and State University

A JOHN WILEY & SONS, INC. PUBLICATION

Published by John Wiley & Sons, Inc., Hoboken, New Jersey
Published simultaneously in Canada

For general information on our other products and services or for technical support, please contact our
Customer Care Department within the United States at (800) 762-2974, outside the United States at
(317) 572-3993 or fax (317) 572-4002.

Wiley also publishes its books in a variety of electronic formats. Some content that appears in print may not be
available in electronic formats. For more information about Wiley products, visit our web site at www.wiley.com.

Library of Congress Cataloging-in-Publication Data:

Gladon, Richard J. 1947-
  Getting Published in the Life Sciences / Richard J. Gladon, William R. Graves, J. Michael Kelly.
    p. cm.
  Includes index.
   ISBN 978-1-118-01716-6 (pbk.)
   1. Biology—Authorship. 2. Authorship—Marketing. 3. Technical writing—Vocational guidance.
  I. Graves, William R., 1960- II. Kelly, James Michael, 1944- III. Title.
  QH304.G53 2011
  808′.06657—dc22
                                    2010049607

eISBN: 9781118041659
oISBN: 9781118041673
ePub: 9781118041666

10 9 8 7 6 5 4 3 2 1

CONTENTS

PREFACE

This book evolved from a course we taught to students in the life sciences at Iowa State University. Most institutions of higher learning do not have a formal course that teaches the processes associated with preparation of a manuscript for a refereed journal. At Iowa State University, students usually take this course when they begin to write their undergraduate research project report or, more commonly, their thesis or dissertation. This book will also be useful for professional scientists who would like to increase their ability to communicate their work to an audience. Ultimately, the main goal of this book is to make it easier for a scientist to write journal articles, and it was developed to guide inexperienced writers through the process of manuscript development and submission to a refereed journal in the life sciences. However, the book should not be limited to writers in the life sciences, and scientists in other disciplines will also find it useful for developing their writing skills.

Part I of this book addresses issues the author(s) must consider before they enter the writing stage, and maybe before they enter the thinking and organizing stage. This section also contains a chapter devoted to ethics in publishing. Part II presents our method for developing and writing the manuscript. Part III recreates the scenario of submission, external peer review, revision, and other miscellaneous events that occur after the manuscript has been written, submitted, and accepted.

Books and manuals are available to assist inexperienced students and professionals with their writing for a refereed journal. These publications are valuable. However, the manner in which the writer typically learns to construct the manuscript often leads to expansion of the scope of the article rather than development of a focus on specific points. In the end, much more has been written than can be accommodated in a typical manuscript, and the writer must reduce the length and content of the manuscript to bring it into compliance with the current standards of the journal of choice. This can be very difficult for an inexperienced writer, especially students who recently finished their thesis or dissertation research, and feel all their data must be presented. Thus, frustrations set in, and productivity wanes. Our approach to organizing, developing, and writing the manuscript is quite different, and it helps to streamline the entire writing process.

Our book differs from others because the focus throughout is on how the writer can unequivocally convey the most salient information to the reader. We call these packages of salient information "take-home messages," and the manuscript is built around them. After the take-home messages have been developed, the writer adds to the manuscript only the information that provides the evidence needed to support, and prove, those take-home messages.

Another unique feature of this book is our liberal inclusion of exercises we have developed while teaching the course. These exercises help novice writers build a solid foundation, and they allow experienced writers to improve their skills in manuscript development. Our core philosophy is to advance science by conveying take-home messages clearly and concisely. Our success rate has been very good. About two-thirds of the manuscripts developed in our course are published, and this is especially significant when one takes into account that our students are usually first-time writers of manuscripts for refereed journals.

We must add a word of caution to this preface. Use of this book, the principles within it, and the exercises within it, cannot cure bad science. Make sure you use good scientific processes and you execute the scientific method to the letter. Bad scientific procedures and practices cannot be repaired; they only can be renovated. A writer cannot compensate for that bad science with an extremely well written manuscript.

ACKNOWLEDGMENTS

We are grateful to our mentors, peers, and students, who have inspired much of the content of this book. For example, Dr George L. Staby introduced the idea of developing take-home messages to his graduate student Richard J. Gladon at The Ohio State University. William R. Graves and Richard J. Gladon both benefitted from a research-writing class at Purdue University taught by Dr Charles Bracker. All three authors appreciate the students in their course at Iowa State University, *Publishing in Biological Sciences Journals*; successful publishing outcomes achieved by these students reinforce to us the value of the methods presented in this book. Thanks are also given to Dr James A. Schrader for his assistance with creating the images found in several chapters and proofreading the entire manuscript. Dr Philip Dixon provided critical review and comments on Chapter 10, and Dr J. Clark Wolf critically reviewed our discussion of ethical issues in publishing in Chapter 3. The American Society for Horticultural Science, and its Director, Michael Neff, generously allowed us to use various publication and reviewers' forms and a society style manual.

PART 1

PRELIMINARY CONSIDERATIONS

CHAPTER 1

THE IMPORTANCE OF AND NEED FOR PUBLISHING

We are all apprentices of a craft where no one ever becomes a master.

—Ernest Hemingway

Press on. Nothing in the world can take the place of persistence. Talent will not; nothing is more common than unsuccessful men (women) with talent. Genius will not; unrewarded genius is almost a proverb. Education alone will not; the world is full of educated derelicts. Persistence and determination alone are omnipotent.

—Ray A. Kroc, Founder of McDonalds

DEFINITION AND NEED

The activities of scientists in the search for new knowledge are focused into a sequence of events we call the scientific method. Dissemination of the information discovered by the scientist is the last step in the scientific method, and publication of the information in written form is one of several vehicles for dissemination. This dissemination of these new discoveries in the form of writing may take place in one of several venues, including books, refereed periodicals, or nonrefereed publications (e.g., trade magazines or popular publications). The process of publication, especially publication of the information in a refereed journal, consists of many activities that must be completed sequentially in a clear and concise manner.

This book focuses on the steps in the process of preparing a manuscript for subsequent submission to a refereed journal in the life sciences. The dissemination of newly discovered information is critical to the advancement of science, and practicing scientists have a duty to complete the scientific method by publishing their information. However, for a variety of reasons, practicing scientists often do not complete this final step in the process efficiently. Science suffers because scientists "reinvent the wheel" when the information they need is locked in the mind or a filing cabinet of someone who has not published important information.

As a practicing scientist, you might start the task of publishing by asking yourself some simple questions. What is it that you enjoy most about being a scientist? How well do you like to communicate your results to other scientists? How well do you like to write? If you are like most of the scientists we have encountered, writing is not what you enjoy most about being a scientist, and the chances are good that you are not particularly eager to write.

Getting Published in the Life Sciences, First Edition. By Richard J. Gladon, William R. Graves, and J. Michael Kelly
© 2011 Wiley-Blackwell. Published 2011 by John Wiley & Sons, Inc.

Often, writing is not on the top-ten list of things that need to be done today, because it is human nature to avoid what we do not enjoy.

For many scientists, the thrilling aspects of science involve developing a hypothesis, conducting experiments, and collaborating with others who can offer new perspectives on and skills for solving the problem at hand. Writing about the research results is well down the list of motivators for many scientists. Yet, writing about our work is essential. And it can be rewarding, perhaps even fun. If you are a practicing scientist not already drawn to the act of writing, then we advise you to learn about writing and to learn to like it. Although you may find this difficult to imagine, your professional career and advancement will depend almost entirely upon your ability to communicate with other scientists. The better you communicate orally and by written word, the more rewarding will be your career as a scientist. Few activities bring more rewards to the career of a scientist than the act of publishing scholarly work.

The essential unit of publishing for a scientist is the refereed journal article, and all work done by a scientist, even the most preliminary experiments, should be conducted with a mindset that a refereed journal article ultimately will be the result of that work. Our overarching goal in this book is to help you develop the techniques and skills that make publishing in refereed journals as pleasant as possible. Let us begin with some basics regarding the contemporary meaning of publishing in refereed journals.

THE EVOLVING DEFINITION OF PUBLISHING

Decades ago, most scientists did not face the pressure to publish as frequently as they do today. A scientist may have worked for several years to examine an intellectual issue from numerous angles and, eventually, a publication might result, but sometimes there was none. The articles that were published, however, tended to be long and very thorough. The term "monograph" was sometimes used to describe such an article, with the implication that just one (mono) article contained all the known information about the topic.

Today, most scientists in the academic arena are under quite substantial pressure to publish at a greater rate. Consequently, contemporary journal articles are often shorter and more restricted in their focus, and, in most cases, the article is focused on just one or a few objectives. The quality of the content of the article remains critical, but the ability of a scientist to produce several shorter articles of high quality rather than one longer diatribe is a critical skill that can make or break a research career. Central to the development of this skill is an awareness of the so-called least publishable unit (LPU). The LPU has been described as the minimum amount of information (data) sufficient for a manuscript to be accepted for publication in a reputable, refereed journal (Broad, 1981).

An LPU must support at least one conclusion that your community of scientists (peers) will consider, and this conclusion should have the following features.

1. It should be *original* (the conclusion has never been drawn before).
2. It should be *important* (the conclusion is likely to have some kind of impact).
3. It should be based on research conducted by *using accepted norms of the discipline* (Broad, 1981).

Another evolving aspect of publishing is the ever-increasing array of venues in which you may choose to submit your manuscript. New journals are being developed and brought to publication at a greater rate than historical journals are being discontinued, and this leads to a net gain in the number of journals that might be an avenue for you to publish your manuscript. This represents a wonderful opportunity for you as an author, because you can select the venue that will be most appropriate for your manuscript. It also means that if your manuscript is not accepted by the first choice of journals, there are several other options. Indeed, perseverance is a key trait of successful scientists.

Another related change in the publication process is the mode of review. Nowadays, very few journals conduct reviews by sending each reviewer a hard copy of the manuscript. Most manuscripts are now submitted electronically, and the editorial office of the journal sends the manuscript for review via electronic mail. In turn, the judgment of the reviewers is reported electronically as well. Questions remain as to whether the electronic movement of manuscripts is affecting the quality of reviews, but there is no doubt that reviews can be done more efficiently and rapidly than ever before. Thus, although your manuscript might be released or rejected, the relatively rapid decision leaves you with more time to improve it and pursue revision and resubmission to the same journal or publication in another journal.

WHY PUBLISH?

Before we address this question of why publish, it is important that each of us critically analyzes ourselves to determine our motivation for publishing. There are several very good reasons why you should publish your work, and Peat et al. (2002) have a good discourse on this subject. We summarize their information here.

1. First and foremost, if you have new information, you have an irrevocable duty as a scientist to disseminate that information to other scientists.

2. Publication of research results permits scientists to study those results and use them to advance science, scientific thought processes, and, ultimately, benefit society via practical utilization of these new discoveries.

3. As an extension of no. 2, with help from our current information retrieval systems, it helps your information get out to a broader audience. This is extremely important, as all disciplines learn from high quality science conducted by people in different disciplines. For instance, many things we know today in plant science have their foundation in animal or medical sciences, and vice versa.

4. Most research today is made possible by funding from many possible entities. These may be large government programs, such as the National Science Foundation, the National Institutes of Health, and so on, state and local sources, and private foundations. It is imperative that you publish the results evolving from the funds the granting agency has invested in your research program.

5. As an extension of no. 4, publications in refereed scientific journals will increase your probability of obtaining continued funding for the same project, or funding to conduct other, related research projects. Success breeds success.

6. Publishing your results can lead to rewards such as promotions and recognitions by professional groups. Publications will strengthen your track record for these promotions and recognitions, and they will add credibility to your dossier. Conversely, not publishing your results can damage an otherwise promising career. Most academic units require a minimum of eight to fifteen publications in refereed scientific journals for promotion (and possible tenure) from one rank to the next highest rank.

Likewise, publications add a strong measure of credibility to the entire research team. A publication, or publications, in a refereed scientific journal is the benchmark by which almost all people measure success in research.

7. In virtually every research organization, except perhaps some industries, publication of research results in refereed scientific journals is the accountability factor used most for decisions affecting the life of that unit.

IN THE END, IT IS REALLY FOR THE SAKE OF SCIENCE

Why is publication so valued? At the simplest level, publications in refereed journals are the vehicles by which science advances, that is, the engine that carries science from one level to the next higher level. Refereed journals represent a repository where, ideally, only new, important, and verifiable data are reported and placed in context. As we will consider later, the referees, your peers, are critical, in fact so critical that the peer review process may frustrate you mightily from time to time. These peers determine the originality, importance, and soundness of your manuscript, and they attempt to "weed out" unoriginal, unimportant, or inadequate work by releasing the manuscript. This release may come in one of two forms. Peer reviewers may release the manuscript but recommend revision and resubmission. On the other hand, one or more of the reviewers may release the manuscript without a recommendation for revision and resubmission. Conversely, the acceptance of your manuscript by peer referees, either as is, or more commonly, with revisions, signifies that your work merits entering the permanent collection of scholarly information on the topic. That collection is available to all other scientists who can use it to shape future research questions and the conduct of future research investigations. Your accepted manuscript thus makes a permanent mark on science and advances our collective state of knowledge. It is a necessary part of science, and also a significant achievement for you, professionally.

We hope it is obvious that you have a responsibility to publish all your work that, at minimum, meets the requirements of the LPU. Publication is essential to science because it is the engine that moves science forward; however, scientists still, sometimes, do not write (Boice and Jones, 1984). It is also essential to you, because it moves you forward professionally. In most instances, only those academic scientists and noncommercial research scientists who demonstrate the ability to publish their data regularly will have stable careers marked by achieving tenure, being promoted, and enjoying a favorable reputation. Publishing skills are critical lessons to learn as early as possible, because those lessons will carry you as far as you can go professionally.

WHY YOU SHOULD BE A GOOD WRITER

High-school students who write well are competitive when seeking entrance to colleges that are highly ranked. Undergraduate students with strong communication skills are recruited actively to become graduate students. A graduate student who has articles accepted during her or his Master of Science degree program will be actively recruited to continue their education and pursue a PhD. PhD graduates with multiple refereed publications are also particularly successful when they search for a postdoctoral, nonacademic professional, or tenure-track faculty position. In addition, a faculty member at a junior academic rank must document their scholarship by publishing in refereed journals, or they will not move to higher ranks (Long et al., 1993).

We can also talk about this concept on a more practical level. Here are some important reasons why you should be a good writer.

1. Manuscripts that are written well are easy to read, and the reader finds them interesting to read and visually pleasing (Peat et al., 2002).

2. Manuscripts that are well written will move through the review process much more quickly. This will lead to it being more likely that the manuscript, which contains a record of all your hard work, will result in a publication (Peat et al., 2002). Reviewers and associate editors are busy people. They do not have time to correct your English and grammar. If your manuscript is written poorly, there is a chance it will not even be sent out for review until you revise it and resubmit it. This process of returning your manuscript, revision, and then resubmission will take several weeks, and that process does not even start until you have recovered from the painful exposure that your writing habits and use of English and grammar are not what they should be. The sooner you can correct these bad habits, the better.

3. If your manuscripts are written well, your peers will take you more seriously (Peat et al., 2002). The last thing a reviewer wants is to receive a poorly written manuscript (maybe from a friend) that is so poorly written they have to reject it on the basis of poor English and grammar before they get to the level of judging the science contained in the article.

4. Continued publication on a topic can also lead to continued funding via successful grant applications (Peat et al., 2002). As you publish more, it becomes apparent that you are becoming an expert (maybe *the* expert) in a given field or discipline. The more you are known for conducting good science and getting it published, the greater is the probability that you will receive continued funding. In addition, good writing skills developed while writing refereed publications will carry over to greater skills in writing grant applications.

5. As you become an expert in a field or discipline, your professional peers will call on you more often to be a reviewer of manuscripts in your area of study. Continued good work by you may lead to you becoming an associate editor, consulting editor, or editorial board member in your discipline. This is the icing on the cake, which will allow you to sharpen your skills even further, and this will advance your career rapidly.

6. The more you review journal manuscripts, the better a writer you will become. During the process of review of submitted manuscripts, you will learn things that you can do to make your writing better on several fronts. However, and maybe even more importantly, you will learn things *not* to do when writing.

7. As you learn to write better, the time required to write and complete a manuscript and submit it for publication will reduce (Peat et al., 2002). This will allow you to become more efficient in your writing. This also will lead to fewer frustrating encounters with the keyboard of your computer as you try to get a manuscript completed and submitted.

8. It is not necessary for you to be verbose to get your point(s) across to the reader. The journal article by Watson and Crick (1953) that explained (i) that DNA existed as a double helix and (ii) how genetic replication could occur was slightly more than one printed page. And we all know they were awarded a Nobel Prize for their efforts.

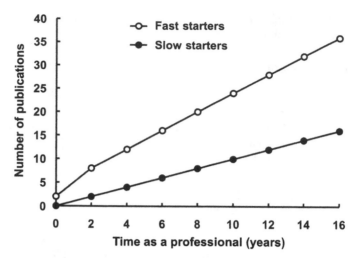

Figure 1.1   Idealized representation of the publication record of a fast starter and a slow starter.

QUICK STARTERS AND CAREER DEVELOPMENT

The quicker a person starts writing and publishing, the faster their professional career develops, and the faster her or his number of publications accumulates (Boice, 1991). In essence, the number of publications the quick starter has over their professional career becomes a "snowball rolling down a hill." In this case, the gap between a slow starter and a quick starter continuously becomes greater throughout each scientist's professional career. This is shown in our idealized representation in Figure 1.1, and it leads to a greater number of doors being opened professionally. The figure depicts the publication record over time of a theoretical fast starter and a slow starter. The slow starter begins a career after graduate school with no publications, whereas the fast starter has published two papers as a graduate student. Note how the difference between the two scientists becomes more pronounced over time.

Evidence supporting the importance of publication(s) in academic careers abounds (Blackburn et al., 1978; Boice, 1991, 1992), and quick starters develop accomplished careers that bring professional rewards that slower starters never attain. Books are available to help young professionals having trouble preparing manuscripts evolving from their scholarly work, and all levels of writers can benefit from this information (Boice, 1990, 1996).

SOME WORDS OF WISDOM

We close this chapter with some advice regarding the entire publication process.

1. Conceptualize every research activity you do in terms of the refereed journal article that could result from it.

2. Design every experiment you conduct, even preliminary or peripheral work, in a statistically appropriate manner so that publishing the work might be possible.

3. Never consider an experiment complete until it is published, whether it was or was not a preliminary or peripheral experiment.

4. If the work you are describing is good, and in your mind it should be published, then never accept release or rejection a manuscript as the final outcome.

5. Do not take negative reviews personally and do not allow them to destroy your confidence and ability to function as a scientist, and for that matter, a writer. Remain positive about your writing ability, persevere to the end, learn from your mistakes, seek the help of good writers, and continuously hone your skills.

6. Learn how to write manuscripts for refereed journals, and then follow through until they are published. The quick starter attains a career level that slower starters do not achieve, and the rewards you will receive will be well worth the time and energy you will need to spend to become a quick starter.

REFERENCES

BLACKBURN, R.T., BEHYMER, C.E. and HALL, D.E. 1978. Correlates of faculty publications. Sociol. Educ. 51:132–141.

BOICE, R. 1990. *Professors as writers. A self-help guide to productive writing.* New Forums Press, Stillwater, OK.

BOICE, R. 1991. Quick starters: New faculty who succeed. *In* M. Theall and J. Franklin (Eds). *Effective practices for improving teaching. New directions for teaching and learning,* no. 48, pp. 111–121. Jossey-Bass Publishers, San Francisco, CA.

BOICE, R. 1992. *The new faculty member. Supporting and fostering professional development.* Jossey-Bass Publishers, San Francisco, CA.

BOICE, R. 1996. *Procrastination and blocking. A novel approach.* Praeger Publishers, Westport, CT.

BOICE, R. and JONES, F. 1984. Why academicians don't write. J. Higher Educ. 55:567–582.

BROAD, W.J. 1981. The publishing game: Getting more for less. Science 211:1137–1139.

LONG, J.S., ALLISON, P.D. and McGINNIS, R. 1993. Rank advancement in academic careers: Sex differences and the effects of productivity. Am. Sociol. Rev. 58:703–722.

PEAT, J., ELLIOTT, E., BAUR, L. and KEENA, V. 2002. *Scientific writing. Easy when you know how.* BMJ Books, London.

WATSON, J.D. and CRICK, F.H.C. 1953. Molecular structure of nucleic acids. A structure for deoxyribose nucleic acid. Nature 171:737–738.

CHAPTER 2

STEPS IN MANUSCRIPT PREPARATION AND GETTING STARTED

In short, the preparation of a scientific paper has less to do with literary skill than with *organization*. A scientific paper is not literature. The preparer of a scientific paper is not an author in the literary sense.

—Robert A. Day and Barbara Gastel

PRINCIPLES

Advanced undergraduate students and graduate students in the middle of their degree programs probably have not been exposed to the steps that constitute the publishing process, simply because they have not yet had the opportunity to publish the results of completed research. In addition, some postgraduate professionals may not have had the opportunity to publish, even though they may have one or more postbaccalaureate degrees. This chapter provides an overview of the publishing process up to the time the manuscript is submitted to a journal. We will present several of the most likely postsubmission events later in this book. We encourage all users of this book to complete the important exercises at the end of this chapter. They set the novice author on a course to learn how to prepare a manuscript for publication.

It is difficult to delineate one universal flow of events that covers all possibilities for the publication process. We present in this chapter an overview of each of the parts of the method we have developed for preparing manuscripts for submission to refereed journals. In Part II of this book, we expand this introduction and teach how to complete each of these parts in a logical, straightforward manner. Use of this process, and especially completion of the associated exercises, lightens the burden of writing a publication on first-time writers because our approach teaches the writer to move on a very focused path from start to finish. It is a unique method for the preparation of manuscripts that is effective because it permits scientists to focus on a few take-home messages and write about them.

As undergraduate students in science classes we were taught to use the scientific method as our basis for constructing our laboratory reports. As such, Knisely (2005) and McMillan (2006) have written excellent student handbooks for writing laboratory reports in biology courses, and again, the basis for preparing the report is the flow of the scientific method. For graduate students, several books or manuals exist for helping the novice writer learn how to synthesize the necessary information into a manuscript for submission

---

Getting Published in the Life Sciences, First Edition. By Richard J. Gladon, William R. Graves, and J. Michael Kelly
© 2011 Wiley-Blackwell. Published 2011 by John Wiley & Sons, Inc.

to a refereed journal (Davis, 2005; Day and Gastel, 2006; Gustavii, 2003; Katz, 1985; Matthews et al., 2000; Woodford, 1968). In addition, there are texts that can aid in writing better scientific reports (Booth, 1993; O'Connor and Woodford, 1975; Tichy [with Fourdrinier], 1988; Woolston et al., 1988).

These books, manuals, and other methods for manuscript preparation focus on a continuous paring of information until the writer arrives at a point where they can start to write the manuscript. Even at this point, there will be continued paring of the information until one reaches the endpoint of the manuscript. This is a difficult approach to master, because it requires a continuously new set of realizations by the author that some of the research she or he has completed is not of enough value for it to be placed in a manuscript. For many researchers, but especially young researchers, this is a difficult pill to swallow, and they fight removal of these data. In the method that is the thesis of this book, we have the writer focus on the major points he or she wants to convey before they start to write any prose. Basically, this allows the writer to focus on the most important points of her or his research, and then write about those points only, expanding the base of information presented as is necessary only for a prospective reader to understand what was done, why it was done, how it was done, and what it means.

Another unique approach we have developed in this book is the sequence with which the manuscript is developed. Most guides to help novice writers through their first several manuscripts have the writer prepare the manuscript in the same sequence in which they read a published journal article (e.g., Day and Gastel, 2006; Gustavii, 2003). In this sequence, the writer first prepares the title and by-line, followed, in order, by the abstract, introduction, materials and methods, results (including tables, figures, and photographs), discussion, and then finishing with the references section. We believe this sequence for producing the manuscript immediately multiplies the troubles encountered by the author producing the manuscript. Getting past the preparation of a good solid title and abstract, as well as appropriate introduction and materials and methods sections, often frustrates first-time writers to the point where they cannot move forward on preparation of the manuscript. First-time writers often get part of the way through these early sections of the manuscript, and in a massive display of frustration, crumple up the manuscript and throw it away so that they can start again with a clean slate. Others begin the sequence of units written with a section from the middle part of the manuscript (e.g., Katz, 1985; Mathews et al., 2000).

We advocate an inside-out approach. By starting the production of the manuscript with the development of the two to four take-home messages that will be the focus of the manuscript, and then a provisional title that encompasses those take-home messages, the writer is now focused on a limited number of items, and this reduces the interferences and blocking of the writing that frustrates so many writers, both first-time and experienced. This first step results in a logical flow to the presentation of the research data in the form of tables, figures, and/or photographs and then the results section text. This allows the writer to start by creating the heart of the paper, which sets the tone for the entire manuscript. This approach has the positive psychological impact of breaking the entire, often daunting, process into a series of smaller, manageable steps, rather than the more laborious process of writing the manuscript from start to finish. In addition, this approach allows the author to begin writing with the end product clearly in mind.

Figure 2.1 presents an overview flow diagram of the method we use in this book. Our hope is that you adopt this sequence and produce high-quality manuscripts with relative ease. Certainly, after the writer has completed several repetitions of this process (i.e., published several journal articles), she or he then can adjust their approach to fit their individual flair.

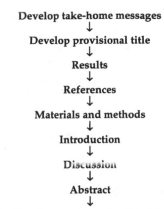

Develop take-home messages
↓
Develop provisional title
↓
Results
↓
References
↓
Materials and methods
↓
Introduction
↓
Discussion
↓
Abstract
↓
Title, by-line, keywords, and authorship footnote

Figure 2.1   Overview of manuscript development.

OVERVIEW OF MANUSCRIPT PREPARATION PROCEDURE

The following paragraphs outline the steps and give key items to which attention must be given during the preparation of the manuscript. This sequence is one that reflects a useful and orderly chronology of activities that will guide authors through the entire process of manuscript preparation. We find this sequence extremely helpful for writers who are experiencing manuscript preparation for the first or second time. We have also found it to be helpful for more experienced writers who have encountered difficulties getting to the point of a finished product ready for submission.

Overview

The production of a manuscript for publication in a refereed journal occurs after the laboratory, field, or clinical research work has been completed. Young scientists new to practicing the scientific method and conducting research often learn of the project from their major professor. A student new to scientific research often assumes the major professor is completely up to date on all relevant literature. The assumption made by the student is that the pile of journal articles they received from their major professor represents a thorough and complete search of the literature. Often, this is completely untrue. The major professor may have missed portions of the relevant literature, and the student new to this research area starts work without a complete review of the pertinent literature. Ultimately, the student is responsible for a complete review of the pertinent literature for their thesis or dissertation research project. In practice, all authors need to be responsible for the review of the relevant literature, and the best start to a research project is not in the laboratory, but rather in the library. The student should take it upon himself or herself to conduct a complete search of the literature and to synthesize it before starting laboratory, field, or clinical work. As the research is conducted and completed, almost without fail, the student will find additional literature that must be cited, and then they realize they should have known about this published research long before now. As this reading and search of the literature is completed, the student should be aggregating these references into a bibliography, part of which will later become the first draft of the references section of the manuscript. At this point, the writer must know, or at least have some notion, about the journal to which the manuscript will be submitted so that the manuscript may be produced in a format that meets the requirements of that journal. To help the writer, especially a novice, in determining the journal to which the manuscript should be submitted, we have included in Chapter 5 information on knowing and choosing your journal. That chapter contains a discourse on factors that influence the decision on which journal to choose and an exercise on knowing and choosing your journal.

Take-Home Messages and Provisional Title

The writer must decide when to start writing and where to publish the manuscript. Immediately after the decision is made to proceed, the writer should begin to identify and articulate the take-home messages that have evolved from the research. The take-home messages are the essence of the publication, and they are the pieces of information you want every reader to remember and understand unequivocally when they have finished reading the article. The writer should now be in position to look at the words used to construct the take-home messages, and from these words, he or she should try and formulate a provisional title that will clarify the author's thought process and establish the focal point(s) and scope of the manuscript.

Authors may also prepare a provisional synopsis or abstract of what will become the contents of the manuscript. This provisional abstract may have been formulated previously as an abstract for a presentation at a scientific meeting. Provisional is an important word to remember at this stage, because it is not the appropriate time to settle on a final title and a final abstract. If the author(s) have not chosen a journal to which they will submit the manuscript, they should begin to do so at this time. Also, the author(s) should have decided on the basic form of the manuscript: will it be a note, a short communication, or a complete research report? Finally, the author(s) should already have collected most of the information needed to complete what will become the major sections of the manuscript.

Results

The purpose of the results section of the manuscript is to relay to the reader the new facts (data) obtained by the conduct of this research. However, these data should be presented only if they support and provide evidence for one or more of the take-home messages of the manuscript. In constructing this section of the manuscript, the author needs to determine what data are critical to building a case for the take-home messages and the conclusions coming from the research. Now is the time to produce rough drafts of the tables, figures, photographs, and any other forms of data that will appear in the manuscript. The author(s) must choose and use their data wisely and objectively, and they should use only enough data (representative and/or informative) to make an unequivocal case for their take-home messages. The writer must use the Système Internationale for all aspects of manuscript preparation. The writer should remember at this point that the tables, figures, and photographs should be chosen and used only for their value as supporting evidence that helps to make the take-home messages easily understood and credible, that provides answers to your objective statement(s) in the introduction section, and that the writer can interpret and place into context easily. Many writers find that tables, figures, and photographs produced at this stage are rough drafts that must be modified later so that the take-home messages can become even clearer to the reader.

If the writer(s) have not done so already, a complete outline of the text of the manuscript should be developed. Writers should never attempt to put text onto the pages of a manuscript without first constructing either a word, sentence, or topic outline for each section of the manuscript. Do not use the IMRAD (introduction, materials and methods, results, and discussion) sequence of sections for developing the manuscript because it will create difficulties developing a linkage between the sections and the take-home messages.

Use take-home messages, an outline of the results section, and the rough drafts of the tables, figures, and photographs to prepare the text of the results section. This will probably be the shortest section of the manuscript, and it will contain only simple, declarative sentences. Usually just one to two typed, double-spaced pages is sufficient for the text of the results, and only information that is supported by the data and provides evidence for the take-home messages should be included. The writer should now reexamine the suitability of the table(s), figure(s), and photograph(s) for the text just written and revise as needed to enhance consistency throughout the results section.

After both the text and supplemental forms of data presentation have been revised and polished, it is a good time for the writer to get the entire results section to the coauthors and maybe even an outside reviewer or two to make sure the writer is setting the correct direction for the manuscript before the writing goes too far. This allows the writer to confirm that the take-home messages truly are the focal point of the manuscript up to this time and that the results presented support and provide evidence for the take-home messages. While the results section is being reviewed, the writer should start on the materials and methods section. Sometimes, if the writer has trouble getting his or her ideas into the results section of the manuscript properly, he or she may want to make some limited progress on the materials and methods section. The results section should be finished before much of the materials and methods section is completed because too much revision of the materials and methods section may need to be done when the results section has not yet been solidified.

References

The overall purpose of the references section is to present the reader with the sources of information used in the present work (appropriate credit to the previous work of the authors and others). It also shows the location of related, extended information on the subject matter covered in the manuscript. This section builds confidence among readers that the author(s) are current in their knowledge of the subject and that the review of literature is complete. There should have been no references to anything presented in the results section, because only the results of the author(s) work were presented. However, at this point, and throughout the remainder of the manuscript, the writer will need to refer to the literature in terms of what has been done previously (the introduction), how the previous research was conducted and used in this research (the materials and methods), and how the research reported in this manuscript compares with previous research (the discussion).

The writer should cite in the text only literature that is of significance to the work in the manuscript, and the references section should not be cluttered with references to theses, dissertations, abstracts, and so on. In addition, information from the World Wide Web may change quickly, so try and not use this source unless absolutely necessary. In the listing of the references, the exact descriptor of the location within the source (volume, page numbers, etc.) must be presented, in the order and style required by the particular journal. The writer or another author should double-check the descriptors of all references for accuracy in their entirety by matching the citation with the resource. Authors must be sure all citations in the text are presented in the references section, and vice versa. If one or more of the authors has accumulated literature and formed a compendium of those sources, then the writer should transfer that compendium into the references section of the manuscript now being produced. When literature searches are conducted, deposit all citation information needed to locate the resource in an accessible file that can be changed readily. As the writer now needs to refer to something they knew from the earlier literature, they may refer to the references in this newly created references file. When the manuscript has been completed, the writer should then be sure to eliminate the citations that were not used. Use the citation system of the journal where you will submit, if it is known during preparation of the manuscript. The Harvard System (name[s] and year) should be used unless it is not allowed by the journal. As any editor of a journal will tell you, no other section of the manuscript will have as many errors as the references section and its associated in-text citations.

Materials and Methods

The overall purposes of the materials and methods section are to describe how the work was conducted, including the experimental design and statistical analysis, and to supply a competent, informed reader with enough detail that he or she will have confidence in the work and be able to repeat it as needed. In some cases, a new procedure or new apparatus will have been developed and used, and describing this is an important part of reporting the research work. In general, the following subsections should be included, usually in this order: source and handling of experimental material; any treatments, and how they were applied; preparation of the experimental material for analysis; chemical analyses; enzymatic analyses; miscellaneous analyses and procedures; and the experimental design and data analysis procedures. The materials and methods section must be able to stand alone, without reference to other parts of the manuscript. Often, this section is the easiest to write because it is simply a descriptive compilation of the operations the researcher(s) completed to accomplish the research. Normally, it will be very factual and straightforward, and nothing should be analyzed and interpreted. Writing should be for readers who are informed about the general subject, so background information should be minimal. The presentation of information should be limited to only what is necessary for the reader to understand how the presented results were generated, and it must be logical, almost to a fault. This section is often two to five typed, double-spaced pages, and it may contain references, tables, figures, or photographs, as necessary, for helping the reader understand what was done and how it was done. Again, the writer must use the Système Internationale for all aspects of manuscript preparation.

Introduction

The overall purpose of the introduction is to set the complete rationale for conducting the research reported in the manuscript. It should begin strongly and end strongly. The introduction supplies sufficient background information to allow the reader to understand why the research was done, and at the same time, allow a reviewer to begin to evaluate whether or not the research merits publication. Normally, the introduction contains three major parts. They are the specific rationale for conducting the research, a review of the pertinent literature, and the statement(s) of objectives that embody the hypotheses the research will test.

The specific rationale is necessary because it sets the stage for the justification of the methods that were used and the data that were gathered by using those methods. The review of literature should not be exhaustive, in a way expected of review articles. The number of citations and the discussion of the pertinent literature should be limited to only that required to demonstrate the need for the research (i.e., the rationale) and to set the tone for what will be presented within the manuscript. References acknowledge previous work of other researchers or one of the authors, and the writer must show how that previous work relates to the research reported in this manuscript. Many times, if the review of literature is written appropriately, the author can create a direct linkage to the discussion section so that they may analyze, interpret, and criticize the existing literature. This allows the writer to show why the previous research needed to be extended. The introduction section should end with a paragraph that contains clear and concise statements of the purpose(s) of the research reported. Depending upon the scope of the research project reported, this could be focused on one very specific objective of the research. Alternately, it could be focused on an overall objective of the research with clarifying specific objectives that, when taken together, make the overall objective more understandable to the reader. This overall objective often relates to the overall conclusion of the results of the work, and the specific objectives will almost certainly correspond to the take-home messages of the manuscript, after both of them have been rearranged, revised, and polished. Some authors and/or some journals will report in the introduction section an overview of the findings of the research. The writer should determine whether this is typical of articles in the journal being targeted and include or avoid an overview of the findings accordingly. Be sure you avoid an unnecessarily long historical review of the published literature, and make sure you avoid loose, scattered, or illogical writing or organizational style. And whatever you do, do not promise more than you can deliver!

Discussion

The overall purposes of the discussion section are to analyze and interpret the data derived from the research and to relate those data to the work of others. There are also several secondary purposes of this section of the manuscript. The discussion drives home the take-home messages, and it shows how this new information is either consistent or inconsistent with our base of previous knowledge. Along with that, the discussion presents the implications of the research, both practical and esoteric, and who should care about that new information. Finally, in many cases, it answers the question that many readers will formulate when they understand the content of the manuscript: what is next?

A good discussion includes, but is not limited to, the following: statements of the take-home messages and conclusion(s); references to acknowledge previously published research and the value of this previous research to the work in this manuscript; comparisons and contrasts between the research in this report and work already published; and a clear indication of how these research results advance the state of knowledge in this specific area of research (i.e., we think you should publish this manuscript because ...).

First-time authors and others with little experience often find the discussion section difficult to write, because the data and conclusions derived from the research must now be analyzed, interpreted, and compared with the work of others. This is usually challenging, but a strong discussion will enhance your paper and its value to others.

Abstract

The purpose of the abstract is to present a synopsis of the entire body of information in the manuscript. The abstract should answer several questions about the research reported. For instance, the abstract should clearly delineate the rationale, or justification, for conducting the research, as well as the objectives of the research. It should describe what new information and conclusion(s) (i.e., take-home messages) can be drawn from this research.

And the abstract should indicate what the new information means to people other than the author(s). An effective abstract normally presents a brief summary of each section of the paper, and the take-home messages and conclusion(s) must be presented unequivocally. The abstract should be a single paragraph, and rarely more than 200 to 250 words; indeed, 100 to 200 words might be a better target. A good abstract must stand alone, because it will be republished by abstracting services, and in this form, it will be read without the supporting evidence contained in the manuscript.

Title, By-line, Keywords, and Authorship Footnote

The title, in its final form, should be an adjustment from the provisional title developed at the earliest stages of the writing process. The title is the most succinct summary of the contents of the manuscript. The title should be brief yet contain high-impact words that tell or indicate to the reader the ultimate take-home message(s) of the article.

The by-line is the name(s) and complete address(es) (for mailing purposes) of all authors (individuals who will take credit and responsibility for the contents of the manuscript). The by-line also designates credit to the institution(s) where the research was conceived and conducted, something that is extremely important in the determination of the status of the institution(s). *Whose name will appear in the by-line, the order of appearance, and who will be the corresponding author are items that must be decided before work on the manuscript commences.* The manner in which the names of the authors are presented (i.e., initials only, complete first name, and initial of middle name, etc.) must conform to the format of the journal. Normally, all coauthors are not involved in correspondence with the editorial staff of the publisher. The person who takes charge of all communications between the authors and the editorial office is called the corresponding author. This individual must "ride herd" on the manuscript to avoid publication delays and to serve the interests of the coauthors.

The keywords are an extremely important part of the manuscript, because they are a major mechanism with which other people can find the published paper. Researchers cannot review tables of content of all journals in their field as each issue is published. Most research scientists will use an abstracting and indexing service that will filter the peripheral articles so that the researcher can quickly find the key articles they need to read and understand to remain up to date in their field. All words of impact in the title become keywords, and the authors can expand this list by selecting words not in the title that have a direct and important relationship with the content of the manuscript. The choice of these keywords should not be taken lightly. They play a critical role in the dissemination of your results and interpretations.

The authorship footnote usually appears at the base of the first page of the published paper. It is a footnote linked to the title, but most journals no longer use a superscripted numeral at the end of the title to designate it. In most instances, the first part of the authorship footnote is the "received for publication" date and the "accepted for publication" date. This footnote may contain an approval for publication by the institution(s) that sponsored the research, which may be in the form of a journal manuscript number assigned to this article. The footnote also may contain reference to funding sources that enabled the work. It may also acknowledge persons who helped with the research in some way, but not to the extent where their involvement warranted authorship. This latter segment of the authorship footnote is sometimes placed in a specific section of the manuscript that is located elsewhere. Be sure to reference the footnoting style of your chosen journal before you submit the manuscript.

Items Between Completion of the Rough Draft and Submission

After you have completed the rough draft (not just the first draft) of the manuscript, put the manuscript away for a period of time so you can distance yourself from it. This time period will give you an opportunity to take a fresh look at the contents of the manuscript, and you will be in a better position to edit, rearrange, and revise the manuscript as necessary. After you have polished the manuscript via several revisions, you should then make certain of the following: data within tables and figures have been imported into the manuscript accurately; the data presented indeed do support and provide evidence for the take-home messages; all citations have been referenced in the references section and the text; and other "house-keeping" items have been checked for accuracy.

At this point, you should send the manuscript to one, two, or three competent reviewers. These reviewers may be in-house, if there are other scientists within your institution who can do an appropriate review. If not, you may have to send your manuscript to a colleague at another institution. You should ask these reviewers to do a thorough review of the manuscript as if they were requested by an editor to make a recommendation on the publication of the manuscript. You should consider carefully the reviewers' comments and make changes as necessary, and after these comments have been incorporated into the manuscript, all of the coauthors should proofread the manuscript one last time before submission. The manuscript should then be arranged in its final form for submission, and all references and table and figure data should be checked for accuracy. Finally, check all requirements of the journal to be sure they have been fulfilled appropriately, and obtain approval to publish from your institution(s), if required. Now make the required number of copies of the manuscript, figures, photographs, and so on, and along with the cover letter, submit the manuscript to the journal you have chosen. Many journals prefer or require electronic submission rather than postal submission of multiple paper copies.

If you are using this book as a text for a course in how to write and publish a journal article, you may have additional events that must be completed during the course. You may have in-class and out-of-class exercises, in-class peer reviews of paragraphs from sections of your manuscript, out-of-class peer reviews of sections of the manuscript, an out-of-class peer review of a rough draft of the entire manuscript, and one or more one-on-one sessions with one or more of the course instructors. These exercises and experiences will help you gain insight into the production of your manuscript, and you should not take these opportunities lightly. These exercises and experiences will help you learn how to produce a better manuscript that will have a greater chance of getting published.

IT'S TIME TO GET STARTED ...

Now is the time to get started on some of the preliminary issues associated with the development of your manuscript. Exercise 2.1, Getting Started, will help you focus on some of the broader issues that will direct you to a point where you can begin writing your manuscript. We are also having you do a preliminary draft of your take-home messages in Exercise 2.2 This exercise is designed to help you begin to crystallize your thoughts about the concise and accurate messages you want to convey unequivocally to every reader of your manuscript. In Exercise 2.3, you will develop your first draft of the title of your manuscript. This is a preliminary draft of the provisional title, which you will use to help keep you on-course when you write the manuscript.

EXERCISE 2.1   Getting Started

1. Please write two statements that identify the single primary issue you were addressing in your research. These statements should say the same thing, but the thesis they represent should be stated differently. You may write each statement as a question or as a declaration of the problem that was addressed/solved. Write your statements for an audience of other scientists who do not necessarily have expertise in your specialized field of research. Practice composing your two statements elsewhere, and then, after you have revised and refined them several times, write the final, refined versions below. You should limit each of your responses to a single question or declarative statement that is written *succinctly and without ambiguity*.

   (a)

   (b)

2. Answer briefly, in one, or at most, two sentences: **(a)** Why is this research important? **(b)** To whom does it matter?

   (a)

   (b)

3. Identify one new question that can now be asked in a future research project. This question should have evolved as a result of the research you have completed, and *it should be one that could not have been asked before you conducted your research and analyzed and interpreted your data.*

EXERCISE 2.2   Preliminary Draft of the Take-Home Messages

After critically considering and analyzing the issues you addressed in your research, and the data you generated, analyzed, and interpreted, complete this exercise. As you do this exercise, package the key information you want the reader to obtain, without question, from your manuscript into a preliminary draft of the take-home messages. In Chapter 7, you will refine and finalize your take-home messages, but we recommend that you begin thinking about them as soon as possible. Thus, we have developed this exercise. You should have two to four of these take-home messages, and each should be written with great care because they will subsequently guide the creation of your entire manuscript. These two to four take-home messages should be synthesized as sentences, or at least phrases, and you should work on drafts of them elsewhere. Allow time to revise and refine this preliminary draft of your take-home messages at least once after letting them "sit and get cold" for at least part of a day. After the messages have been refined, complete the entries below. Finally, for each take-home message, write one or more sentences or phrases to show how you will support the take-home message with a set of data you collected, analyzed, and interpreted.

Take-home message #1: _____

_____

Supporting data: _____

_____

Take-home message #2: _____

_____

Supporting data: _____

_____

Take-home message #3: _____

_____

Supporting data: _____

_____

Take-home message #4: _____

_____

Supporting data: _____

_____

EXERCISE 2.3   Preliminary Draft of a Provisional Title

1. Look at the words in your take-home messages and any keywords you have discovered about your manuscript contents as you have worked with them to this point. List these words on the lines below, but restrict yourself to a maximum of 12 important words you feel will be the essence of a title that describes your work.

_____   _____

_____   _____

_____   _____

_____   _____

_____   _____

2. Now, organize, arrange, rearrange, and string together these words to develop a concise and unambiguous first draft of a statement that summarizes your work in the manuscript (i.e., your provisional title of the manuscript). Limit yourself to 25 words. Let this draft get "cold" for a period of time before you proceed to (3).

3. After you have let this first draft of a provisional title get "cold" for a few days, return to it and try to "tighten" it to a maximum of 15 to 18 words.

REFERENCES

BOOTH, V. 1993. *Communicating in science. Writing a scientific paper and speaking at scientific meetings.* Second edition. Cambridge University Press, New York, NY.

DAVIS, M. 2005. *Scientific papers and presentations.* Second edition. Academic Press, San Diego, CA.

DAY, R.A. and GASTEL, B. 2006. *How to write and publish a scientific paper,* sixth edition. Greenwood Press, Westport, CT.

GUSTAVII, B. 2003. *How to write and illustrate a scientific paper.* Cambridge University Press, New York, NY,

KATZ, M.J. 1985. *Elements of the scientific paper. A step-by-step guide for students and professionals.* Yale University Press, New Haven, CT.

KNISELY, K. 2005. *A student handbook for writing in biology,* second edition. Sinauer Associates, Inc., Sunderland, MA.

MATTHEWS, J.R., BOWEN, J.M. and MATTHEWS, R.W. 2000. *Successful scientific writing. A step-by-step guide for the* biological and medical sciences. Cambridge University Press, New York.

McMILLAN, V.E. 2006. *Writing papers in the biological sciences.* Fourth edition. Bedford/St. Martin's, Boston, MA.

O'CONNOR, M. and WOODFORD, F.P. 1975. *Writing scientific papers in English.* American Elsevier, New York, NY.

TICHY, H.J. [with S. Fourdrinier] 1988. *Effective writing for engineers, managers, scientists.* Second edition. John Wiley & Sons, New York, NY.

WOODFORD, F.P. (Ed.) 1968. *Scientific writing for graduate students. A manual on the teaching of scientific writing.* The Rockefeller University Press, New York, NY.

WOOLSTON, D.C., ROBINSON, P.A. and KUTZBACH, G. 1988. *Effective writing strategies for engineers and scientists,* Lewis Publishers, Inc., Chelsea, MI.

CHAPTER 3

ETHICAL ISSUES IN PUBLISHING

O what a tangled web we weave when first we practice to deceive.

—Sir Walter Scott

Ethics are matters of making good decisions about questions of appropriate conduct regarding your work and that of others.

—Martha Davis

DEFINITIONS AND NEED

Although few scientists acknowledge that scientific misconduct in the form of unethical behavior exists, examples abound. There are numerous cases of plagiarism in the literature, and squabbles over authorship rights and privileges surface frequently (Wilcox, 1998). What is appropriate behavior in some parts of the world may be taboo elsewhere (i.e., things are culturally relative). This relativism can be extended to differences between scientific disciplines such as the life sciences and the natural sciences (e.g., chemistry or physics). It is important for writers to understand that behaviors accepted in one discipline may not be acceptable in another discipline.

Let us first ask the question, what is ethics? Ethics is a branch of philosophy, which itself is defined as the search for wisdom (Mish, 2004). A definition of ethics may include, depending upon the focus of the arguments, a code of societal behavior, a system for decision-making and problem solving, or a method to resolve ought/should questions. Ethics and ethical behaviors should be dissociated from religion, and one should not permit oneself or another person to back out of an issue of ethics on the basis of religious convictions.

Individuals interested in further study within the areas of ethics and moral theory should consult some of the textbooks used for teaching ethics and moral theory on college campuses (Comstock, 2002; Rachels and Rachels, 2007; Rollin, 2006; Timmons, 2006). Also, there are several excellent resources focused on ethics in the biological and life sciences, and they should be consulted for a deeper understanding of these ethical issues in the arena of publishing (Bulger et al., 1993; Korthals and Bogers, 2004; Resnik, 1998). In particular, an entire section in each of the books by Korthals and Bogers (2004) and Bulger et al. (1993) is devoted to responsible authorship and communication in the biological sciences. Our treatment of ethics is solely for the purpose of communicating to prospective authors that they need to understand that ethical considerations must be addressed during the process of publishing a journal article. This includes publication of

Getting Published in the Life Sciences, First Edition. By Richard J. Gladon, William R. Graves, and J. Michael Kelly
© 2011 Wiley-Blackwell. Published 2011 by John Wiley & Sons, Inc.

research results in refereed journals in the life sciences, and as with our legal system, ignorance is no excuse for not knowing what is right or wrong when publishing a manuscript.

Within the entire scope of conducting research and then publishing it, there are numerous opportunities for unethical conduct. More broadly, this unethical conduct is called professional scientific misconduct. In this chapter, we will focus on the three major forms of misconduct associated with publishing: issues of scientific misconduct, plagiarism, and authorship responsibility issues.

ETHICAL ISSUES RELATED TO SCIENTIFIC MISCONDUCT

LaFollette (1992) presents an excellent overview of scientific misconduct and its many related issues in her seminal work on the topic. In some forms of disagreements over issues in scientific publishing, distinctions can be made easily. For instance, *illegal acts* are those forbidden by an organization granted the authority to formulate and distinguish between appropriate and inappropriate behaviors (Reagan, 1971). *Immorality* is behavior that violates the moral standards of a society (LaFollette, 1992; Reagan, 1971). *Improper professional conduct* in publishing is abuse of the proper norms of behavior of authors and other professional communicators, as defined by a professional society or association (LaFollette, 1992). *Fraud, or deception*, occurs when these forms of misconduct occur with the intention of false representation to gain an unfair advantage, or to hurt someone or a group deliberately (LaFollette, 1992). This *fraud* in scientific publishing consists of false representation of the authenticity of data, text, authorship, or a publication decision, and it includes theft of another's ideas (LaFollette, 1992). A key issue here is that the false representation had to have been made with the express *intent* to advance one's position or harm the position of another person. Although referees and editors that participate in the publishing process can also misbehave ethically, we will focus our attention only on the scientific misconduct of authors, especially as related to the production of manuscripts for publication in refereed journals in the life sciences.

On being a scientist: Responsible conduct in research (Committee on Science, Engineering, and Public Policy, 1995) devotes several sections to about ten forms of scientific misconduct. Several of these forms of misconduct are not related specifically to writing and the publishing process. These are honest errors, errors caused by negligence, misallocation of credit for something done by another person, cover-ups of misconduct in science, reprisals against whistleblowers, malicious allegations of misconduct in science, and violations of due process in handling complaints of misconduct in science. A major form of professional scientific misconduct is a category broadly labeled "deception," which can manifest itself in several forms of poor judgment. Three of these infractions are related directly to publishing and are particularly common:

- *Fabrication*, which is defined as the complete synthesis, or making up, of data or results that never have existed.
- *Falsification*, which is the changing or misreporting of data or results that exist in a form different, but true, from the data reported by the author(s).
- *Plagiarism*, which is the use of the data, ideas, words, and so on, of some other person(s) without giving appropriate credit (attribution) to the originator(s).

Although fabrication and falsification are extremely important, they usually occur before, or at the latest, during the process of writing the first few drafts of the manuscript.

A separate form of falsification has come onto the scene more recently. Scientists, especially molecular biologists, often do work with DNA, RNA, and proteins, and that work often results in the formation of an electrophoretic gel that was used to separate the components in question. In many cases, a digital photograph of the gel is taken and included in a journal manuscript. However, in the interlude between taking a photograph of the gel and submitting it for publication, the author could manipulate the image digitally by any of several means. Only certain manipulations are acceptable, and the author must determine what, if anything, she or he can do to enhance the photograph without overstepping ethical boundaries. As this has occurred several times recently, many journals have developed a policy statement of what is and is not acceptable practice in making a digital image ready for publication.

These acts of scientific misconduct undermine progress in science, but more importantly, they undermine the set of values on which scientific enterprise rests. Scientists who engage in these practices put their scientific careers at risk, and few, if any, scientists found guilty of scientific misconduct have recovered to continue their careers fully. In Marcel LaFollette's book *Stealing into print*, she states that even the most minor of infractions, which usually lead to some form of disagreement, can end up with the scientist in error, whether it be purposeful or inadvertent, being severely punished. Disagreements that lead to severe punishment(s) are often not merely rhetorical quibbles, but rather are frequently a reflection of important differences in the way the following items are viewed and held by the participants: value systems, codes of professional conduct, ambiguities in institutional policies, and ambiguities in institutional procedures.

LaFollette (1992) has organized and grouped this information on deception, misrepresentation, fraud, and other forms of unethical conduct in scientific and technical publishing as authors doing the following:

- Describing data or artifacts that do not exist (fabrication);
- Describing documents or objects that have been forged (falsification);
- Misrepresenting real data, or deliberately distorting evidence or data (falsification);
- Presenting another person's ideas or text without attribution (plagiarism), including deliberate violation of copyright;
- Misrepresenting the privilege of authorship by omitting a contributing author;
- Misrepresenting the privilege of authorship by including a noncontributing author or authors;
- Misrepresenting publication status.

PLAGIARISM

We wish to help inexperienced writers know what is and what is not plagiarism so that it does not occur during manuscript preparation. We also hope to help these writers identify when credit needs to be extended for the ideas and information of others, and in some cases, appropriate credit for the words of another person. It may seem we are using scare tactics to drive home our points about plagiarism. However, the plain and simple fact is that we cannot emphasize strongly enough how important it is that authors are free of any concern about whether they have plagiarized. Writers must be sure they are not plagiarizing, because it could be devastating to their professional career. It is of utmost importance that authors understand what is common knowledge, and more importantly, that authors know what are the bounds of common knowledge in their discipline. Authors, and especially inexperienced authors, should err on the side of safety until they know unequivocally where the lines have been drawn by their discipline. In some instances, these boundaries of common knowledge will vary from person to person within the discipline. Thus, it is important that the author err on the side of a conservative approach.

Definition of Plagiarism

Let us now focus our effort on plagiarism, as this form of misconduct is the type that occurs most frequently during writing and revising a scientific manuscript. *Plagiarism is the misappropriation (or theft) of the contribution(s) of another author.* This definition includes theft of the following items: idea(s); information; and writing, in terms of the use of unusual phrases or terms, sentences or significant parts of sentences, and sentence or paragraph organization.

Here are several issues that writers, especially inexperienced writers, should know about plagiarism (Charlotte Bronson, 2007, personal communication).

- An act of plagiarism can be committed by accident. However, this will never hold up as an excuse for the act.
- Citation standards vary between disciplines. What constitutes plagiarism in one discipline or professional society may not be viewed as plagiarism by another discipline or professional society. It is of the utmost importance to understand your disciplinary culture so that you may remain within its bounds.
- There may be culturally different standards for the appropriate use of other people's material. Again, it is important to understand your disciplinary culture so that you may remain within its bounds.
- There are "risk factors" associated with plagiarism. People are more likely to commit plagiarism when they are writing in a second language, when they are pressed for time, or when the original source in question is, or is thought to be, obscure so that the person thinks they are less likely to get caught.
- The consequences associated with getting caught are often catastrophic, and people found guilty of plagiarism never regain the professional status accorded them before they were caught.

Rules for Avoiding Plagiarism

Plagiarism occurs when an individual, either purposefully or inadvertently, uses the ideas, information, or writing of another author. Authors should do their best to avoid instances where plagiarism might occur, and the following several rules will help authors avoid these occurrences. Keep in mind that a citation, such as Johal et al. (1995), gives credit for the information and ideas that are contained within the resource, but if the writer is quoting or paraphrasing the work of another author, then the citation, itself, does not give sufficient credit for the writing the author has done. In other words, one among the "et al." may be the person who deserves the acknowledgment, rather than the senior author of the publication.

- Authors should *cite all ideas and information* taken from another author.
- Do not quote unless *absolutely necessary*. If you must quote, make sure the quote is exact (verbatim), use quotation marks, and accurately cite the source. This

recommendation is quite discipline-specific, as some professions maintain that the easiest way to avoid plagiarism issues is to quote the person's ideas directly.

- Do not paraphrase, because there are very few situations in scientific writing where it is necessary or appropriate.

How to Write Well *and* Avoid Plagiarism

When you are producing a manuscript and want to avoid plagiarism issues, you should remember that your first goal is to have good insights into what the author has written, and at the same time, respect their right to ownership of what they have said. Although you must have good clarity and understandability in your writing, your goal at this point is not necessarily good grammar but rather appropriate accreditation of the work of another author. In this case, ideas must come first and the words you use and how you use them must come second. Until you are able to develop your own mechanism or protocol for avoiding plagiarism, you might want to follow this set of rules for avoiding plagiarism while you are writing your manuscript (Charlotte Bronson, 2007, personal communication).

- Read the original work (writing) of the author(s) and make sure you understand everything about what they have written.
- Think about what the author(s) have written and the ideas they want to get across to the reader. Ask yourself these questions, and get answers to them before proceeding: What did the authors say? How does what they said relate to the topic of the manuscript now being written?
- Put aside the resource that contains the information you want to incorporate into your manuscript.
- Use *your own* language and style (whatever it is) and write out what you want to say.
- Add proper citation, or citations, depending upon the structure of what you wish to accredit. Check what you have written for accuracy against the original writing of the author you wish to acknowledge.
- After you have written what you wish to say and checked it for accuracy, then translate it into good English usage and fix the grammar, syntax, and so on, so that it reads properly.

ETHICAL ISSUES PERTAINING TO AUTHORSHIP

We will now focus on another aspect of ethics in publishing. Years ago, many researchers worked singly, and independently, and issues of authorship rights, privileges, and responsibilities arose infrequently. More recently, an individual scientist produces few research projects independently and, therefore, few manuscripts. Therefore, authors must deal with issues of author rights, privileges, and responsibilities to an extent not present 25 to 50 years ago. The pressure to publish has continued to spiral upward. Ethical issues related to authorship that never arose years ago are surfacing now that these pressures have become more commonplace. Abuses of authorship rights, privileges, and responsibilities are also a form of scientific misconduct, and this general lack of responsibility in authorship issues is a major form of this scientific misconduct. In the booklet *On being a scientist: Responsible conduct in research* (Committee on Science, Engineering, and Public Policy, 1995), the authors devote several sections to about ten major forms of scientific misconduct. One is a category broadly labeled deception. There are many forms of poor judgment considered deception, several of which revolve around unethical conduct in authorship issues. Several of these infractions can occur during the publishing process, most commonly:

- Misrepresenting authorship by omitting a contributing author;
- Misrepresenting authorship by including a noncontributing author;
- Misrepresenting publication status; and
- Presenting ideas, text, or words of other authors without appropriate attribution (plagiarism), which includes deliberate violation of copyright by plagiarizing one's own work.

Absence of a Code of Ethics

We will approach authorship rights, privileges, and responsibilities in publishing mostly by describing what misconduct is and how to avoid committing it. It is easier for us to explain what not to do/what is wrong, than to describe what to do/what is right. Unfortunately, there are almost no written codes of ethics in disciplines in the life sciences. Therefore, scientists such as yourself are basing your conduct on traditions and unwritten values, rather than rules, codes, and laws that govern the behavior of scientists in a particular discipline (Pigman and Carmichael, 1993). Some professional societies have recognized this lack of direction in their membership, and committees and boards have been established for the express purpose of developing a code of ethics for the professional society.

Publish or Perish—The Pressure to Publish

Why is there so much pressure to publish? What is its origin? Originally, there was "mild" pressure to publish. Scientists everywhere felt, rightfully so, that unless the information was communicated so that it could stand the tests of time and scrutiny, the research work was not complete. However, publications have become one of the mechanisms for evaluating scientists, their productivity, and their professional progress, which has led to an incredible shift in the focus on publishing.

Scientists quickly realized the rules of the game had changed. In response, they reduced the size of the article to that of a "least publishable unit" (Broad, 1981). Often, two to five least publishable units today might have made up the content of one paper several decades ago.

This pressure to publish is not necessarily all bad. It has certainly stimulated scientists to greater levels of achievement, but there also came abuses of the system under which all scientists operate. Whether it is good or bad, we are certainly stuck with it, at least for the time being. Thus, today's scientist is feeling two forms of pressure to publish. One is from the tradition of communicating their work to other scientists as the final step in the scientific method. The second one is from themselves and the administration at their home institution. Both sets of individuals want to record greater numbers of publications, grants, and so on, so that the scientist and the institution can gain stature in her or his field and get "easier access" to funding.

Justified and Unjustified Authorship

There are two ways fraudulent, unjustified authorship can occur. First, a deserving author is not included on the manuscript. This probably occurs most frequently when the omitted author would not have been happy with the content of the manuscript. This unhappy coauthor might "stonewall" or otherwise slow down the publication process, maybe to the point that the manuscript would never be submitted. Omitting a deserving author might also occur when one or more authors may be able to secure their own personal advancement over the omitted author if the manuscript is published without the omitted author being a part of the publication. It has been argued that much unjustified authorship is due to tenure policies that reward the number of papers published rather than their quality, and this leads to wasteful publication (Huth, 1993).

Our second form of unjustified authorship comes in the form of a scientist being included as an author on a publication when they are not deserving. In many cases, at best these individuals should appear in the acknowledgments section only. What makes a potential author deserving? Excellent guidance is provided by the International Committee of Medical Journal Editors (ICMJE, 1997), who developed the Vancouver Guidelines on Authorship. They state that anyone listed as an author should have participated in the work sufficiently to take public responsibility for the content of the article and its work. Thus, along with credit for the work comes accountability for its quality, originality, reproducibility, and validity.

Many problems have arisen from changes that are occurring in the research world, with more emphasis being placed on team-based rather than individual approaches to research. Today, teams of authors publish most articles, and it has become difficult to determine whether a given author has or has not contributed to the work to a level where they deserve to be named coauthor. Here we list some of the general obligations that are part of the right and privilege of being called a coauthor of a manuscript.

- Authors take responsibility for the overall quality of the manuscript, in terms of content, originality, and reproducibility.
- They have responsibility for being aware of and reporting prior work by both the coauthors and other researchers. Has all or part of the work already been done?
- Authors accept criticism of both the written word and the scientific content of the manuscript, and understand that disagreement is an important and necessary event that may be expected in research endeavors. They must expect that their work can and should be subject to scrutiny by their peers and that they must be able to answer those criticisms, when appropriate.
- Authors understand and respect the property rights of scientists and their work. This is important, because there have been instances where an associate "pirates" the work of her or his colleagues and passes it off as their own, without acknowledging the scientist who may have done the majority of the original work contained in the manuscript.
- Authors must be prepared for the publicity that may come from the manuscript. Scientists seem increasingly interested in publicity and public notice, perhaps because of the potential impact on funding.

There are also several issues that come into focus when multiple authorship must be considered:

- Senior authorship—who will be the first author?
- The order of the other names (a.k.a., the junior authors) after the senior author has been designated.

- Should certain administrators be listed in the by-line, simply because they are in charge of the laboratory, department, college, or other unit where the research occurred? Likewise, we may ask if financial supporters also should be considered as coauthors.

- Should a graduate student have a right of authorship on a manuscript in which the work was not their own thesis/dissertation work, but rather work they were required to perform in return for an assistantship? In the same manner, should a technician/technical assistant have a right to authorship that has evolved from her or his daily work? In both cases, the issue is one of when does "work for hire" begin and end.

- In addition, should a graduate student automatically become the senior author on a publication that evolved from his or her research, especially in cases where the student has not completed his or her degree program or provided an appropriate contribution to the conduct of the project.

- If the research was a group project, who has done enough to warrant authorship?

- What is the right/privilege available to the person who has taken the lead in the preparation of the manuscript? How does that amount of work correspond to the effort that has been extended by the person(s) revising, reviewing, and editing the manuscript or functioning as the corresponding author?

Out of this definition come three types of irresponsible authorship:

- *Unjustified authorship.* This occurs when a person makes technical suggestions only, does not participate sufficiently in the reported work, did not help write the paper, and did not see or judge the final version of the manuscript. In general, these individuals cannot take public responsibility for the work, and they cannot defend the content of the article. Often, they should receive their attribution in the form of an acknowledgment.

- *Incomplete authorship.* This comes when one or more people are not listed as coauthors on the paper, and they have had responsibility for the critical content of the paper. Generally, this is considered rare, but it has certainly happened many times.

- *Wasteful publication* (Huth, 1993). This occurs when there is an abuse of the least publishable unit. At times, researchers will take what should have been in one or two publications, and will stretch it out into four, five, or six papers. This is also called divided publication. There is also *repetitive publication*, which occurs when authors republish essentially the same content with only minor additions.

Inexperienced writers need to gain an appropriate understanding of who should and should not be an author. To this end, we have incorporated two exercises that will help inexperienced writers select appropriate coauthors for the manuscript. An exercise in Chapter 4 will help lay the groundwork for the choice of coauthors, and it is structured on a more or less general basis. At the end of Chapter 19, after the manuscript has essentially been completed, this issue of appropriate authorship will be revisited. In this instance, the privilege of authorship is evaluated much more stringently and with greater scrutiny. Hopefully, this will ensure that properly contributing authors will be included and noncontributing authors will be removed.

Postscript

We cannot emphasize strongly enough that you, as a writer, must be cognizant of these ethical issues related to publishing your work. Most scientists who have been caught in the act of scientific misconduct, or have been accused of it, whether they were found guilty of the charges or not, have not recovered sufficiently to resume a productive and successful career as a scientist. As with our legal system, ignorance is no excuse for not knowing what is right or wrong when publishing a manuscript.

OVERVIEW OF THESE EXERCISES[1]

The overall purpose of these exercises is to help the reader understand what is and is not plagiarism and, through this identification process, learn to identify when credit needs to be extended to previous authors for their ideas, information, or written words. The reader should take home with them an understanding that citation is needed for accreditation of these ideas and information, but proper accreditation of the words and sentences of another author occurs only when the original author is quoted verbatim and quotation marks are used around the words and sentences of the previous author.

---

[1]Exercises 3.1 and 3.2 have been developed and used by Dr. Charlotte Bronson of the Department of Plant Pathology, Iowa State University, Ames, Iowa, and she has given complete permission for them to be used in exercises that teach plagiarism to students.

EXERCISE 3.1   Writing Summary Statements Without Plagiarizing

Part 1 is a take-home exercise, and Part 2 will be completed in-class at the next class meeting time. Please read the following passage that has been quoted from G.S. Johal, S.H. Hulbert, and S.P. Briggs (1995). Disease lesion mimics of maize: a model for cell death in plants. BioEssays 17:685–692. Please make sure that you understand fully the information that the authors are trying to get across to the readers of this article.

> "A class of maize mutants, collectively known as disease lesion mimics, display discrete disease-like symptoms in the absence of pathogens. It is intriguing that a majority of these lesion mimics behave as dominant gain-of-function mutations. The production of lesions is strongly influenced by light, temperature, developmental state and genetic background. Presently, the biological significance of this lesion mimicry is not clear, although suggestions have been made that they may represent defects in the plants' recognition of, or response to, pathogens. '...' In this paper we argue that this might be the case..."

1. Please write one or two sentences with a total of 30 words or less that summarize, in your own words, the take-home message of the quoted passage above. You will probably want to practice on a separate piece of paper, and then write, very legibly, your response on the lines below.

_____

_____

_____

_____

_____

_____

_____

2. When the class meets next time, you will be divided into groups of two, three, or four. Each member of your group will read the sentence(s) written by the other members of the group, discuss them, and determine whether or not plagiarism has occurred.

EXERCISE 3.2   When Does Plagiarism Occur?

Part 1 of Exercise 3.1 is a take-home exercise, and Part 2 of it will be completed in-class at the next class meeting time. Exercise 3.2 should be completed immediately after Part 2 of Exercise 3.1 has been completed. Please read the following passage, which has been quoted from G.S. Johal, S.H. Hulbert, and S.P. Briggs (1995). Disease lesion mimics of maize: a model for cell death in plants. BioEssays 17:685–692. Please make sure that you understand fully the information that the authors are trying to get across to the readers of this article. (This is exactly the same quotation that was used in Exercise 3.1.)

> "A class of maize mutants, collectively known as disease lesion mimics, display discrete disease-like symptoms in the absence of pathogens. It is intriguing that a majority of these lesion mimics behave as dominant gain-of-function mutations. The production of lesions is strongly influenced by light, temperature, developmental state and genetic background. Presently, the biological significance of this lesion mimicry is not clear, although suggestions have been made that they may represent defects in the plants' recognition of, or response to, pathogens. '...' In this paper we argue that this might be the case..."

1. Please read and understand each statement recorded below. Judge for yourself whether or not you feel plagiarism has occurred because the writer of the statements below has failed to give proper credit for the ideas, information, and/or writing of Johal, Hulbert, and Briggs. Place a check on the line following each choice. The group judgment section of this exercise will occur during the next class meeting time, immediately after Part 2 of Exercise 3.1 has been completed.

   (a) Currently, the biological significance of lesion mimicry in plants is not known, although suggestions have been made that they may represent defects in the plants' recognition of, or response to, pathogens.

   Your judgment: Plagiarism occurred _____; Plagiarism did not occur _____ .

   Group judgment: Plagiarism occurred _____; Plagiarism did not occur _____ .

   (b) Currently, the biological significance of lesion mimicry in plants is not known, although suggestions have been made that they may represent defects in the plants' recognition of, or response to, pathogens (Johal et al., 1995).

   Your judgment: Plagiarism occurred _____; Plagiarism did not occur _____ .

   Group judgment: Plagiarism occurred _____; Plagiarism did not occur _____ .

   (c) Currently, "the biological significance of lesion mimicry in plants is not known, although suggestions have been made that they may represent defects in the plants' recognition of, or response to, pathogens" (Johal et al., 1995).

   Your judgment: Plagiarism occurred _____; Plagiarism did not occur _____ .

   Group judgment: Plagiarism occurred _____; Plagiarism did not occur _____ .

   (d) The biological significance of lesion mimicry in plants is currently not known, although some researchers believe that they may represent defects in the ability of plants to recognize or respond to pathogens.

Your judgment: Plagiarism occurred _____; Plagiarism did not occur _____ .

Group judgment: Plagiarism occurred _____; Plagiarism did not occur _____ .

(e) The biological significance of lesion mimicry in plants is currently not known, although some researchers believe that they may represent defects in the ability of plants to recognize or respond to pathogens (Johal et al., 1995).

Your judgment: Plagiarism occurred _____; Plagiarism did not occur _____ .

Group judgment: Plagiarism occurred _____; Plagiarism did not occur _____ .

(f) Lesion mimicry has been proposed to be due to mutations in genes controlling the ability of plants to detect and respond to pathogens.

Your judgment: Plagiarism occurred _____; Plagiarism did not occur _____ .

Group judgment: Plagiarism occurred _____; Plagiarism did not occur _____ .

(g) Lesion mimicry has been proposed to be due to mutations in genes controlling the ability of plants to detect and respond to pathogens (Johal et al., 1995).

Your judgment: Plagiarism occurred _____; Plagiarism did not occur _____ .

Group judgment: Plagiarism occurred _____; Plagiarism did not occur _____ .

(h) Disease-like lesions in plants may be due to mutations in genes controlling the ability of plants to defend themselves against pathogens (Johal et al., 1995).

Your judgment: Plagiarism occurred _____; Plagiarism did not occur _____ .

Group judgment: Plagiarism occurred _____; Plagiarism did not occur _____ .

2. After you have made your judgments, you will be divided into groups of two, three, or four. Your group then will determine whether or not plagiarism has occurred.

REFERENCES

BROAD, W.J. 1981. The publishing game: Getting more for less. Science 211:1137–1139.

BULGER, R.E., HEITMAN, E. and REISER, S.J. (Eds) 1993. The ethical dimensions of the biological sciences. Cambridge University Press, Cambridge, UK.

Committee on Science, Engineering, and Public Policy (National Academy of Sciences). 1995. *On being a scientist. Responsible conduct in research.* National Academy Press, Washington, DC.

COMSTOCK, G.L. (Ed.) 2002. *Life science ethics.* Iowa State Press, Ames, IA.

HUTH, E.J. 1993. Irresponsible authorship and wasteful publication. *In*: BULGER, R.E., HEITMAN, E. and REISER, S.J. (Eds). *The ethical dimensions of the biological sciences*, pp. 134–137. Cambridge University Press, Cambridge, UK.

ICMJE (International Committee of Medical Journal Editors). 1997. Uniform requirements for manuscripts submitted to biomedical journals. Ann. Intern. Med. 126:36–47.

JOHAL, G.J., HULBERT, S.H. and BRIGGS, S.P. 1995. Disease lesion mimics of maize: A model for cell death in plants. BioEssays 17:685–692.

KORTHALS, M. and BOGERS, R.J. (Eds) 2004. *Ethics for life scientists.* Springer, Dordrecht, The Netherlands.

LAFOLLETTE, M.C. 1992. *Stealing into print. Fraud, plagiarism, and misconduct in scientific publishing.* University of California Press, Berkeley, CA.

MISH, F.C. 2004. *Merriam-Webster's collegiate dictionary.* Eleventh edition. Merriam-Webster, Inc., Springfield, MA.

PIGMAN, W. and CARMICHAEL, E.B. 1993. An ethical code for scientists. *In*: BULGER, R.E., HEITMAN, E. and REISER, S.J. (Eds) *The ethical dimensions of the biological sciences*, pp. 117–124. Cambridge University Press, Cambridge, UK.

RACHELS, J. and RACHELS, S. 2007. *The elements of moral philosophy*, fifth edition. McGraw-Hill, Inc., Boston, MA.

REAGAN, C.E. 1971. *Ethics for scientific researchers*. Second edition. Charles C. Thomas, Springfield, IL.

RESNIK, D.B. 1998. *The ethics of science. An introduction*. Routledge, New York, NY.

ROLLIN, B.E. 2006. *Science and ethics*. Cambridge University Press, Cambridge, UK.

TIMMONS, M. 2006. *Conduct and character*. Fifth edition. Thomson Wadsworth, Belmont, CA.

WILCOX, L.J. 1998. The coin of the realm, the source of complaints. J. Am. Med. Assoc. 280:216–218.

CHAPTER 4

CHOOSING YOUR COAUTHORS

If you have coauthors, problems about authorship can range from the trivial to the catastrophic.

—M. O'Connor

PRINCIPLES AND NEED

Before you begin to write, you must determine who your coauthors will be and what will be the preliminary order of the authors in the by-line of the article. Several scientists working in concert accomplish most research, but most journals and authors have not codified a mechanism to determine which scientists should be coauthors and in what order their names should appear in the by-line. Few things cause as much strife among colleagues as issues related to authorship on refereed publications.

One way to reduce this strife is for the lead writer(s) to determine early on, before any research has been conducted, and certainly before any writing is attempted, who will be the coauthor(s) and in what order they will appear. On the one hand, it is unethical to include someone as a coauthor when they have not made a significant contribution to the whole of the research [conceptualization, experimental design, conduct of the work, data analysis and interpretation, or writing part(s) of the manuscript]. Conversely, someone who has made an appropriate contribution to the work should have his or her name appear in either the by-line of the manuscript or at least the acknowledgments. Again, it is unethical not to include someone who has made a substantial contribution to the work. We have included here a short discourse on what is generally required of an individual if they are to be considered a coauthor on a manuscript. Also it should be recognized that the order of authorship may evolve as the manuscript is developed. Any changes in the order of the appearance should be thoroughly discussed with all coauthors. We finish this topic with an exercise you need to complete before you start on your manuscript.

ITEMS TO CONSIDER IN CHOOSING YOUR COAUTHORS

Years ago, before there was such a great level of technology associated with the conduct of research, scientists often worked alone on their particular area of research. At its earliest stages of development, research often involved the thought processes of only one individual. That person conceived there was a problem and sought to discover a solution to that problem. In due time, that scientist would then submit a manuscript that explained the problem, how the problem was solved, and what that solution meant to people interested in that information. The manuscript describing those events would have only one author, as no-one else had worked on the problem and its solution.

Getting Published in the Life Sciences, First Edition. By Richard J. Gladon, William R. Graves, and J. Michael Kelly
© 2011 Wiley-Blackwell. Published 2011 by John Wiley & Sons, Inc.

Science, and all its associated processes, has become more collaborative than ever before (Weltzin et al., 2006). When the manuscript describing the research is written, all contributors must be granted some form of appreciation for their effort. Instead of one author, there now may be two, three, four, or many more coauthors on the manuscript (Cronin, 2001; Regalado, 1995). Weltzin et al. (2006) studied the rise in the number of coauthors in articles published in the journal *Ecology*. They found the average number of coauthors per article rose from 1.1 in 1925 to 3.3 in 2005. In addition, they found the maximum number of coauthors per article varied from two to seventeen. And herein lies the dilemma: how does the lead writer draw the line between thanking a contributor by acknowledging their help versus extending to them the reward and responsibility of coauthorship on the manuscript? A satisfactory answer to this dilemma is not easy to develop, and disagreements accumulate to the point of broken friendships, broken hearts, and colleagues no longer communicating with one another.

An excellent review of the problem of authorship in terms of attribution, accountability, and responsibility has been published, and we will use much material from that report (Weltzin et al., 2006). Additional information on definitions associated with authorship is available (Claxton, 2005; Davidoff, 2000; Huth, 1986). In many instances, journals and professional societies have not defined the essence of authorship and again, in many cases, they have not yet developed a code of ethics about authorship issues. An exception is the medical profession, which has a code for authorship issues (ICMJE, 1997). In instances where there are no clear lines that define authorship, one or more lead authors of the manuscript will invoke their individual criteria (preferences) for authorship, and they will invite individuals to share in authorship with them (Rennie et al., 2000). Thus, whether an individual is listed as an author and the order in which the authors appear in the by-line gives uninformed readers of the article little true information about the contributions of any individual listed in the by-line.

Probably the best, most general, definition of an author is an individual who made a significant contribution to some segment of the research conducted, from conceiving the problem to writing the manuscript for publication. Some journals and professional societies, such as the Ecological Society of America, go a little bit farther. They segment authorship so that the individual can make their contribution via one or more of the following: conceiving the idea to conduct the research, conceiving the experimental design, actively executing the work, analyzing the data, interpreting the data, or writing the manuscript (Weltzin et al., 2006). The most difficult issues to resolve in deciding on authorship are the individual interpretations of the words significant and contribution, and who makes these interpretations. In some disciplines, such as medical research, the International Committee of Medical Journal Editors, in their *Uniform Requirements for Manuscripts Submitted to Biomedical Journals*, states that an individual must contribute to all of the segments listed in the uniform requirements, rather than just one or more segments (ICMJE, 2005).

There are no sets of rules or any conventions that can be used to guide the author(s) of the manuscript to a logical, acceptable statement of who should be an author and what will be the order of coauthors on the manuscript. However, some professional societies are working toward the development of a code for authorship, and the Vancouver Guidelines (ICMJE, 1997) have come closest to broad acceptance. At times, this can be an issue of ethics, and a part of Chapter 3 is devoted to ethical considerations of authorship issues.

The Vancouver Guidelines on Authorship suggest that a person be considered as a coauthor when, and only when, they actively contribute to some predetermined, sufficient level in at least one or more parts of the entire project. Thus, this does not comprise just the writing of the manuscript, but rather more activity associated with the entire project, so that they take full responsibility for the content of the manuscript (ICMJE, 1997). This includes the planning, design, execution, analysis, interpretation, and reporting of the experimental work described in the manuscript. Another way of defining this is that each coauthor must take intellectual responsibility for the results that are being reported, and they must therefore take full responsibility for the entire content of the manuscript. A third way to look at this is that each coauthor must make some "important or substantial contribution" to the conduct of the research.

The Vancouver Guidelines cover the following three facets of the entire project, and according to the guidelines, all three must be met in full (ICMJE, 1997). This is a rather idealistic goal to have each coauthor participate fully in all three areas, and many organizations that have adopted these guidelines have opted to reduce this requirement to one or two of the conditions. Any part(s) of the manuscript critical to its main conclusions must be the responsibility of at least one of the authors. Authorship credit should be based only on the following:

- substantial contributions to idea development and experimental design, or analysis and interpretation of the data;
- writing the manuscript or revising it critically for its intellectual content; and
- final approval(s) of the manuscript to be submitted.

The issues of ghost authors (a writer [usually paid a fee] whose name does not appear on the manuscript) and guest contributors (a writer whose name is placed on the manuscript because they have made a minor contribution, rather than had a major role in the work) have been studied (Peat et al., 2002; Rennie and Flanagin, 1994). Honorary or gift authors (those accorded authorship due to status in the profession) have also been studied (Flanagin et al., 1998; Peat et al., 2002). Incorporation of one of these types of coauthors does not do justice to the work that has been accomplished by others, and use of these types of authors constitutes a serious breach of ethical behavior. Some societies, such as the Ecological Society of America (ESA, 2006), also address the ethical issues of authorship. Some societies have authorship guidelines, which, although quite nebulous, are concerned with adding or deleting coauthors, and most importantly, each coauthor taking responsibility for the entire content of the article (ESA, 2006).

There are some additional, general responsibilities, or obligations, that may be required for one to be considered a coauthor on a manuscript. We suggest that the most important thing the lead author(s) of the manuscript can do is to start dialog about who will be the coauthors, as soon as is practicable, perhaps even before the research work commences. We suggest that, in addition to the above characteristics, to be considered a coauthor, an individual must take a full share of responsibility in all these areas:

- overall quality of the paper, in terms of content, originality, and reproducibility;
- direct responsibility to be aware of and report prior work by any of the coauthors or other researchers, provided those earlier works are relevant to the conduct of the research now being reported;
- accepting criticism of both the written word and the scientific content of the manuscript and understanding that disagreement is an important and necessary event that may be expected in research endeavors (authors must expect that their work can and should be subject to scrutiny by their peers and that they must be able to answer those criticisms, when appropriate);
- understanding and respecting the property rights of scientists and their work (there have been instances where an associate "pirates" the work of his or her colleagues and passes it off as his or her own, without acknowledging the scientist(s) who may have done the majority of the original work).

When more than one scientist contributes significantly enough to the work contained in a manuscript, the presence of and the order of appearance of the coauthors becomes an issue. While you, as lead author, should be sensitive to the wishes and egos of your coauthors, you must also facilitate discussions that ultimately lead to a fair and ethical arrangement of the authors' names in the byline. An early attempt at determining whether an individual should be granted the status of coauthor has been developed (Hunt, 1991).

Hunt's system was refined and strengthened into a two-stage system by Galindo-Leal (1996). The first stage consists mainly of communication among the coauthors during the entire conduct of the research and a self-regulated appraisal of each person's contribution to the research. In the second stage, each contributor self-assigns, or a lead author on the manuscript assigns to each contributor, a score from 0 to 20 points, in increments of five points, in five activity areas that pertain to the conduct of the research and the dissemination of the information. These activity areas are planning, executing, analyzing, interpreting, and writing (Galindo-Leal, 1996); each of these areas must therefore have a maximum total of 20 points, with a grand total of 100 points. The assigned score in each of these five areas leads to a total score for each contributor. Coauthorship is then granted when an individual scores greater than some predetermined amount, usually 20, 25, or 30 points.

In addition, the order of appearance of the coauthors in the byline can be constructed by listing the sequence of coauthor names in descending order of total score (Galindo-Leal, 1996). Others who have studied this facet of publishing agree that persons who make the greater contributions to the research reported should be placed first, second, etc., in the byline, and lesser contributors should be added in descending order of their contribution (Rennie et al., 1997). This scoring system of Galindo-Leal (1996) is quite good for shorter-term, well-defined research work by a small number of researchers, but it becomes less useful for research projects that are longer termed, with many different authors making more complex contributions (Weltzin et al., 2006). Our adaptation of the system developed by Galindo-Leal (1996) should work well for scientists using this book, and Exercise 4.1 should be completed by the lead author(s) of the manuscript.

Some journals have now imposed limits on the number of coauthors on a manuscript. This is due largely to the recent explosion in the number of coauthors on articles. In 1975, over all journal articles published that year, there was an average of 2.2 coauthors per article, and by 1995, this number had climbed to 4.5 (Peat et al., 2002). The cause of this may be that interdisciplinary approaches and collaborative research are more commonplace these days, especially across institutions. However, there may be issues with coauthors trading favors; for example, "I will put you on this manuscript, if you will put me on your next manuscript."

Make sure you know if there are regulations governing the number of coauthors of your manuscript if you will have many of them. Peat and colleagues (2002) suggest that the maximum number of coauthors should be about eight to nine for regular journal articles, four or five for letters to the editor, and three or four for review articles. Some journals impose limits on the number of coauthors. For instance, *Thrombosis and Haemostasis* allows a maximum of eight, *Chest* allows a maximum of seven, and the *New England Journal of Medicine* allows 12. With this last example, articles with more than 12 coauthors have the byline truncated at the first 12, and the remainder of the coauthors are placed in a footnote.

Rennie and colleagues (1997) proposed a system that placed a high value on accepting responsibility and accountability for the results reported. This acceptance leads to credit, as an author, for the individual's contribution to the work reported. They proposed a system of "contributorship" that recognizes the contribution of each individual to the whole of the research and the manuscript, rather than blanket use of the word author. The "word and concept of an author" has been replaced by the "word and concept of a contributor" (Rennie et al., 1997). In exercising this system of contributorship, each person is recognized for their contribution to the entire project, and the accountability of that contributor is designated via disclosing the segments of work for which they were responsible and accountable in a separate byline box (see Panel 1, Weltzin et al., 2006). Thus, each author, or better yet, contributor, publicly accepts responsibility and accountability for their individual contribution to the work reported.

In an attempt to avoid conflicts and other unpleasant scenarios such as these, some journals are moving toward incorporation of a credit line or a contributor's line that presents and discusses each coauthor's contribution to the article (Gustavii, 2003; Peat et al., 2002). Several major journals have now either "strongly encouraged" scientists to provide a "contributorship" endnote to the manuscript (*Nature*, 2006; Anonymous, 1999) or mandated an online-only authorship footnote to the manuscript (*PNAS*, 2011). Other possibilities for handling this contributorship issue are available (Klein and Moser-Veillon, 1999; Rennie et al., 1997; *PNAS*, 2011).

The first author on the byline list is called the senior author, regardless of age, rank, title, and so on, and the others are called coauthors or junior authors. For some people and some institutions, the order of the authors is critically important for promotion and tenure purposes. Sometimes it is important simply to satisfy egos.

For many years, it has been assumed that the senior author on the manuscript has full responsibility for, or guarantees, the validity of the research results and their analysis and interpretation. Other circles of scientists may consider the corresponding author (defined later in this chapter) as the individual ensuring the validity of the entire research. It would be much better if readers did not have to make an assumption about a guarantee of the validity of the research. It would be helpful if at least one author could be designated publicly as the guarantor of the content of the manuscript. The role of this guarantor would be to ensure the integrity of the manuscript as a whole, and this role would have developed out of their contribution to all phases of both the research work and the preparation of the manuscript (Weltzin et al., 2006). Most importantly, the selection, or designation, of a guarantor ensures that an individual who was part of the research team publicly accepts full responsibility and accountability for the entire project (Weltzin et al., 2006). The name(s) of the guarantor(s) would be designated in the same byline box, endnote, or authorship footnote discussed above.

Where is the line drawn between a person earning the privilege of authorship and the person being acknowledged appropriately for their contributions? It has been suggested that a major contributor who accepts full responsibility and accountability for the entire content of the manuscript should be credited as an author (Weltzin et al., 2006). If this individual could be credited with only a small portion of the entire content of the published article, then the lead author(s) of the manuscript must determine whether this justifies authorship. In addition, if this person believes she or he cannot accept responsibility and accountability for the entire content of the manuscript, then she or he should be credited only in the acknowledgments. Again, open discussion may avert conflict and discontent.

We encourage you to complete the exercise at the end of this chapter, and then later, as you are finishing your manuscript, you will complete a similar exercise in Chapter 19. These exercises should help you begin to define the process you will adopt for designating authors on your manuscripts. Consider this example. At Harvard University, there is an Office of the Ombudsman. During the 1991–1992 academic year, 2.3% of the issues adjudicated by the Ombudsman were focused on authorship issues. By the 1996–1997 academic year, this figure reached 10.7% (Wilcox, 1998). Thus, guidelines on how to select coauthors and the order of coauthors need to be developed as soon as possible for major scientific disciplines. Objectivity should rule these issues, and systems based on subjective opinions such as grace and favor should be eliminated (Peat et al., 2002).

One coauthor on the manuscript must be designated as the corresponding author. The coauthor that fills the role of the corresponding author should be chosen before or soon after writing of the manuscript commences. The corresponding author is the person who takes responsibility for the initial submission and for all aspects of getting the manuscript published, including revisions and submissions that occur after the initial submission of the manuscript. This individual is the single line of communication that represents the interface between the representatives from the journal and the coauthors of the manuscript. In some journals, especially international ones, this is a position of honor, and this person is listed last in the byline. In some cases, the corresponding author is considered more important than the senior author.

The corresponding author should be the only one to communicate with the Associate Editor or adjudicator of the manuscript so that the adjudicator does not have several people calling her or him and asking when the review will be finished, what needs to be done, and so on. In most cases, the corresponding author also assumes responsibility for all changes in the manuscript and the costs of publishing it. The corresponding author is also the point of contact for communication by scientists who need to discuss some aspect of the content of the published article in the future.

Normally, the corresponding author also acts as the source of reprints of the published article. In most technologically advanced countries, researchers will simply make a photocopy of the article out of an issue of the journal, or obtain the article via the Internet. However, many developing countries do not have an extensive library and computer systems, and the researchers in that country may have to rely on receiving reprints from the corresponding author. Often, the corresponding author is designated by an asterisk, or some other form of alert mechanism, that points an editor or reader to the source of answers about the manuscript.

In many cases, almost to the point of it being an unwritten rule, a graduate student preparing a manuscript from his or her thesis or dissertation work will automatically become the senior author on the manuscript. Because of the incredibly wide range of possible events that can occur as a student moves through a degree program, neither the graduate student nor the major professor should consider this an unequivocal rule. There may be times when a graduate student finishes his or her degree program, but "walks away" from converting the thesis or dissertation work into a manuscript. Under this set of conditions, it may be entirely appropriate for the major professor to write the manuscript and place himself or herself in the position of the senior author. Again, depending upon circumstances, the graduate student may or may not be included as an author on the manuscript. In this scenario, all parties must be careful because breaches of ethical behavior can occur, and stigmas may be attached to either the graduate student or the major professor, and these stigmas may stay with the individual throughout their entire career.

THE SPECIAL CASE FOR STATISTICIANS

In many cases, scientists know just enough statistics to be dangerous, but the good news is that many of them readily acknowledge they are statistically challenged. To alleviate this problem, researchers often employ the services of a statistician, and the statistician may play up to several roles in the course of the research project. When the researchers decide it is time to write the manuscript for the work just completed, there is often angst related to whether the statistician should be included as a coauthor.

TABLE 4.1   Points Awarded for various Contributions of Statisticians to be Considered as Coauthors

Statistician contribution	Points for involvement
Study Design	
Substantive input to overall design and protocol development	4
Writing one or more sections of grant application(s)	2
Review(s) of grant application(s) before submission	1
Implementation	
Appropriate participation in regular research meetings	4
Implementation of data collection and database management	2
Advice/consultation with only special problems	1
Analysis	
Planning and directing all data analyses	4
Preparation of written reports summarizing data analyses	2
Conducting data analyses	1
Maximum Possible Points	**21**

To try and quantify the involvement of the statistician, at least in medical research, Parker and Berman (1998) developed a system of assigning points for various roles the statistician may play. In this system, the role of the statistician has been allocated a maximum of 21 points, and these are spread over the three main areas of study design, implementation, and analysis. Table 4.1 summarizes this system, as it appeared in Peat et al. (2002). A total of five points or fewer does not warrant coauthorship, six to seven might warrant coauthorship, and eight or more points indicates certain coauthorship (Parker and Berman, 1998). The key issue here is that the contribution of the statistician meets at least some of the Vancouver Guidelines (ICMJE, 1997). If additional questions arise about whether a statistician should be granted coauthorship, consult Parker and Berman (1998).

We will close this chapter with some additional items that the lead author(s) must consider before coming to an understanding and acceptance of who should be a coauthor and what should be the order of the coauthor(s).

SOME ADDITIONAL ITEMS TO CONSIDER IN CHOOSING COAUTHORS

- Sometimes, to avoid bitter confrontation(s), the coauthors will agree before, during, or sometimes after, the research has been completed, to list the names of the authors alphabetically, regardless of the degree of contribution of each author. Let us caution you that the ethics of a decision such as this are questionable, and this type of activity should take place only when there are no other ways to resolve the issues surrounding the confrontation(s). Nonetheless, who will be a coauthor and what will be the order of the coauthors should be established as soon as possible (Matthews et al., 2000; O'Connor, 1991). If issues arise with this process, resolution may come by consulting the Instructions to Authors document, the style manual, or recent issues of the journal of choice, as explicit statements are sometimes included to help writers come to an equitable resolution to this problem.

- If necessary, as a result of some strange set of circumstances, a coauthor or coauthors can be eliminated at a later time. Or, there may be a late addition to the list of coauthors. In addition, the order of appearance of the coauthors in the byline can be changed up until the time the paper is submitted for publication in its original form. Extreme care must be used when any of these three scenarios are followed, because there may be hurt feelings, or an undeserving person may have been added late. All three scenarios are fraught with the potential for breaches of ethical behavior, and care should be taken when any of them become a reality.

- Early in your career, you should choose how you want to label yourself, and then stick with it. This entire issue may be rendered even more complex by occasions such as a marriage. Women scientists will need to determine whether they will retain their original surname (i.e., maiden name), hyphenate their married name onto their original surname, or change their surname to that of their partner. In addition, men may need to make similar considerations when they marry. Certainly, you must use your entire surname, however you have decided to present it, but your other names may vary. It is in your best interest to decide if you would like to use your entire given name or just the initial of your given name followed by a period. The same can be said about your middle name, if you have one, or you may not wish to use your middle name at all. Many cultures in the world do not bestow middle names on their children. In general, the preferred designator is the complete given name, followed by the initial of the middle name followed immediately by a period, followed by the complete surname (Day and Gastel, 2006). All this said, certain journals may require you to present your name in the way they want it, and you may not have a choice in the matter.

Several additional issues come into focus when multiple authorship must be considered:

- Senior authorship—who will be the first author?

- Should (must) certain administrators be listed in the byline, simply because they are in charge of the equipment used, the laboratory, department, college, or other unit where the research occurred. Likewise, we may ask if financial supporters should also be considered as coauthors.

- Should a graduate student have an automatic right of authorship on a manuscript in which the work contributed was not in conjunction with his or her thesis or dissertation? Should the graduate student be an author if the work contributed was in return for an assistantship that was separate from a thesis project? In the same manner, should a technician or technical assistant have an automatic right to authorship that has evolved from their daily work? In both cases, the issue is one of when does "work for hire" begin and end.

- If the research was a very large, long-term group project, who has done enough to warrant authorship?

- What is the right or privilege available to the person who has taken the lead in preparation of the manuscript? How does that amount of work correspond to the effort that has been extended by the person(s) conducting the research, analyzing and interpreting the data, or reviewing, revising, and editing the manuscript?

There really are no clear-cut answers to many of these issues (i.e., there is no system available that resolves these challenges in their entirety), and you should invest time and effort in deciding who will be coauthors. This is not a trivial concern, because decisions on authorship are intimately related to promotions, tenure, and ethics. When making authorship decisions, be sure you have not breached an ethical issue by erring to either the conservative side, by not including someone deserving as an author, or the liberal side, by granting authorship when not deserved. Please see Chapter 3 for more details about ethics and ethical issues related to authorship. Please complete the Determining the Privilege of Authorship exercise at the end of this chapter.

EXERCISE 4.1   Determining the Privilege of Authorship

Consider the manuscript you plan to prepare, and complete the grid below as a preliminary guide to authorship according to the scale proposed by Hunt (1991), but modified by us. Please use this scale to assign a point score for each of these activities to yourself, as the writer, as well as to each potential coauthor of your manuscript. The scale below is in five subdivisions, but any number may be allocated to a person in any activity, as long as the total for the activity has a sum of 20 points no matter how many potential coauthors there are. At this point, there may be many more potential coauthors than there will be when the manuscript has been completed. This entire process will be revisited after your manuscript has been completed to ensure your appraisals have not changed significantly during the writing process.

Start this exercise by assigning the initials of possible coauthors to each column, including the one for you. Please complete only as many columns as appropriate for your situation. If there are more than ten possible coauthors, then duplicate the grid and add the additional columns to make a grid that encompasses all possible coauthors. It is likely you will need at least two columns, as few units of research are completed by only one person. *The sum for each of the five activities across all potential coauthors must be a total of 20 points, so be sure your point allocations add up properly. The maximum each author could achieve is not bound the same way, except that it would be impossible, using the scale below, for one person to achieve more than 100 points.* In general, a possible coauthor should have contributed at least 10 to 15 points worth of effort to warrant authorship, especially if there are many potential coauthors. If there are fewer potential coauthors, say three or four, then a potential coauthor should accumulate at least 20 to 25 points worth of effort. If an individual accumulates less than 10 to 15 points, or 20 to 25 points when there are fewer potential coauthors, you must consider seriously whether or not this potential coauthor deserves to be accorded the privilege of coauthorship. Be sure you have not breached an ethical issue by erring to either the conservative or liberal side when making authorship decisions. Please see Chapter 3 for more details about ethics and ethical issues related to authorship.

Are you comfortable or uncomfortable with the results? Why? Let us assume you have completed this authorship grid, and essentially you are confident of, if not also pleased by, the results of this exercise. You now have a list of which individuals you should include as coauthors, and you now can list the order of the authors in the byline by ordering them into position as a result of the total score for each individual, from highest to lowest. Although the assignment of points for contributions to these activities is subjective, at least there is some level of objectivity to this decision, rather than it being completely subjective, and based on favor and grace.

Contribution	Points
None	0
Minor	1 to 6
Moderate	7 to 13
Major	14 to 19
Entire	20

Activity	Yourself	Potential Coauthor 1	Potential Coauthor 2	Potential Coauthor 3	Potential Coauthor 4
(Initials)	( )	( )	( )	( )	( )
Planning	_____	_____	_____	_____	_____
Executing	_____	_____	_____	_____	_____
Analyzing	_____	_____	_____	_____	_____

Interpreting	_____	_____	_____	_____	_____	
Writing	_____	_____	_____	_____	_____	
TOTAL	_____	_____	_____	_____	_____	

Activity	Potential Coauthor 5	Potential Coauthor 6	Potential Coauthor 7	Potential Coauthor 8	Potential Coauthor 9	Activity Total
(Initials)	( )	( )	( )	( )	( )	
Planning	_____	_____	_____	_____	_____	20
Executing	_____	_____	_____	_____	_____	20
Analyzing	_____	_____	_____	_____	_____	20
Interpreting	_____	_____	_____	_____	_____	20
Writing	_____	_____	_____	_____	_____	20
TOTAL	_____	_____	_____	_____	_____	100

Comments and musings: _____

_____

Are you comfortable/uncomfortable with the results? Why? _____

_____

List of coauthors as they should appear in the byline: _____

_____

REFERENCES

ANONYMOUS. 1999. Policy on papers' contributors. Nature 399:393.

CLAXTON, L.J. 2005. Scientific authorship. Part 2. History, recurring issues, practices, and guidelines. Mutat. Res. 589:31–45.

CRONIN, B. 2001. Hyperauthorship: A postmodern perversion or evidence of a structural shift in scholarly communication practices? J. Am. Soc. Inf. Technol. 52:558–569.

DAVIDOFF, F. 2000. Who's the author? Problems with biomedical authorship, and some possible solutions. Science Editor 23:111–119.

DAY, R.A. and GASTEL, B. 2006. *How to write and publish a scientific paper.* Sixth edition. Greenwood Press, Westport, CT.

ESA (ECOLOGICAL SOCIETY OF AMERICA). 2006. *Ecological Society of America Code of Ethics.* Adopted August 2000. www.esapubs.org/esapubs/ethics.htm. Accessed 29 January 2006.

FLANAGIN, A., CAREY, L.A., FONTANAROSA, P.B. et al. 1998. Prevalence of articles with honorary authors and ghost authors in peer-reviewed medical journals. J. Am. Med. Assoc. 280:222–224.

GALINDO-LEAL, C. 1996. Explicit authorship. Bull. Ecol. Soc. Am. 77:219–220.

GUSTAVII, B. 2003. *How to write and illustrate a scientific paper.* Cambridge University Press, New York, NY.

HUNT, R. 1991. Trying an authorship index. Nature 352:187.

HUTH, E.J. 1986. Guidelines on authorship of medical papers. Ann. Intern. Med. 104:269–274.

ICMJE (INTERNATIONAL COMMITTEE OF MEDICAL JOURNAL EDITORS). 1997. Uniform requirements for manuscripts submitted to biomedical journals. Ann. Intern. Med. 126:36–47.

ICMJE (INTERNATIONAL COMMITTEE OF MEDICAL JOURNAL EDITORS). 2005. Uniform requirements for manuscripts submitted to biomedical journals: Writing and editing for biomedical publication. www.icmje.org/index.html. Accessed 7 December 2005.

KLEIN, C.J. and MOSER-VEILLON, P.B. 1999. Authorship: Can you claim a byline? J. Am. Diet. Assoc. 99:77–79.

MATTHEWS, J.R., BOWEN, J.M. and MATTHEWS, R.W. 2000. *Successful scientific writing. A step-by-step guide for the biological and medical sciences*. Second edition. Cambridge University Press, New York, NY.

Nature. 2006. *Publication policies*. www.nature.com/nature/authors/policy/index.html. Accessed 29 January 2006.

O'CONNOR, M. 1991. *Writing successfully in science*. HarperCollins Academic, London, UK.

PEAT, J., ELLIOTT, E., BAUR, L. and KEENA, V. 2002. *Scientific writing. Easy when you know how*. BMJ Books, London, UK.

PNAS (PROCEEDINGS OF THE NATIONAL ACADEMY OF SCIENCES OF THE UNITED STATES OF AMERICA). 2011. *Information for authors*. www.pnas.org/site/misc/iforc.shtml. Acccessed 20 March 2011.

PARKER, R.A. and BERMAN, N.G. 1998. Criteria for authorship for statisticians in medical papers. Stat. Med. 17: 2289–2299.

REGALADO, A. 1995. Multiauthor papers on the rise. Science 268: 25.

RENNIE, D. and FLANAGIN, A. 1994. Authorship! Authorship! Guests, ghosts, grafters, and the two-sided coin. J. Am. Med. Assoc. 271:469–471.

RENNIE, D., YANK, V. and EMMANUEL, L. 1997. When authorship fails: A proposal to make contributors accountable. J. Am. Med. Assoc. 278:579–585.

RENNIE, D., FLANAGIN, A. and YANK, V. 2000. The contributions of authors. J. Am. Med. Assoc. 284: 89–91.

WELTZIN, J.F., BELOTE, R.T., WILLIAMS, L.T., KELLER, J.K. and ENGEL, E.C. 2006. Authorship in ecology: Attribution, accountability, and responsibility. Front. Ecol. Environ. 4:435–441.

WILCOX, L.J. 1998. The coin of the realm, the source of complaints. J. Am. Med. Assoc. 280:216–218.

CHAPTER 5

CHOOSING YOUR JOURNAL

I don't mind your thinking slowly, but I do mind your publishing faster than you think.

—Wolfgang Pauli

PRINCIPLES AND NEED

Before you write much of your manuscript, learn as much as possible about the journal you are targeting. It is best if the decision of where to submit the manuscript is made before writing commences. The key issue is one of matching the topic or content of the manuscript with the appropriate journal and its audience (Matthews et al., 2000). Often, perusing the list of resources you plan to cite in your literature review will give you an excellent idea of where similar research should be published. If you know to which journal, or at least two or three journals, you might submit the manuscript, then you can spend an hour or two perusing several of the most recent issues that have been published.

One goal of this perusal is to determine some of the basic requirements that this journal has listed in a document titled "Instructions to Authors." This document will usually be found in the first issue of each volume, sometimes in each issue, or better yet, on the website of the journal. In the specific case of biomedical journals (about 3500 titles), the editors of these journals have developed a common set of instructions titled the Uniform Requirements for Manuscripts Submitted to Biomedical Journals. This set of instructions is called the Vancouver Document and is available at ICMJE (1997) or www.icmje.org (ICMJE, 2005). Another goal of this perusal is to determine whether your journal of choice has a particular "personality." Sensitivity to this will help you write your manuscript in a style that better resembles how your manuscript will appear if it is published in that journal. Journal personality traits include the degree of speculation permitted, the depth of the review of literature, the depth of presentation of the materials and methods, and so on. When you discover this personality, next examine several recently published articles to determine how closely the content and results of your work match that of the prospective journal.

We cannot overemphasize how important it is to match your manuscript with the appropriate journal. Failure to do this will waste months of time in reviews that will not lead to publication. Once you find the appropriate journal, determine whether your methodologies, data interpretations, and so on, match those of the prospective journal. Also consider asking colleagues to render an opinion on where they think your manuscript should be submitted.

Getting Published in the Life Sciences, First Edition. By Richard J. Gladon, William R. Graves, and J. Michael Kelly
© 2011 Wiley-Blackwell. Published 2011 by John Wiley & Sons, Inc.

ITEMS TO CONSIDER IN CHOOSING YOUR JOURNAL

As you complete a course on writing a journal article manuscript or proceed through this book, you and the coauthors will need to make a decision about where your manuscript will be submitted for publication. Do this as soon as possible. You might ask yourself what factors should play a role in your decision of where, why, and how you will publish the paper. We present a list of possible factors you should consider before you make a final decision. Certainly, there may be other issues that may factor into your decision, but here we present at least a starting point. Answers to many of these questions are often found in the "Instructions to Authors" of the particular journal. The best place to look for those answers is in the statement(s) of purpose or the statement(s) of the scope of the journal.

Most disciplines have one to several flagship journals, followed by more journals that continue to separate the content of the discipline into smaller and smaller divisions, until we get to a journal that is especially designed for one small segment of the given discipline (the specialty journal). Most manuscripts are best suited for a specialty journal in the discipline associated with the scientists publishing the work. These journals are focused into a narrow spectrum of interest in a particular field. Often, a specialty journal will quickly return a misdirected manuscript to the author(s). The editor may not even review it if the content of the manuscript is inappropriate. Within a given discipline, there will also be levels of perceived quality of various journals. While quality can be defined quantitatively, it is nonetheless subjective. Strive to make your manuscript as strong as possible, regardless of where it is to be submitted.

Try and make the most educated decision about the choice of journal that you can, given the information you can access. The more closely the scientific content of your manuscript and its organization and style match the prospective journal, the more quickly your manuscript will be published. Conversely, if your journal of choice does not meet content, organization, or style considerations, you will probably waste several months and endure a lot of frustration before you get your manuscript resubmitted and published elsewhere. Please complete Exercises 5.1 and 5.2 after you have read and understood the information below.

Overall Quality or Prestige of the Journal

In one way or another, many of these things relate to the "quality" or "prestige" of the journal. This can be very difficult to determine, but it may be a very real issue for one or more of the coauthors. Here are some components that measure the quality or prestige of the journal where you will publish your manuscript:

- If you have narrowed your choice to several closely related journals, then one way to compare journals within a given discipline is to compare their impact factors (Garfield, 1999). The greater the impact factor, the higher the perceived quality of the journal. There may be dozens of journals in a given disciplinary area, and the impact factor quickly arranges them in descending order. The best papers in the discipline usually find their way into the journals of the greatest quality.

- Look at the Editorial Board/Associate Editors. Are they "giants" in their respective fields? Have they published for a long and productive career?

- What is the release/acceptance rate of the journal? Usually, good journals have a 50% or more release rate, and some may release closer to 70% of the manuscripts submitted to them. Certain journals, such as *Science* or *Nature*, have a release rate that is greater than 90%. Normally, journals with a high impact factor will have a high release rate.

Submission to a journal with a high release rate requires a calculated risk by the author(s), because fewer manuscripts get published.

- Who are the likely peer reviewers? Are they well known? Will they give your manuscript a thorough review that is constructive, or will they not take the job of reviewing seriously or provide only unwarranted criticism?

- What is the quality of the printing process? What is the quality of the paper used for printing the journal? Will it resist deterioration? What is the quality of the binding? Will pages become detached when someone spreads the issue out to make a photocopy? What is the quality of the reproductions of the photomicrographs? Are they too large or too small? Can a normal reader clearly see the needed detail such that they can interpret the information on the photomicrograph?

- Probably the most important of all of these is what is your and the other coauthors' personal opinion about the quality of the particular journal? Will you be proud of the fact that your research article has been published in that journal?

Reputation of the Journal

In certain disciplines, one journal is more highly regarded than another journal, and you should be aware of the standing of your journal within your discipline. As you publish more, this will become evident to you.

Scope of the Journal

You must be sure to check on the scope of your chosen journal, because it may waste several months of your time, and the time of the editorial staff of the journal, if you are submitting a mismatch. Most journals are quite specific, and the statement(s) of purpose or scope of the journal will help you understand what the particular journal wants to publish.

Similar Articles

If similar articles have been published previously in your journal of choice, then you probably have a good chance of not getting your article returned to you because it does not match the scope of the journal you chose. Often, this is an excellent benchmark to use in making your decision on the choice of a journal.

Type of Article

There are several types of articles that any journal may publish. The three types below should be considered representative of many journals, but other journals may have other ways of classifying articles. Some journals do not print all of these types of articles, and you must be aware of what type they do publish so that you do not expend time preparing the manuscript, only to discover the publisher does not publish a type that matches the article you have prepared. A *research note* or *brief communication* is a short article, usually about one printed page in length. Some journals have a word-count or character-count to restrict the article to only one page. This type is used for rapid communications of very novel procedures or discoveries, and the turn-around time for these articles can be very short. However, be aware that various journals define notes in slightly different ways, so it is best to check the instructions to authors to be certain that a note in your journal will achieve your desired objective(s). The regular *research report* article that is seen in most journals will average about three to five printed pages. Some journals now require longer, more complete, articles, reaching eight to ten printed pages, with many tables and figures in the article, and an extremely thorough treatment of the subject. Other journals will allow these longer articles, but they do not require them, and may not really want them. Often the largest article is a *review article*. These may comprise 20 or more printed pages. These articles are focused on one very narrow topic, and the article will delve into great detail about all aspects of what has been published about that subject. Two other types, much less used, are the case history and the case-series analyses (Matthews et al., 2000).

Type of Audience

The type of audience could probably be a facet of the scope of the journal. In many cases, each journal has a specific type of audience, such as molecular biologists, applied botanists, applied horticulturists, forest silviculturists, forest biometricians, horticultural physiologists, breeders, geneticists, and so on. Again, be sure before you submit that you have determined the type of audience so that you do not fall into the trap of a mismatch. For some authors, the size of the audience is very important, and most authors will trade away other issues to get the article into a journal that has wide circulation.

Accessibility by the Audience

Accessibility to the journal by the audience is an important factor, although it is becoming less so because more journals are becoming available online. More journals each year are becoming "open access journals" (Day and Gastel, 2006), in which the entire content of the journal is available, free of charge, online.

Personal Opinion

Sometimes, this is the only thing that you can use to help you make this decision. If you find yourself in this predicament, then you certainly need to do more background searching and ask more people about the quality of the journal before you make your final decision. Also, be absolutely sure that you are taking into account the opinion(s) of your coauthor(s), if there are any.

Membership Requirements

Some journals only allow publication if the article was first presented at the annual meeting or a conference of the society that publishes the journal. In addition, some journals may require that one or more authors be members of the society that is publishing the journal.

Page Charges and Reprint/Offprint Charges (and Who Will Pay Them)

Years ago, this was not an issue, because the institution where you worked absorbed these costs, in most cases. With the huge budget cuts some institutions have experienced, there has been a movement away from the institution paying the costs to publish the manuscript, and the researcher(s) must absorb the costs. More recently, researchers have made all forms of publication fees part of the grant request that funded the research. Page charges are usually $100 to $200 per printed page, and part thereof, and more can be charged for color photographs and other special features. Most articles will be about three or four printed pages, but sometimes there will be a part of a page or there will be more pages. Thus, the page charges could easily be $500 to $1000 for the article. The reprint or offprint charges are usually $100 to $300 for the first 100 units, usually with a reduced cost for additional reprints/offprints. Offprints must be ordered before the article is printed, and sometimes, the journal will make you pay for them before they are printed. Certain journals, usually those not published by a professional society, will have no page charges, and/or they have no reprint/offprint charges. Be sure you know this situation before you commit to something. Usually, when there are no charges for pages and/or reprints/offprints, the publisher will send only 25 or 50 reprints to you. Some journals do not have any reprints available to the authors, and the authors simply provide photocopies to those who request reprints. We would suggest that you do not get too many reprints because most people who want a copy of your article will simply photocopy it themselves or get it over the Internet. Many journals will waive, or reduce, page charges, and maybe reprint charges, to authors from third-world countries when it is known there are dire economic circumstances in the author's country. Authors from third-world countries should make this request upon submission so that time is not lost later, when the journal demands payment, or it will not publish the manuscript.

Time from Submission to Publication

This will vary with the journal. However, you should probably count on four to six months from submission to acceptance, and a total time of six to 15 months from submission to publication. This can be drawn out even more if the editor requires revisions and resubmissions. Most journals are trying to reduce this time lag, but publishing a paper is a long, drawn-out process. Remember, when you are preparing your dossier for a tenure or promotion process, a crisis situation on your part (read I should have submitted these manuscripts months ago) does not necessarily translate to a crisis situation on the part of the journal editors and reviewers. If you are considering a particular journal, you can estimate the time to publication by looking at the received for publication or accepted for publication date versus the time the article appeared in print for a sample of several articles. This may be a telling story.

The Review Process

This process varies with the journal, and it would be difficult to group journals that have similar procedures. Some journals have blind reviews, whereas others have open reviews in which the reviewer(s) know the author(s) of the article. In addition, the number of reviewers may vary from journal to journal. Some have an Associate Editor review the article along with one or two outside reviewers, and other journals have an Editorial Board that makes the final decision without seeking outside reviewers or maybe seeking one or two outside reviewers. The level of scrutiny by the reviewers also varies with the journal. Some journals use reviewers that are notorious for their intense scrutiny of the article, and other journals use reviewers that do not worry about more mundane issues and focus on the science the authors are trying to execute.

Format Considerations

Be sure the format you used in your manuscript is compatible with the journal to which you will submit the article. Some journals do not want to publish northern or Southern blots, and if your paper includes any of these, then you will have a mismatch. Format can also be related to how the journal structures the flow of the sections of the paper. Some journals now have the Materials and Methods section at the end of the article, and others may reduce the size of the font used for some part(s) of the article. Almost all journals will have a specific style used for their in-text citations and the listing of the resources in the References section of the manuscript. In most cases, journals have designated styles for the font of headings and subheadings, if they are allowed.

Miscellaneous

There are numerous issues that may come into consideration. Such items as the interest(s) of the readership, the total circulation, and the end use (individual subscribers versus libraries only) are important here. In addition, some journals "separate themselves" from other journals by citing the number of times writers cite articles that journal has published, the "impact factor" of the journal (Garfield, 1999).

EXERCISE 5.1   Narrowing Your Journal Choices

Instructions:

You may need to go to the library or go online to complete this exercise, as many questions cannot be answered without talking to someone who knows about the journals in your field or obtaining that information yourself. To answer each question below, list about five journals that meet the criteria of each question. In some cases, there may not be five journals that will meet the criteria, and in that case, only complete the question to that extent. The concept of this exercise originated in O'Connor (1991), and all these questions came from that source.

1. Which journals are read most frequently by the target audience of your manuscript?

    A.

    B.

    C.

    D.

    E.

2. Which journals publish articles of the type and length you wish to publish?

    A.

    B.

    C.

    D.

    E.

3. Which journals are the best established in the field of your manuscript?

    A.

    B.

    C.

    D.

    E.

4. Which journals in your field publish high-quality articles (scientifically) but have a moderately low (about 30% of submissions) rejection rate?

    A.

    B.

    C.

    D.

    E.

5. Which journals have the best Editors and Associate Editors in your field?

    A.

    B.

    C.

    D.

    E.

6. Which journals provide the most prompt, fair, and helpful review process?

    A.

    B.

    C.

    D.

    E.

7. Which journals are published often enough that your paper may be out in printed form within about six to nine months from initial submission?

A.

B.

C.

D.

E.

8. Which journals print the highest-quality photographic artwork, if that is necessary for your manuscript?

A.

B.

C.

D.

E.

9. Which journals use a literature citation system that is compatible with the system you are using?

A.

B.

C.

D.

E.

10. List any journals *not covered* by the main abstracting and title-listing services.

A.

B.

C.

D.

E.

11. List any journals that require you, or at least one author, to be a member of its parent society in order to publish in its journals.

A.

B.

C.

D.

E.

12. List any journals that have page charges and you do not have the resources to meet that commitment.

A.

B.

C.

D.

E.

13. List the journals that have reprint or offprint charges and you do not have the resources to meet that commitment.

A.

    B.

    C.

    D.

    E.

14. Do any of your journal choices require payment of a submission fee just to submit the manuscript for publication?

15. Do all of your journal choices provide enough reprints or offprints to meet the needs of your coauthors?

EXERCISE 5.2 Knowing and Choosing Your Journal

Now that you have read through the information that presents and discusses the factors that might influence your choice of a journal for publishing your research findings, it is time to put that information into practice. Please answer these questions *briefly*. You may enlist the help of your major professor(s) and the other coauthors to complete this exercise, if you so desire. Or, you may want to work through this exercise yourself first, and then work with these other coauthors on a second edition of it.

1. List five (5) factors that have influenced, or you think might influence, your decision of the most appropriate journal *for your particular manuscript*. List them in decreasing order of importance (A most important and E least important).

   A.

   B.

   C.

   D.

   E.

2. List the complete name of five (5) journals you think could be the most appropriate journal *for your particular manuscript*. List them in decreasing order of appropriateness (A most appropriate and E least appropriate).

   A.

   B.

   C.

   D.

   E.

3. Either go to the library, or go on the Internet, and obtain copies of the "Instructions to Authors," or a similarly titled document, for each of these five (5) journals that are options for your manuscript. As you read through these documents, circle or highlight with a marker key items that relate directly to your choice, or rejection, of that given journal as your vehicle for publishing your manuscript. For each item you circled or highlighted, place a note next to it indicating whether it was a factor for choosing or rejecting the particular journal.

4. Name the journal that was your first choice:

5. In one to two sentences, justify to yourself and your other coauthors why you have chosen this particular journal for your manuscript.

6. For each of these items below, characterize *in one sentence your perception* of the journal you have chosen. This exercise has functioned as a first attempt at choosing your journal, but you should not be surprised if it changes one or more times before you submit your manuscript.

   A. Reputation (relative to other possible journals):

   B. Audience:

   C. Overall quality:

   D. Review policies:

   E. Circulation:

REFERENCES

DAY, R.A. and GASTEL, B. 2006. *How to write and publish a scientific paper*. Sixth edition. Greenwood Press, Westport, CT.

GARFIELD, E. 1999. Journal impact factor: A brief review. Can. Med. Assoc. J. 161:979–980.

ICMJE (INTERNATIONAL COMMITTEE OF MEDICAL JOURNAL EDITORS). 1997. Uniform requirements for manuscripts submitted to biomedical journals. Ann. Intern. Med. 126:36–47.

ICMJE (INTERNATIONAL COMMITTEE OF MEDICAL JOURNAL EDITORS). 2005. Uniform requirements for manuscripts submitted to biomedical journals: Writing and editing for biomedical publication. www.icmje.org/index.html. Accessed 7 December 2005.

MATTHEWS, J.R., BOWEN, J.M. and MATTHEWS, R.W. 2000. *Successful scientific writing. A step-by-step guide for the biological sciences*. Second edition. Cambridge University Press, New York, NY.

O'CONNOR, M. 1991. Writing successfully in science. HarperCollinsAcademic, London, UK.

CHAPTER 6

PRINCIPLES AND CHARACTERISTICS OF GOOD SCIENTIFIC WRITING

A man (woman) of true science uses but few hard words, and those only when none other will answer his (her) purpose; whereas the smatterer in science thinks that by mouthing hard words he (she) proves that he (she) understands hard things.

—Herman Melville

DEFINITION AND NEED

Writing manuscripts for refereed scientific journals (a.k.a., scientific writing) is not the same as science writing, nor is it the same as creative writing. Day (1995) has an excellent discourse on these differences. Scientific writing is done by scientists for an audience of scientists, and the desired result is the unequivocal transfer of information from the writer, one scientist, to the reader, another scientist. On the other hand, science writing may be done by scientists, journalists, or nonscientists, but the purpose of the writing is generally to transfer information about science to an audience of nonscientists (Day, 1995). This is an extremely important distinction, and the "vocabulary, tone, and complexity" differ dramatically between the two types of writing (Day, 1995). Creative writing is useful for literature, but the basic principles of creative writing are mutually exclusive to scientific writing. Scientific writing keeps the focus on newly discovered data and the analysis, interpretation, meaning, and impact of those data on our base of knowledge. In contrast, creative writing might emphasize the novel use of the English language, which is rare in scientific writing.

Three basic purposes of scientific writing are important to remember, because they underpin the work of scientists who write (Day, 1995). The first purpose of scientific writing, and therefore, scientific publications, is to record what was done, and get that information into an archive (i.e., a scientific research journal). The second purpose is to inform our peers of what has been done, which occurs as they read scientific research journals. The third purpose is to educate future generations of scientists about what has been done previously so they do not repeat previous work unnecessarily.

Our goal in this chapter is to introduce basic principles and characteristics of how to use English in simple ways so that language use is not what draws attention to your work. Your writing will be better received if the data and their interpretation, meaning, and impact draw the attention of the reader, rather than your English usage. If your manuscript is written poorly, the reviewer will not understand it, and you will have almost no chance of getting the manuscript published. Reviewers or referees should be free to adjudicate the scientific content of the manuscript, without being burdened with the challenges caused by poor use of English. Reviewers should not accept a poorly written manuscript, because

Getting Published in the Life Sciences, First Edition. By Richard J. Gladon, William R. Graves, and J. Michael Kelly
© 2011 Wiley-Blackwell. Published 2011 by John Wiley & Sons, Inc.

there is a chance the poor use of language will lead to misinterpretations during the review. From our experience in writing scientific articles, we have identified some broadly based principles and characteristics of good scientific writing, and we will describe them below. Adherence to these principles and incorporation of these characteristics into your writing will help you over the hurdle of appropriate English usage. In one of the chapters of Luellen's (2001) book, he presents an excellent discourse on what makes good writing and a well written manuscript. The take-home message from that chapter is that writing moves from poor to fair to good to excellent as fewer of the essential components of the manuscript are no longer an issue—whether it is data presentation, interpretation of the data, presenting how the work was completed, the grammar, the punctuation, and so on.

PRINCIPLES TO CONSIDER

Make Time for Writing

Yes, this really is a principle of good writing! Successful scientific writers designate time to write regularly in discrete sessions, not as a marathon. The author knows to devote a certain amount of time and energy each day to producing written work, and it stays on the writer's daily docket. This work time for writing can come in several forms. As you develop into a mature writer, you will find your correct mechanism for providing this writing interval. We recommend you start by first deciding you will buy into this concept. If you do, then you should choose one of these mechanisms for creating a period for writing as a starting point until you can determine what works for you. This daily period devoted to writing prevents marathon writing sessions, which are often less focused and, therefore, unproductive.

There are several ways a writer can develop the structural basis for their daily writing period. Some authors daily set aside a one- to three-hour block in which the only thing they do is write. For some, this is first thing in the morning, sometimes only at a home office so there are fewer distractions. Others go to the library or close their office door to the public so that distractions are minimized. These writers designate their writing period in the form of time. Other authors designate their writing period by word count. For instance, they have a goal to write 250, 500, 750, or 1000 words each day, and as soon as they reach that limit, they stop writing. Writing 1000 words for a scientific publication takes more time than you might imagine, so start with a more modest goal if you are new to writing. Editing and polishing that written piece is done at a different time or on another day. Perhaps a more common approach is to write some previously designated and consistent number of pages of manuscript daily. Parini (2005) has proposed this as his "two pages a day" concept. Parini, just like the great author Ernest Hemingway, always used as a stopping point a place where he knew exactly what he would begin writing the next day. In this way, he eliminated one of many writers' greatest blocks toward getting started—knowing what is the starting point and where you will go with it. Two pages per day will correspond to about 500 words, as most manuscripts written double-spaced with normal margins will be between 225 and 250 words per page. For many writers, this works well because they see progress daily; it is not an over-burdening marathon session and, best of all, they can readily give themselves permission to stop at this limit because they know they will resume their writing the next day at a known point. Let us do some math here to illustrate why this works. There are about 230 working days (Monday through Friday) per year. If we allow for 10 holidays, five sick days, and 15 vacation days, we are left with about 200 working days. If we do two pages a day, we produce $200 \times 2 = 400$ pages of text per year. This could be a small book, about 15 to 20 journal manuscripts, or some combination thereof.

Know Your Limit—and Know the Cause of the Limit

Virtually all writers endure phases in which they are unproductive. When you have become unproductive, by all means, stop trying to force yourself to write. If you continue to push yourself, you will gain nothing except the scars of frustration and a loss of confidence in your writing abilities. We all know how we have been told repeatedly throughout our schooling not to let writing projects go until the last minute. This is not a 500- or 1000-word theme for English composition that you can manufacture in one evening and part of the night. This just does not work well for someone trying to write a larger document, especially one as complicated as a manuscript for a refereed journal. It is important that you know yourself well enough that you can determine what is causing the stoppage, and then do something about it.

Some individuals get fatigued as they write. The fatigue may be associated with the use of a computer and looking at its screen. It may also be a simple problem of trying to work at the task at hand for too long, and everything runs together. In these instances, you may return to your work on the manuscript after a short period of time, say one or two hours. Your outlook toward the task at hand will improve immensely, and productive writing will resume. Another problem may be that you are simply tired of dealing with the set of facts associated with one section of the manuscript. In this case, try switching to another section that needs attention, and when you feel sharper, return to the section you left earlier.

On the other hand, some individuals suffer from various forms of "writer's blocks" (Boice, 1990, 1996; Boice and Jones, 1984). Actually, there is some debate as to whether writer's block really occurs, as opposed to rampant excuse-making to justify why progress has not been made (Matthews et al., 2000). Matthews and colleagues (2000) present some solutions that should be considered by writers experiencing problems moving forward. These individuals must take more drastic measures to overcome these issues of blockage. If, after you have prepared several manuscripts, and maybe have published a few, you still have a lot of problems getting started on your writing, then you might consult some of these resources to begin to work at overcoming these blocks to writing. Many a professional or academic career has been ruined by undiagnosed writer's blocks that have not been cured in time for a promotion decision (Boice, 1991; Boice and Jones, 1984).

Use Reference Books Liberally

Before you begin writing, you should equip yourself with several reference books that will be important to your ability to write well. Keep a good dictionary close as you work. Almost any dictionary, including a small, pocket dictionary, should meet your needs. However, as you may need to clarify your understanding of a word of greater complexity so that you are using it correctly, we might suggest a desk-size dictionary (Mish, 2004). In all of your writing, strive to use the simplest word possible that does not allow the reader to misunderstand what you mean by your use of a given word. This discourse reminds us of a famous quotation from Sir Winston Churchill:

> "Old words are best, and old words when short are best of all."

A second reference book is the style manual developed by your profession, your professional society, or the publishing house where your manuscript will be submitted, if they have prepared one. Many scientific societies have style manuals. These manuals are more detailed than the "Instructions to Authors" documents you found to complete the "Knowing and Choosing Your Journal" exercise in Chapter 5. These "Instructions to

Authors" documents are excellent starting points for the writer, but they do not contain much detail about what the journal editors expect from authors in regard to style. Style manuals will cover such items as how to format tables, what to include in figure captions, what style of in-text citation should be used, what style should be used for listing the resources in the references section, and so on. However, because the "Instructions to Authors" is a condensed version of the most important items to help guide the author, it should be read thoroughly, and maybe often, once the author(s) have selected which journal will receive the manuscript submission. Certainly, it should be read with great attention before any work toward production of the manuscript has begun.

Another excellent resource is the Council of Science Editors (CSE) Style Manual (CSE, 2006). The CSE Style Manual transcends manuals of a profession, a professional society, or a journal-publishing house. It has been written and edited by professionals who edit journals throughout all of the natural sciences. This massive tome contains an answer to nearly every question an author may have about their manuscript, but it is too mundane and too detailed for the author to read all of it. However, like a dictionary, when a particular item needs to be queried, the answer almost always will be there.

The last reference item needed is another book on style, but this time it is a book on style in the English language, rather than a specific profession or journal. Writers have several choices in this area, each of which may fill a specific need of the writer. When information about more advanced writing in English is needed, it is hard to beat *The Elements of Style* (Strunk and White, 2000). This small book used by professional writers is focused on the items necessary for the writer to achieve a high quality document. There are also several books used by teachers of English composition that are quite good. At least one of them should be in the library of every writer. *The Little, Brown Handbook* (Fowler, 1986) and *The St. Martin's Handbook* (Lunsford, 2003) are excellent resources. Another useful resource is a book focused on English word usage for scientific writing (Day, 1995). This book clarifies basic principles for written English, in terms of style, grammar, and structure, all in the context of a scientific manuscript. For scientists who have published several articles and want to develop their skills more thoroughly, we suggest Luellen's book on fine-tuning your writing (Luellen, 2001). This book is extremely valuable, but it should not be used until the writer has mastered several basic skills needed for the preparation of a manuscript.

Develop a Purpose and Convince Your Readers to Believe It—Be Persuasive

Develop a strong set of take-home messages that you can defend with strong data. Your messages should lead to an overarching purpose or theme for your manuscript. Make clear in the various sections of your manuscript the need for the work you have done, the methods used to accomplish your goals, your results, and the meaning and impact of those results. Like a successful politician running for office, keep on message, be convincing, and be persuasive. Maintaining a consistent sense of purpose throughout your manuscript is an element of good scientific writing that fosters success.

Appropriate Structural Format of the Manuscript (IMRAD)

For eons, humankind has been recording events and observations by nonscientists, whose records were not structured as researchers now write. Early scientists recorded their observations as they saw them, but neither the scientific method nor a structural format for reporting these items was available to them. The first scientific journals appeared in 1665, about 350 years ago, but scientists recording these events still did not have a structured format for reporting their observations (Day, 1995).

During the nineteenth century, the eminent scientist Louis Pasteur began to conduct experiments in microbiology and life sciences. Some of his incredible results (e.g., germ theory for explaining human diseases) seemed so far-fetched to the other scientists of that period that they wanted to duplicate Pasteur's studies so that they could be sure of his results. Over a period of many years, the scientific method evolved, so that these imitating scientists knew how things were done previously, and could perform the same experiments. A structural format evolved that allowed subsequent scientists to read, understand, and confirm the work of others. Thus was born the system for formatting reports of scientific investigations, and today we call this structural format the IMRAD system. This IMRAD system for reporting research consists of **I**ntroduction, **M**aterials and methods, **R**esults, **A**nd **D**iscussion. You must know your journal so that you can conform to the structure of that journal, if it does not use the IMRAD system. Several disciplines in the natural sciences and social sciences use alternate systems.

CHARACTERISTICS TO CONSIDER

Organization

This subsection of the characteristics to consider is so important it warrants its own quotation:

> "Scientific writing is primarily an exercise in organization. A scientific paper is not literature."
>
> —Robert A. Day

Scientific journals force a certain degree of organization on you by way of the section headings they use, that is, the IMRAD system for structurally formatting the manuscript. However, within those sections, the burden rests on the writer to keep the reader oriented. The key issues with organization are to present information in a logical order and group similar ideas and topics. The writer must organize the flow of information for the reader, because the reader does not know what the writer will be telling the reader next, and it is the responsibility of the writer to remember that the purpose of publishing the work is to inform present and future scientists working in the discipline. Chapter 8 on Organizational Outlining will aid your development of a path for writing the manuscript. The writer must also make smooth transitions between topics, and good topic sentences must be constructed and used to start each paragraph. The remaining sentences within the paragraph are there to define and support the information delivered in the topic sentence. Those sentences should be complete enough that the argument you are presenting is developed completely and unequivocally to a logical conclusion. Each paragraph must evolve from the general statement made in the topic sentence to the specific facts that define and support the information in the topic sentence. This can be done almost seamlessly by use of parallel structure of sentences and statements within each paragraph. Thus, the writer must use appropriately structured paragraphs to develop a logical flow of events that the writer wishes to report. The writer needs to develop a clear progression track that takes the reader fluidly from newly discovered knowledge into what is known, without the reader realizing the transformation has taken place (Lebrun, 2007). Many journals permit authors to use subheadings within the major sections of the manuscript. The development and use of concise and logical sub-headings can be a simple way the author can stay organized, and in the end, keep the reader oriented throughout the entire manuscript. Appropriate organization of the manuscript is probably the key ingredient in getting the work published, and it remains the responsibility of the author(s) to ensure it occurs. Use headings, subheadings, topic sentences, an organized flow of words, and other organizational tools to follow the formula of many an orator: Tell them what you are going to tell them, tell them, and then tell them what you told them.

Grammar, Syntax, and Punctuation Matter

It is important to understand that proper use of grammar, syntax, punctuation, and punctuation marks is a critical component of the manuscript. Most editors, associate editors, and reviewers will return a manuscript without completing a proper review if the manuscript is fraught with spelling, grammatical, or punctuation errors. If you are having problems with grammar and syntax issues, you should consult some of the reference books we discussed in an earlier part of this chapter. Alternately, you might enlist a writing specialist who can help you with these problems. Issues with punctuation will probably be resolved more easily than those with grammar or syntax. Luellen's (2001) and Kirkman's (2006) books are excellent resources for proper use and avoidance of abuse of the principles of good grammar, proper syntax, and punctuation.

Accuracy and precision

Accuracy is defined as a measure of the truth associated with an event, for example, how close the arrow is to the bullseye of the target. Precision is defined as a measure of the repeatability of the event. Thus, if an archer manages to shoot four arrows within one centimeter of one another, that archer could be called very precise. On the other hand, if all four of those arrows landed within one centimeter at the outside edge of the target, the archer would be very precise, but not very accurate. This analogy can be extended to writing and word usage. Accuracy results from the proper choice of words assembled accurately and precisely to convey only the intended meaning. Here is an important quotation to remember:

> "Do not write so that you can be understood, write so that you cannot be misunderstood."
>
> —Epictatus

You should check all sentences to ensure they cannot be misconstrued. We suggest that you strive to be specific rather than general, definite rather than vague, and compose concrete rather than abstract expressions. It is more meaningful, and therefore more informative, to say "air temperature in the incubator was 23 to 28 °C" than it would be to say "temperature in the incubator generally was maintained within an acceptable range."

Correct Spelling and Correct Word Usage—Appropriate Language

Every word in your manuscript should be checked against a spell checker, which is probably a component of the software you use. When the spellings do not match, some mechanism to inform you that they do not match alerts you to that fact. Sometimes words spelled backwards (or palindromes) are correct spellings, and the software does not interpret them as words that have been spelled incorrectly in the intended context (e.g., me versus em [a printer's measure]). *There is no excuse for the presence of misspelled words in your manuscript.*

In addition, be sure that the words you have chosen carry the correct meaning. Words such as insure versus ensure, there versus their, and many others have specific meanings that are not interchangeable. Some spell-checking functions in your software package will do this for you, but you cannot count on the software to give you unequivocal usage. If you have any doubt about the correctness of your word usage, then look up the definitions of the word in a dictionary. *As with spelling, there is no excuse for the presence of incorrectly used words in your manuscript.*

Clarity

It is your duty to write about your science clearly, unequivocally, and succinctly. It is not the duty of the reader to resolve confusing text. Introductions should include a succinct statement of the purpose for conducting the research to orient the reader from the very beginning. Sentences throughout the manuscript should be simple and declarative to minimize the risk of inaccuracy or confusion. Simplicity in writing is a direct sign of clarity in thinking during the research work and an immeasurable reflection of the depth of thought and understanding of the author(s). Words that have multiple meanings should not be used in scientific writing. Likewise, words that have more than one spelling should be used only when no other word is available. Be clear in all you state. If a reader is presented with unclear writing, he or she may suspect that the researcher was unclear about what they did and how they did it.

Recall the words of Confuscious:

"In language, clarity is everything."

Economy of words

Only after you have produced a manuscript that leaves the reader with only one possible interpretation should you begin to work on economy of words. Economy of words is important, but economy must not be achieved at the expense of either the organization or the clarity of the manuscript. Economy of words can take on many forms. Uneconomical word usage comes from unnecessary word usage ("outside of" versus "outside"), stacked modifiers, dangling modifiers, inappropriate word usage ("previous to" or "prior to" instead of "before"), and other vagaries of the English language. All these extra words can act to obscure your message, and that is not good. Inexperienced writers should not dwell on economy early in the production of a manuscript. You have more important issues to resolve with preparing a rough draft of the manuscript. Chapter 20, "Polishing Your Writing with Good Word Usage," should help you reduce your word count without loss of meaning. Chapter 20 will also address such topics as use of the active voice rather than passive voice, syntax (word order), choice of words, and use of informative, rather than indicative, words, phrases, and statements, among many other items. These topics are important but are understood and appreciated better when you have a strong draft of your manuscript ready to polish.

Writing Objectively

Scientists must write objectively. Be sure that you present all sides of every argument, so that the reader can critically evaluate the strengths and weaknesses in your statements. In addition, do not extend your arguments beyond the facts available. When subjectivity creeps into the writing, a disservice is done to future generations of scientists who may not have known the writer was not relaying an unbiased report of his or her activities. It is important that writers neither overestimate the importance of their own work nor allow themselves to let their hypotheses get in the way of their data. The best scientific writing will be that which balances the need for objectivity with enough persuasion to show the reader the information in the manuscript is credible.

Writing with Consistency

Authors should strive for consistency in all aspects of production of the manuscript. This includes such items as spelling consistently (e.g., colour versus color) and ensuring that the format for spelling is consistent with what is required by the journal in which the manuscript is to be submitted. Another form of consistency is in the structure of the sentences, paragraphs, and sections of the manuscript. Parallel structure should be used throughout the manuscript, as once the reader adjusts to the parallel structure used by the author, the remainder of the manuscript almost unfolds by itself as the reader progresses through the sections of the article. Additional forms of consistency may be focused on capitalization issues, proper grammar issues, and proper punctuation issues.

Make your Writing Fluid

Let us start this section with a quotation that is extremely meaningful for writers who try and make their writing seamless in the eyes of the reader. It comes from one of the greatest authors of American literature.

> "The greatest possible merit of style is, of course, to make the words absolutely disappear into the thought."
>
> —Nathaniel Hawthorne

There are several things you can do to make your writing more fluid, so that the reader seamlessly moves from sentence to sentence, paragraph to paragraph, and section to section of your manuscript. Alley (1996) has an interesting chapter on making your writing fluid, and we will summarize that information here. There are two major ways in which less fluid writing can be transformed into more fluid writing. The first major way is for the writer to find ways to vary the rhythm of the piece so that the material presented does not become boring or monotonous. All written works have a rhythm, whether it was or was not planned by the author. As the reader becomes more and more bored with the written words, the speed of reading and comprehension decrease irretrievably.

The first way to change the rhythm is to vary the beginning of each of the sentences. This can be done by starting sentences with one of the following: a simple subject and verb, a prepositional phrase, transition words, an introductory clause, an infinitive phrase, a participial phrase, or in some specific cases, a question. These, and other openers not addressed here, can be varied as needed to make the reading flow. The second way is to vary sentence length. Most sentences should have a length of 20 words or less, with an optimum of around 12 to 15 words. You should try and vary the length, but stay in the teens, change the length every two or three sentences, and occasionally use a short (less than 10 words) or long (greater than 20 to 25 words) sentence. Be cautious in using many short sentences in a row, as that makes the writing seem choppy. The third way to change rhythm is to vary sentence structure. There are three basic sentence structures, the simple sentence, the compound sentence, and the complex sentence. The simple sentence is usually a declarative sentence, and it has one independent clause. The compound sentence contains two, or maybe more, independent, but related, clauses linked by a conjunction such as or, and, or but. The complex sentence contains one independent clause and one, two, or rarely more, dependent clauses. The fourth way to vary rhythm is to vary paragraph length. This has little to do directly with either the number of words in the sentences or the number of sentences in the paragraphs. It has to do with the number of lines there are in each paragraph when the piece is printed. When paragraphs are consistently short, less than about seven lines, there is too much stop and go to the reading, and the reader becomes bored. Likewise, when the paragraphs are too long, more than about 14 lines, the paragraphs are too long, and this also leads to boredom, because of there being too much to digest, as well as intimidation issues for the reader. Thus, most paragraphs should be between seven and fourteen lines long, with an occasional long and/or short paragraph between.

The second major way to make the writing fluid is to eliminate discontinuities. Basically, these are the transitions between sentences and, separately, between paragraphs. A paragraph is a grouping of related ideas in the form of sentences. There are three ways in which discontinuities can be eliminated. The first way is to not permit the transition from one idea (one sentence) to the next idea (another sentence) to get separated by too many sentences in between. The linkage between these ideas should be in close proximity, and not the 20, 30, or 40 words of several intervening sentences. The transitional words used by writers fall into three categories. (i) Continuations continue to move ideas in the same direction, and they are represented by words such as "also" and "moreover," and numerical ordering (like these examples). (ii) Pauses will stop the reader for a short period of time only, and these are represented by "for instance" and "for example." (iii) Reversals turn the flow around, and they are represented by such words as "however" and "conversely." The second way is to eliminate needless typography. Needless typography is characterized by abbreviations, capitalizations, numerals within sentences, and other things. These items are discontinuations because they intimidate readers and cause them to stop reading and concentrate on the item. However, much scientific writing cannot exist without the use of numbers and abbreviations, so the writer must strike some form of balance between these items. Abbreviations are particularly troublesome because many of them end with a period, and the reader must stop to determine whether or not they have come to the end of a sentence (i.e., the idea) or the end of an abbreviation. Capitalization is a useful tool if the item described is common, such as ATP, DNA, RNA, or other common chemicals. This problem arises because people do not read each letter in a word, and when all the letters are capitalized, the reader becomes confused if the word is not recognized readily. Actually, most readers recognize a word not only by its letters, but also by its shape, which is now distorted. The third form of needless typography is the placement of numerals in the text. Most readers come to a complete stop when they see a numeral, because they are sure the number has great significance, even when it does not. This third way also includes the use and misuse of equations within the text. At times, this is an excellent way to get information across to the reader, but it will occur at the expense of a complete stop by the reader, and the writer must take this into account. Conversely, it may take numerous words to get across the same point that an equation makes in less than one line of text. The mathematical content of the equation must be unequivocal, or all is lost.

Make Your Writing Interesting

Although we have stressed in this book that your writing should be well organized, clear, and economical, that does not mean you must develop a style that is "stuffy." Under the correct circumstances, a description of an interesting, and maybe even humorous, way the development of the research occurred may be quite timely in presenting your work. However, writers must be careful they do not sacrifice the clarity and understanding of the message for the sake of adding a little levity to the manuscript. Through use of tables and visual items such as photographs, drawings, and figures of data, make the manuscript attractive to the eyes of the reader.

EPILOG

Through understanding and using these principles and characteristics of good scientific writing, you should be able to give your manuscript a greater chance to get published with fewer rejections and editing issues. These principles and characteristics will reappear in many places throughout this book. It is good for the writer to look for how these are applied in various forms while reading the chapters of this book and completing the exercises in the back of several of the chapters.

REFERENCES

ALLEY, M. 1996. *The craft of scientific writing.* Third edition. Springer-Verlag New York, Inc. New York, NY.

BOICE, R. 1990. *Professors as writers. A self-help guide to productive writing.* New Forums Press, Stillwater, OK.

BOICE, R. 1991. Quick starters: New faculty who succeed. p. 111–121. *In* M. THEALL and J. FRANKLIN (Eds). *Effective practices for improving teaching. New directions for teaching and learning.* No. 48. Jossey-Bass Publishers, San Francisco, CA.

BOICE, R. 1996. *Procrastination and blocking. A novel approach.* Praeger Publishers, Westport, CT.

BOICE, R. and JONES, F. 1984. Why academicians don't write. J. Higher Educ. 55:567–582.

[CSE] COUNCIL OF SCIENCE EDITORS, STYLE MANUAL COMMITTEE. 2006. *Scientific style and format: the CSE manual for authors, editors, and publishers.* Seventh edition. The Council, Reston, VA.

DAY, R.A. 1995. *Scientific English: A guide for scientists and other professionals.* Second edition. The Oryx Press, Phoenix, AZ.

FOWLER, H.R. 1986. *The Little, Brown handbook.* Third edition. Little, Brown and Co, Boston, MA.

KIRKMAN, J. 2006. *Punctuation matters.* Fourth edition. Routledge, New York, NY.

LEBRUN, J.-L. 2007. *Scientific writing. A reader and writer's guide.* World Scientific Publishing Co., Singapore.

LUELLEN, W.R. 2001. *Fine-tuning your writing.* Wise Owl Publishing Co., Madison, WI.

LUNSFORD, A.A. 2003. *The St. Martin's Handbook.* Fifth edition. Bedford/St. Martin's. Boston, MA.

MATTHEWS, J.R., BOWEN, J.M. and MATTHEWS, R.W. 2000. *Successful scientific writing. A step-by-step guide for the biological and medical sciences.* Second edition. Cambridge University Press, New York, NY.

MISH, F.C. 2004. *Merriam-Webster's Collegiate Dictionary.* Eleventh edition. Merriam-Webster, Inc., Springfield, MA.

PARINI, J. 8 April 2005. The considerable satisfaction of 2 pages a day. The Chronicle of Higher Education 51(31): 31. http://chronicle.com/weekly/v51/i31/31b.

STRUNK, JR, W. and WHITE, E.B. 2000. *The elements of style.* Fourth edition. Allyn and Bacon, Boston, MA.

PART II

DEVELOPMENT OF THE MANUSCRIPT

CHAPTER 7

DEVELOPING THE TAKE-HOME MESSAGES AND A PROVISIONAL TITLE

Everything should be as simple as it can be, yet no simpler.

—Albert Einstein

DEFINITION AND NEED

The take-home messages are the central ideas, or pieces of information, you want the reader to understand unequivocally by the time they have finished reading your published article. Most published journal articles are about 1500 to 7000 words in length. It is important for you as an author to remember that the readers of your manuscript will retain only a few select ideas from the mass of text you write. Thus, the only chance the author really has to get information across to the reader is in the form of take-home messages that essentially "hit the reader between the eyes." The writer must craft the presentation of these relatively few words, sentences, and ideas to make unequivocal sense to the reader, or the discoveries made by the researcher will be lost. These take-home messages will surface throughout your manuscript, and your goal in writing your manuscript is to get each one across to the reader with no chance for misinterpretation.

The title is an accurate and economical description of the entire contents of your manuscript. The title is extremely important and necessary because it "pushes" the interested reader to continue her or his next step in their perusal of the published article. An appropriately interesting title may also lure an ambivalent reader into reading the article, thereby allowing your messages to reach an ever-broader audience. Because the title is such an important part of the manuscript, we are having you start on the development of a good title early in the writing process so that you have much time to revise and refine it. Your first attempt at developing a title is your formulation of a provisional title for use in guiding your writing of the first few drafts of the manuscript. At a later point, the provisional title will be revised into the final title of the article. Do not be surprised if the final title on your submitted manuscript does not look anything like your provisional title developed in this chapter.

Getting Published in the Life Sciences, First Edition. By Richard J. Gladon, William R. Graves, and J. Michael Kelly
© 2011 Wiley-Blackwell. Published 2011 by John Wiley & Sons, Inc.

PRINCIPLES BEHIND AND DEVELOPMENT OF TAKE-HOME MESSAGES

Principles Behind the Take-Home Messages

After a reader peruses the title of your article, he or she will make a decision about how much more of the article will receive his or her attention. If you have been successful in convincing someone to read further, then the abstract will probably be read next. If these two sections of the article have piqued interest satisfactorily, then the tables and figures will be reviewed. Finally, all or parts of your entire published article will be read. When a reader has completed perusal of the article, it will be pushed aside. It is then the reader will reflect and discern, consciously or unconsciously, what was learned. As a writer, the most important thing you can do for the readers of your article is to provide them with two, three, or four of what we call take-home messages. These should be presented in a manner that makes them unequivocal in the reader's understanding of the contents of the article. Thus, the take-home messages are really the central messages or ideas you want the reader, every reader, to grasp at the time they have finished reading the article. In some instances these two to four individual take-home messages may be interrelated to such a degree that they may be woven together into an over-arching message that represents the essence of the research project in one simple, straightforward idea. An individual take-home message or an over-arching combination of several take-home messages may be called a conclusion coming from the work. These over-arching take-home messages or the more simple, single take-home message will then become the focal points of several sections of the manuscript. They will appear prominently throughout the body of the manuscript.

Development of the Take-Home Messages

A take-home message may be formed in one of several ways. The structural form of a take-home message may be somewhat amorphous, but in the end, it usually takes the form of a very simple, straightforward, declarative sentence about a fact or group of facts. The fact(s) may have evolved from analysis and interpretation of the numerical data or some other form of observation that occurred during the conduct of the research. A take-home message also may appear as a phrase or group of words that describe some event, but in most written forms, it takes the shape of a declarative sentence.

The researcher who is finishing a research project will probably have the concepts or ideas associated with a given take-home message or two "pop into their heads" some time after the data have been taken or maybe analyzed and interpreted. When these concepts or ideas come to mind, be sure you write them down or log them in some way. At a later time, you will need to refer to them in the formulation of the take-home messages of the manuscript. The writer forms the take-home message by transforming one or more facts from the research into a concept or idea. Simple, declarative statements of fact are then created to represent these messages.

If the reader of your manuscript or article is a well trained scientist, then he or she will not accept what you have to say about the topic unless you provide evidence that supports your claim(s) made in the take-home messages. Thus, at this point, you must provide unequivocal evidence in the form of your data so that the reader of your manuscript or, later, your article, will understand what you want her or him to gain by reading your article. Therefore, for each of the take-home messages you develop as central points of your manuscript, you must provide factual data in the form of a table, figure, photograph, or some other form of information that helps the reader accept the take-home message. In some cases, one table or one figure will suffice. However, in other instances, you may need a combination of several tables only, figures only, or a mixture of tables and figures to make the reader understand, in no uncertain terms, what your take-home message is saying.

Examples of Take-Home Messages and Their Corresponding Use of Data

Example Take-Home Message #1   Ethylene must be present within tomato fruit tissue for the tomato to turn red during ripening.

How data can be used to provide evidence for example take-home message #1   The pattern (as a figure with time on the x-axis and ethylene production rate on the y-axis) of the ethylene production rate corresponds to, and precedes, lycopene (red pigment) production in ripening tomatoes. A northern blot (as a photograph) of the mRNA for the gene that solely controls the synthesis of lycopene shows that the expression of the gene is up-regulated in response to the presence of ethylene.

Example Take-Home Message #2   Holding germinated impatiens seeds in 1 to 2% oxygen, balance 99 or 98% nitrogen, for 12 to 24 hours reduces radicle growth to less than 1 mm, but the radicle and shoot will resume normal growth when the germinated seed is returned to air.

How data can be used to provide evidence for example take-home message #2   Measurements of radicle length will be taken immediately after removal from the low-oxygen environment, and those data will be presented in a table. The measurements of radicle growth (length) after exposure to low concentrations of oxygen (1 to 2% for 12 to 24 hours) show that the radicle grows less than 1 mm, whereas concentrations of oxygen greater than 2%, and for times greater than 24 hours, caused radicles to grow longer than 1 mm. During the seven days after the seedlings were returned to air, no abnormal seedlings developed, and the radicles continued to grow (presented either in tabular form or within the text of the results).

PRINCIPLES BEHIND AND DEVELOPMENT OF A PROVISIONAL TITLE

Principles Behind the Provisional Title

The title of your manuscript is the most succinct summary of your work (Day and Gastel, 2006). However, most writers will change the title several times as they develop their manuscript. New insights develop as one synthesizes the facets of the research into a manuscript, and this often causes a change in the existing title. It should not alarm the writer that the title first brought forward changes. This is a quite normal sequence of events as the manuscript is being developed. You should expect it!

The goal in formulating a provisional title is to allow yourself to synthesize the contents of your take-home messages into a single summary phrase, or in some cases, a sentence, that best relays to the reader the entire content of your manuscript. Having a provisional title will help the writer to stay on the appropriate course until the manuscript is finished (O'Connor, 1991). The combination of the take-home messages and the provisional title can give the writer a set of limits that will act as a guide in producing the manuscript. For writers, and especially first-time writers, it is important to have a guiding light that helps keep them focused on the entire story they want to relay to the reader. In this way, the author can keep the direction or flow of the information in the manuscript focused on solving the question that originally brought about the research. In addition, it has been suggested that writers also develop a provisional abstract, which, along with the provisional title and the take-home messages, will help to keep the reader on-track (O'Connor and Woodford, 1975). An abstract may already have been developed for a presentation at a scientific or professional meeting, and this can act as a provisional abstract during development of the manuscript. The final title and final abstract (in the finished manuscript) guide the reader, but the take-home messages, provisional title, and provisional abstract are used by the writer to guide his or her efforts during the development of the manuscript.

The most important thing to keep in mind is that your title will be read by thousands of scientists either in the original journal or through an indexing or abstracting service. However, fewer people will read the abstract, and fewer yet will read the entire article or even parts of the article. Thus, you must do a good job of preparing a title that will catch the attention of prospective readers and cause them to feel they must read your article.

A title may be either *indicative* or *informative*, but the best titles will be ones that are *informative* (Luellen, 2001). Indicative titles do not come with a straightforward statement of what the research determined, that is, its conclusion(s). The indicative title indicates what is in the article, but the reader must go into the article to find out what were the conclusion(s) that arose from the research. All scientists are busy, and if time is an issue for a scientist, they may not go into the article to discover the conclusion(s). You have now lost a potential reader of your information. Informative titles are meaningful on first reading, and they relay information to the reader immediately upon reading the title. At that instant, a scientist can decide whether to read more of the article.

Development of a Provisional Title

To formulate a provisional title, the writer should look at the recently developed take-home messages, and their supporting data, and develop a title for the manuscript that encompasses the essential information in the manuscript. The development of this provisional title can often be done by extracting keywords from the take-home messages and statements of results and stringing them together, in an appropriate manner, to form a provisional title. Once the appropriate keywords have been gathered, the writer should rearrange the words into an active (rather than passive), declarative statement or sentence that informs the reader about the contents of the manuscript on first reading. Often, authors will have to rearrange these keywords several times before they arrive at a sensible title, but we have found that the take-home messages are an excellent source of keywords that may appear in a title. (These keywords coming out of the take-home messages, but not being used in the provisional title, are often an excellent source of the other keywords for the entire article. You will need to develop a list of keywords at the time of submission, and these keywords will help immensely at that time.)

Some writers may find it easier to develop the provisional title first, and then develop the take-home messages subsequently. Our recommendation is to first develop the take-home messages, and then out of these take-home messages develop the provisional title. We believe the best manuscript will be developed by first formulating the take-home messages and then developing all segments of the manuscript out of these take-home messages. In any event, these two very important parts of the manuscript should be developed at nearly the same time. The exercises near the end of this chapter will help you develop a provisional title that later, in Chapter 19, will be revised into the final title. We would suggest that the provisional title is developed for *you, the writer of the manuscript*, as a means to guide you through the writing process, and the final title is developed for the reader of the article.

Please Consider These General Rules (Luellen, 2001; Day and Gastel, 2006).

- The title should be meaningful to the reader on first sight. There should be no doubt in the mind of the reader regarding the content of the manuscript.

- The title is not necessarily a sentence, but it is a carefully chosen set of words, and these words must make the title completely understandable. Some journals do not allow sentences, and others permit them if they are appropriate. In reality, this set of words should not just make the title understood; it should make it so that the title cannot be misunderstood.

- A short title may be appropriate; however, make sure it is descriptive and informative to the reader. Most titles are too long.

- Avoid lengthy titles because most journals have a word limit (maybe 12 or 15 words), a character limit (maybe a total of 75 or 80 characters), or a number of lines limit. This will vary with each journal.

- There should be no abbreviations, chemical names, or jargon in the title.

- The title should facilitate a quick retrieval of the published article during a search by an indexing service.

- Avoid the use of serial titles—those with Roman numerals linking a long series of articles on the same topic. These are also known as title–subtitle units. Normally, each article will have the same title, and they will be distinguished from one another by a Roman numeral followed by a subtitle specific to that article. If the quality of the science in one of them, say number VII, is poor, then the lack of publication of the information in number VII may preclude subsequent papers from being published, as there is a "hole" in the series.

- Avoid the use of "hanging titles." These titles contain a colon in the middle of the title, and the purpose of the colon is to try and link two dissimilar items together into one title. Simply delete the colon and rearrange the words so that a simple, declarative title emerges.

- The most common fault of a bad title is faulty syntax, or word order, leading to an incorrect understanding of the content of the article. Syntax problems appear regardless of the length of the title.

- The second-most common fault of a bad title is the use of low-impact words such as "effect of . . ." and "response of . . . ," use of "that," and use of articles such as "a" and "the."

Do not be afraid to spend a lot of time on the development of your take-home messages and title. If you do not develop clear, succinct, and direct take-home messages, the development of your manuscript will be time-consuming and the manuscript will ramble because you have no purpose and direction in what you are trying to write. Likewise, spend lots of time developing a good title, because that is the first thing your reader sees, and you do not have a second chance to make a good first impression.

EXERCISE 7.1 Developing Take-Home Messages— The Rough Draft

Background

The first purpose of this exercise is to define the take-home messages you consider most important for your manuscript. The second purpose is to make you think critically about the information you convey to the reader and how strongly that information is supported by the evidence you will provide (i.e., your data). In most cases, a take-home message will relate to one, or more, of the objectives or hypotheses of your research project. Thus, your objectives or hypotheses from your research proposal might serve as a basis for construction of your take-home messages. Another basis, probably even more direct, for your take-home messages will be the results (i.e., your data) that you present in the manuscript. In addition, you may have a separate take-home message that is related to methods development. You should not feel insecure about making a take-home message about a new technique you have developed or, indeed, about a published technique that did not work exactly, requiring you to make a substantial improvement. Manuscripts for most journals should have two to four take-home messages, and you should keep this upper limit of four in mind as you develop them. If the information you want to include in your manuscript develops into five or more take-home messages, then you should consider breaking the body of research results into two manuscripts.

Instructions

For each of your take-home messages, write a *clear and direct statement or sentence* that represents the message in an appropriate balance of conciseness and detail. In *one or more clear and direct sentences, for each take-home message*, explain how a portion of your data can be used to support or prove each of your take-home messages. In some cases, more than one table, figure, or photograph may be needed as your evidence for a given take-home message. You should not feel that each take-home message must be supported by only one set of data. Please practice writing your take-home messages and their supporting statements of evidence elsewhere. When you are satisfied that they represent what you want to relay to the reader, use the space provided for writing and defending, by use of your data, *up to four take-home messages* that will become the basis of your manuscript. Allow some period of time between each time you refine and improve each take-home message. This may be an hour or two, or perhaps a day or two. Let this rough draft of the take-home messages "get good and cold" for a few days before you move to the final draft (Exercise 7.2) that you will prepare for use in writing your manuscript.

Take-home message #1: _____

_____

_____

Data needed to support/prove message #1: _____

_____

_____

Take-home message #2: _____

_____

_____

Data needed to support/prove message #2: _____

_____

_____

Take-home message #3: _____

_____

Data needed to support/prove message #3: _____

_____

_____

Take-home message #4: _____

_____

_____

Data needed to support/prove message #4: _____

_____

_____

EXERCISE 7.2   Developing Take-Home Messages— The Final Draft

Background

Previously, you completed the rough draft of this exercise (Exercise 7.1), and you have had an opportunity to let those take-home messages get "cold" for at least a few days. We have found that many writers, especially first-time writers, have not had a thorough enough "experience" with processing, organizing, analyzing, and interpreting their data. Therefore, most writers were not in a position to write two to four good take-home messages that will form the basis of their manuscript. In addition, classmates or instructors may have edited the rough draft of those take-home messages. Thus, between your additional thought about what messages you want to relay to the reader, your greater involvement and experience with your data, and the comments of your peer classmates and instructors, you should be in a position to refine your take-home messages and write them into this final draft of this exercise. These refined take-home messages will surely be better than those in your rough draft. Because you will start writing the text of your manuscript soon, you also need to draw the line and say "Now I must determine what guidelines and limitations I need to place on what I will write." Thus, the purpose of this exercise is to provide you with an opportunity to refine the take-home messages you consider most important for your manuscript before you get very far into writing it.

Instructions

Start this exercise by refining your rough-draft take-home messages, incorporating any appropriate comments and criticisms received from peers, instructors, reviewers, and yourself. For each of your refined take-home messages, write a *clear and direct statement or sentence* that represents the message you want to get across in an appropriate balance of conciseness and detail. In *one or more clear and direct sentences for each take-home message*, explain how a portion of your data can be used to support or prove each of your take-home messages. In some cases, more than one table or figure may be needed as your evidence for a given take-home message, and you should not feel that each take-home message must be supported by only one set of data. Please practice writing your refined, final-draft take-home messages and their supporting statements of evidence on a separate piece of paper. When you are satisfied they represent what you really want to relay to the reader, and what guidelines and limitations you want to place on your writing, complete the form below by writing and defending, by use of your data, *up to four refined take-home messages* that will become the basis of your manuscript.

Refined take-home message #1: _____

_____

_____

Data needed to support/prove message #1: _____

_____

_____

Refined take-home message #2: _____

_____

_____

Data needed to support/prove message #2: _____

_____

_____

Refined take-home message #3: _____

_____

_____

Data needed to support/prove message #3: _____

_____

_____

Refined take-home message #4: _____

_____

_____

Data needed to support/prove message #4: _____

_____

_____

EXERCISE 7.3   Developing A Provisional Title

Part A

Look at the words in your take-home messages, supporting statements, and any key words you have discovered so far in the content of your manuscript. List these words on the lines below. Restrict yourself to a maximum of 14 important key words you believe describe the essence of your work.

_____   _____

_____   _____

_____   _____

_____   _____

_____   _____

_____   _____

_____   _____

Part B

Now, organize, arrange, rearrange, and string together these words to develop a concise and unambiguous first draft of a statement that summarizes your work in the manuscript. Limit yourself to 20 to 22 words.

_____

_____

_____

Part C

After you have let this first draft of a provisional title get "cold" for at least a few days, return to it and try to "tighten" it to a maximum of 12 to 15 words.

_____

_____

EXERCISE 7.4   Would You Read this Article, Given This Title?

Instructions

In Exercise 7.3, you were asked to formulate a provisional title for your manuscript. In this exercise, we would like for you to critique several titles that we have developed ourselves. Similar titles may be out there, but to protect the innocent, we are claiming these as titles derived for the purpose of this exercise. (We think you can read between the lines on this one.) Study each title, and after some reflection on it, please cite what you feel is wrong with it, and then try and fix it. Please refer to some of our comments about titles near the end of the text of this chapter. There really are no right or wrong answers for this exercise, but the critical thinking associated with your analysis of the title should help you come to a good final title as you come to completion of your manuscript.

1. Human growth in response to human growth hormones

What is wrong?

Your revised title:

2. Tissue-culture media components affect HeLa cell growth and development

What is wrong?

Your revised title:

3. Global warming

What is wrong?

Your revised title:

4. Propagation by stem cuttings of 54 species of trees and shrubs as affected by time of year, growth-regulator treatment, and bottom heat: Overcoming genetic barriers

What is wrong?

Your revised title:

5. Pollinator activity near flowers of *Rhamnus caroliniana*

What is wrong?

Your revised title:

6. Toward a new green revolution: Genetic engineering to make flower petals photosynthetic

What is wrong?

Your revised title:

7. Membrane-mediated transport

What is wrong?

Your revised title:

8. A new method to isolate antioxidants from bovine cell suspensions

What is wrong?

Your revised title:

9. Effects of five atmospheric pollutants on the activity of nine invasive insects in chaparral habitats in Spain

What is wrong?

Your revised title:

10. New evidence regarding models to predict herbivory: An ecosystem approach

What is wrong?

Your revised title:

11. How are signals that initiate tumor formation in mice transported?

What is wrong?

Your revised title:

12. *FRL08* is responsible for sex expression in banana slugs

What is wrong?

Your revised title:

13. Investigations on the long-recognized but poorly understood global phenomenon of accelerated aging among coastal bird species exposed to elevated CO_2

What is wrong?

Your revised title:

Instructions

In the five spaces below, enter poor (maybe ridiculous) titles you have seen as you have read published literature. Describe what is wrong with the title, and revise the title so that it relays to a reader the true content of the article.

14.

What is wrong?

Your revised title:

15.

What is wrong?

Your revised title:

16.

What is wrong?

Your revised title:

17.

What is wrong?

Your revised title:

18.

What is wrong?

Your revised title:

EXERCISE 7.5   Peer Critiques of Provisional Title and Take-Home Messages

Instructions

You have completed Exercise 7.2 on a final draft of your take-home messages and Exercise 7.3 on the formulation of a provisional title. Please cut and paste electronic copies of your final answers for those two exercises into a new document, and place on that document your name and identify it as Exercise 7.5. Make three copies of this document, and if you are using this book as a textbook for a course, make another copy for each instructor. Distribute the document to three other individuals and each instructor. Ask them to complete a review of your provisional title and provide suggestions for its improvement. Ask them also if they would look at your final draft of your take-home messages and determine whether or not they seem to fit with the provisional title. Finally, ask them to render a judgment on the suitability of your proposed data as support/proof of your take-home messages.

We cannot emphasize strongly enough the need for writers to have others review and critique her or his written work. We all become so "close" to our writing projects that we often do not "see" that what we thought was clear is, in fact, very ambiguous to another reader. Through the detection of these faults in writing, whether it is a title or a statement of fact (or so we thought) about the results, we can all benefit greatly from having others, scientists and nonscientists, review our material. There is no shame in having a friend point out a problem to you before it is too late. It is especially helpful if inexperienced writers with little editorial experience have others review and edit their work often during the production of a manuscript, because errors *will* enter into the manuscript; the sooner they can be eliminated, the better for all involved.

REFERENCES

DAY, R.A. and GASTEL, B. 2006. *How to write and publish a scientific paper.* Sixth edition. Greenwood Press, Westport, CT.

LUELLEN, W.R. 2001. *Fine-tuning your writing.* Wise Owl Publishing Co, Madison, WI.

O'CONNOR, M. 1991. *Writing successfully in science.* HarperCollinsAcademic, London, UK.

O'CONNOR, M. and WOODFORD, F.P. 1975. *Writing scientific papers in English.* American Elsevier, New York, NY.

CHAPTER 8

ORGANIZING AND OUTLINING YOUR MANUSCRIPT

If a man (woman) can group his (her) ideas, he (she) is a good writer.

—Robert Louis Stevenson

IMPORTANCE OF ORGANIZATION

The goal of the writer is to position the reader to understand unequivocally the messages contained in the manuscript and then the subsequent published article. The best way to ensure this occurs is to present the reader with text that flows clearly and logically from start to finish. To complete this task, the writer must organize the content of the manuscript so that the coauthors, then the reviewers, then the editor(s), and finally, the future scientist-reader can follow the flow of information without becoming distracted. The main benefit of this organization is that it certainly helps the reader understand the content of the published article. However, it also helps keep the writer on the correct track to write only what is necessary. An important factor to note at this point is the assumption that you have conducted the research within the proper framework of the scientific method. *If your work has not been completed within the proper framework of the scientific method, this book cannot help you correct that problem. You must return to the laboratory, field, or clinic to conduct your experiments again, because no amount of proper writing can cure badly conducted science.*

Most journals organize the presentation of information into the IMRAD structure (**I**ntroduction, **M**ethods, **R**esults, **A**nd **D**iscussion) for communicating the information. When you have chosen your journal, be sure to confirm that your journal uses the IMRAD structure, or if it does not, make sure your manuscript conforms to the structure required by that journal. Before the development of IMRAD, all journal articles were considered descriptive, and Day and Gastel (2006) provide an excellent discourse on the development and use of the IMRAD structure for journal articles. Use of this structure is good, because it positions the reader to obtain answers to the following questions (Day and Gastel, 2006). What problem was studied? How was the work conducted? What did the research discover? What is the significance of the research? These structural units of the manuscript may also be called the main headings or the skeletal outline, as suggested, of course, by a biomedical researcher (Katz, 1985, 2006), and their role is to make the manuscript more understandable to both the writer and the reader (O'Connor, 1991).

Getting Published in the Life Sciences, First Edition. By Richard J. Gladon, William R. Graves, and J. Michael Kelly
© 2011 Wiley-Blackwell. Published 2011 by John Wiley & Sons, Inc.

Certain disciplines (e.g., descriptive field science, theoretical sciences, and reports on new methodology; Day and Gastel, 2006; O'Connor and Woodford, 1975; O'Connor, 1991) and, more recently, certain journals, have deviated from the IMRAD structure. Certainly, there is nothing inherently wrong with these other structures. However, they are not the norm for most journals, and a quick check of recent issues of your journal of choice or reference to the "Instructions to Authors" or "Instructions to Contributors" may clarify the structure before writing commences and the structure must be changed. Nonetheless, IMRAD remains the method of choice for structuring a manuscript to present scientific results in the clearest, most logical flow of information.

After the writer has come to a sense of appropriate overall organization of the manuscript, she or he can proceed to outline these pieces into a logical flow of information. The writer might look at it this way, as she or he prepares a manuscript: the overall organization is the road map, and the outline is the vehicle that carries the author from Point A to Point B on the map.

Organizing his or her thoughts about the structure and content of the manuscript will cost the writer some time at the start of the writing process. However, it saves much time in the long run (Matthews et al., 2000). Few inexperienced writers realize that much thought goes into the process of organizing the manuscript before the fingers touch the keyboard. Most inexperienced writers think the first step in writing a manuscript is the formulation of a title, but this is far from true. An excellent start to the writing process is to define what you want to say, organize how you want to say it, and then plan the flow of writing the manuscript using the outline. Scientists realize quickly that the essence of this last sentence is the application of problem-solving strategies (Matthews et al., 2000). We now apply that problem-solving process to writing the manuscript. The first part, defining what is to be said, is represented by the take-home messages you developed in Chapter 7. In this chapter, you will develop the overall organization of how you want to get your message across, and this will be followed by a section in which you plan the flow of the manuscript—your outline.

PHASE I: OVERALL ORGANIZATION OR STRUCTURE

The process of preparing to write a manuscript should be divided into two phases. In the first phase, the author(s) should think about what to include. This must be followed by a second phase in which an outline that guides the writer is formulated. This "thinking" part must come first, and it often constitutes 50% or more of the energy spent on getting ready to write (Matthews et al., 2000). Most importantly, the author(s) should not be developing an outline at this point because it becomes too restrictive regarding the "thinking" stage. The imagination of the writer(s) must not be hindered by the restrictions of an outline at the start of the writing process.

At this thinking stage, the author(s) should mostly be brainstorming, which is defined as the development of a random-topic list (Matthews et al., 2000). Of course, this topic list must be focused on the content of the manuscript. For the purpose of organizing your manuscript, an excellent approach to brainstorming is to select the most straightforward take-home message, and then brainstorm about it. Subsequently, the writer should continue to move through all take-home messages sequentially. The brainstorming writer should restrict or focus his or her energy on the take-home message to pursue at this point, but should not restrict herself or himself to any particular section of the manuscript. Thus, random thoughts about why the research needed to be done, how it was done, what the results were, and what those results mean should be logged on to a piece of paper or on to a word-processing document. The writer should let the thoughts flow as they evolve and move from brain to paper. This is a case of quantity over quality (Matthews et al., 2000). Exercises 8.1 through 8.4 will provide experience with brainstorming the organization of your manuscript. The approach used in these exercises is necessitated by the fact that the entire organization and development of the manuscript will occur around the take-home messages. The brainstorming and organization of the manuscript is not done around the IMRAD sections, because the take-home messages are the thread holding the manuscript together rather than the order of the sections. Normally, the next step after brainstorming, after some time away from the writing project, is to begin to cluster the ideas into groups that will facilitate formation of an outline of the brainstormed ideas (Matthews et al., 2000). Katz (2006) offers a similar, but slightly different, approach to this brainstorming– clustering–outline development scenario, and authors may wish to see if one approach is better for his or her particular case.

There are several structures for organizing the material to be written, and the writer should consult additional sources if help is needed in the organizational process (Matthews et al., 2000). The most prevalent structure, and probably the most expedient one for someone writing a journal manuscript, is to follow the IMRAD method for organizing the manuscript. Other main structures for organizing the material include chronological, geographical/spatial, functional, importance, possible solutions, specificity, complexity, pro/con, and causality (Matthews et al., 2000). In some cases, these other structures may work well as a method for organizing a section of the manuscript after its primary organization using the IMRAD method.

After some time, the brainstorming process for this particular take-home message will wane, and the writer(s) should move on to the next take-home message. The brainstorming process should be repeated for each take-home message, and the random thoughts about each take-home message should be documented in one place. Most writers will move to another take-home message, but thoughts about the second take-home message may conjure up thoughts related to a previous take-home message, or a take-home message yet to be addressed. This is OK. Get the spurious thoughts about these previous and future take-home messages into their respective document(s) and return to the take-home message you are addressing at this time. It is not advisable to ignore these other thoughts that enter your thought process for the given take-home message. You should not simply discard those thoughts because they do not relate to the take-home message at hand. Once they are lost, they may never be retrieved and used in your manuscript, and this may not be a good thing.

Brainstorming and clustering of ideas are more efficient at the thinking stage, and outlining is more efficient at the writing stage, which, perforce, must come after the thinking stage (Matthews et al., 2000). At some point, this process will become less productive or efficient within each of the take-home messages. Now, the writer should move to "clustering" those random thoughts into segments that will lend themselves to outlining the manuscript. Sometimes, this operation of clustering leads to the development of what is called a concept map, but it also may lead to other forms of organization (Matthews et al., 2000). We suggest you cluster this information into a format that follows the IMRAD structure for writing manuscripts, as this will be your next step in the writing process. Exercises 8.5, 8.6, 8.7, and 8.8 will help you achieve a structured approach to this next step in organizing your manuscript.

If you are having problems getting started with formulating your outline, you might try developing what is called an "issue tree" (Flower, 2000). An issue tree looks like the root system of a tree, and it helps you double-check the balance of topics and the relationship of these issues to one another. Both clustering and the formulation of issue trees are an intermediate step between brainstorming and the development of the outline that guides the writing (Matthews et al., 2000). Some writers may choose to jump over clustering or the formulation of issue trees, and go straight to the outline. In the end, each writer must use "whatever works for them" to develop an outline that will guide herself or himself through writing the manuscript.

PHASE II: OUTLINES AND OUTLINING

An approximate definition. One definition of an outline is "a preliminary account of a project" (Mish, 2004). For our purposes in writing a manuscript, we will define an outline as a patterned, organized, or in some way systematized arrangement of topics or elements that allows both the writer and, subsequently, the reader to understand better the development of the manuscript.

An outline has several important, but distinct, purposes, and it presents the writer with several advantages (Tichy, 1988). In its most basic sense, the outline helps the writer (i) organize his or her thoughts and (ii) organize the words to be used (Matthews et al., 2000; O'Connor, 1991). Here are some benefits of a thorough, well designed outline that come from an organized approach to developing the manuscript.

- The most straightforward benefit is that it improves the logical thinking and critical analysis skills of the writer, who can quickly tell when something is wrong in the body of the manuscript (Tichy, 1988). It helps organize thoughts about the writing project (research results) and what you want and do not want to say about your topic.

- A second advantage is that a lack of coherence is very evident in the outline, even more so than in the final written piece (Tichy, 1988). It gives you a sense of direction that allows you to keep focused on the task at hand, which is writing the manuscript around your take-home messages, and it provides guideposts for making sure you are sticking to these take-home messages of the research.

- Third, an outline allows the writer to place the proper level of emphasis on the topic, and ensures it will be done at the correct point in the manuscript (Tichy, 1988). It will also help you ensure you have covered all the topics you wish to cover, and none that you did not want to cover.

- Another advantage is that the development of an outline can be stopped and restarted easily by writers who have frequent interruptions (Tichy, 1988).

- Another extremely important advantage in this day of multiauthor papers is that the lead author can communicate to the other coauthors very quickly and accurately the content and flow of the manuscript by presenting them with an outline (Tichy, 1988).

- Probably the most important advantage of the outline is that it allows the author to identify and remove unnecessary information from the manuscript (Tichy, 1988), which can be a very real problem for first-time writers who are having difficulty accepting the fact that some of their data will not appear in the manuscript. As you develop your outline, your mind will develop ideas and begin to group those ideas into meaningful units that create a logical sequence of words in the manuscript. Outlining helps develop and keep these meaningful units and logical sequences of words moving in the correct direction, which is the development of the manuscript.

- If you have done a good job of producing an outline, and you have stuck to using it, then you will find a good correlation between the topics of units of the outline and the eventual paragraphs, sentences, and even words of your manuscript.

- The outline not only helps the writer, but it also will help the reader understand the flow of the article after it has been published (O'Connor, 1991).

One very important reason for developing an outline before writing commences is that many writers are reluctant to change the basic structure of a manuscript once they have completed a rough draft of it. Most writers are so relieved to have finished the rough draft that they will not make major changes to the manuscript, because the rough draft represents the work as a whole, and the author is reluctant to tinker with it. The importance of the outline also cannot be overemphasized for the sake of the reader. Some journals now require that an outline of the published article is located either just before or just after the abstract in review articles and longer research reports so that the reader has an understanding of the basic structure of what they will soon read (Day and Gastel, 2006). Basically, the outline functions as a table of contents for these longer works (Day and Gastel, 2006).

There are several outline formats from which the writer may choose, and two of the most usable ones are shown on the next pages. Either format should yield a satisfactory outline. Format Type I is useful for shorter pieces such as a long paragraph within a section or a section of the manuscript, whereas Format Type II may be better suited for an entire manuscript. These two formats are shown in an alphanumeric system, which features alternating numbers and letters. The relative degree of importance of the item is shown by its position of indentation and/or its movement from greater-sized numbers and letters to those that are smaller. Another system is based on numbers only and/or decimals to separate and stratify the entries. In this latter system, entries may be indented to various degrees, or they may be kept flush left with no indentation. Although this may be a generalization, the alphanumeric system is usually associated with scientific work, whereas the numeric decimal system is often associated with government, military, or industrial work. These systems may be observed in the works of Matthews et al. (2000) and O'Connor (1991), and either one of the formats may be used according to the wishes of the writer. The alphanumeric system seems to be more popular with a wider audience of writers than the number/decimal system, because it is easier to grasp (Matthews et al., 2000). Some computer software systems have outline packages, and writers may find them useful. However, some of these programs are locked into a specific format, and it is impossible to vary the program to meet the needs of the writer.

Outlines may be formatted as either topic outlines, sentence outlines, or paragraph outlines, and inexperienced writers should see one or more references for an excellent treatment of the topic of outlines and outlining (Fowler, 1986; Lunsford, 2003; O'Connor and Woodford, 1975; O'Connor, 1991; Tichy, 1988). The use of one format over another is determined by the preference of each author. A topic outline is composed of only words, clauses, or phrases throughout all parts, and end punctuation is not used. This format is especially useful and more efficient when only the original author will work with the outline. It is an excellent choice for conciseness and brevity, especially at an early stage of the writing project. It also has advantages and disadvantages (Tichy, 1988). A second format, the sentence outline, as the name implies, uses complete sentences throughout all divisions and subdivisions of the outline. In nearly all cases, the sentence outline evolves out of the topic outline (O'Connor and Woodford, 1975). It is the format of choice when a person other than the original author, or author of the outline, must work with the material, as would be the case when the writing load is spread over several individuals. Complete sentences must be used, and there should be end punctuation throughout the outline. The order and meaning are clearer and more exact when a sentence outline format is used. There are both advantages and disadvantages to use of the sentence outline (Tichy, 1988). The third type is the paragraph outline. In this case, each division of the outline is a summary of an entire paragraph. This type is good for short pieces, but it is not a good choice for longer pieces such as an entire manuscript (Tichy, 1988). It may find use when the writer is composing one section of the manuscript at a time, such as the materials and methods section, but this type does not lend itself well to the development of thought relationships (Tichy, 1988). Often, these thought relationships are the thread that sews together the entire manuscript.

Many writers use the best parts of sentence outlines and topic outlines, and they combine them into one format. This variant allows the writer to use the best aspects of both types. Normally, sentences are used for the major topics, and clauses, phrases, or words are used for secondary, tertiary, etc., topics (Tichy, 1988). Parallel construction of the outline (see below) and the writing can be a problem when these two types are mixed. Please keep in mind that the outline is a tool, not an end unto itself (Lunsford, 2003). Parallel structure and wording should be used throughout the outline, without regard to the type or format of outline used by the writer. All forms of grammatical units (e.g., nouns, verbs, prepositions, prepositional phrases, clauses, etc.) should exhibit parallel word usage (Lunsford, 2003). Nonparallel word usage only leads to confusion on the part of the reader and, sometimes, even the writer. An example of parallel and nonparallel wording is shown in Table 8.1 (Fowler, 1986; Lunsford, 2003).

Several miscellaneous issues must also be addressed. The item, or items, you most want to emphasize should be placed toward the end of the listing so that the writer ends with a bang. It has been suggested that every division within the outline must have at least two parts, regardless of format type or outline type (Fowler, 1986; Lunsford, 2003). Thus, there is no A unless there is at least a corresponding B. This issue has received mixed reviews by some writers, because they do not understand the reasoning behind the need for two corresponding parts. We will not enter this fracas. However, you will notice that the two format types we have illustrated below each have two corresponding parts. On the other hand, our sample outline (also below) contains several instances in which there is only one entry under a given division. Statements of comparison (often made in the discussion section of a manuscript) should be linked in parallel fashion by using the appropriate *coordinating conjunctions or comparison words* (e.g., and, but, or nor) and *correlative conjunctions or pairing words* (e.g., either . . . or; neither . . . nor; not only . . . but also) (Lunsford, 2003). Appropriately balanced sentences are an especially powerful tool for persuading readers to accept points the writer is arguing (Lunsford, 2003). Finally, include all necessary words (e.g., articles, prepositions, adverbs) that permit clarity, parallel structure, and grammatical correctness in what the writer desires to communicate (Lunsford, 2003). A lack of articles or prepositions is probably the most common cause for a lack of clarity becomes evident.

TABLE 8.1 Parallel and Nonparallel Wording

Parallel wording—correct	Nonparallel wording—incorrect
How to Ride a Horse	*How to Ride a Horse*
I. Mounting the Horse	I. Mounting the Horse
II. Starting the Horse	II. To start the horse
III. Controlling the horse	III. Control of the horse
IV. Stopping the horse	IV. How to stop a horse

OUTLINE FORMAT TYPES

Format Type I—Useful for Shorter Outlines
Manuscript Title

I.
  A.
    1.
      a.
        (1)
          (a)
            i)
            ii)
          (b)
        (2)
      b.
    2.
  B.
II.

Format Type II—Useful for Longer Outlines
Manuscript Title

I.
  A.
    1.
      a.
        i.
          A)
            1)
              a)
                i)
                  (A)
                    (1)
                      (a)
                        (i)
                        (ii)
                      (b)
                    (2)
                  (B)
                ii)
              b)
            2)
          B)
        ii.
      b.
    2.
  B.
II.

A SAMPLE OUTLINE

Here is a sample topic outline we used to write a manuscript about how we teach Publishing in Biological Science Journals at Iowa State University. (However, there are a few sentences in it.) The many revisions, additions, and deletions needed to get this outline took almost an entire week of man-hours to produce, but, subsequently, the entire rough draft of the manuscript was written in about 20 hours.

Publishing in Plant Science Journals: A Course that Jump-starts Professional Careers

I. Course development, structure, and need
  A. Course development
    1. Charles Bracker's course at Purdue served as the model
    2. Prepare students for the real world ("We are in the publishing business.")
  B. Course structure and pertinent information
    1. Two semester credit hours
    2. Spring semester
    3. Graduate-level course; also undergraduate honors students
    4. Textbooks and manuals
      a. *How to Write and Publish a Scientific...* (Day)
      b. *Scientific English*, Second Ed. (Day)
      c. *ASHS Publications Manual*
      d. *Style Manual for the Journal of Publishing in Plant Science Journals*
  C. Need for this course
    1. Professionals (both researchers and educators) have a duty/responsibility to disseminate their results and information to all segments of society
      a. Final step in the research process
    2. Needed in an ideal situation?
      a. No—Responsibility/Duty of the Major Professor
      b. Real answer—Yes—Major Professor often cannot do because of time constraints; sometimes/many times they are not good writers and do not feel comfortable teaching writing; also many times do not have the basic skills necessary to train graduate students how to do a journal article
    3. Influence of publication(s) in professional development
      a. Academic—promotion, tenure decisions, and professional ethos
        i. Known that professionals that start well early maintain that over entire career (1978 and Quick Start. arts.)
        ii. Known that publications and grants received are extremely important for career development in almost all disciplines
      b. Research Institute—climbing administrative ladder
      c. Industry—climbing the corporate ladder
II. What We Teach—The Course Content
  A. Miscellaneous introductory topics
    1. The publication process
      a. Publishable units
        i. Least
        ii. In excess of least (larger, longer articles for *Plant Physiol.*) for certain journals

        b. The review process

           i. Synopsis of the flow before submission

           ii. Synopsis of the flow after submission to publication

      2. Ethics

        a. Plagiarism issues

        b. Authorship responsibilities and issues

  B. Parts of the paper I: Results

      1. Begin with results as the focal point of the paper

      2. Use of text vs. tables or figures for reporting results

      3. Construction of tables and figures

      4. Limitations on text content (data/facts only)

  C. Parts of the paper II: Materials and Methods

      1. Overall purpose

      2. Structure (possible use of subheads)

      3. SI system

  D. Parts of the paper III: Introduction

      1. Stimulate interest

      2. Convince readers of importance of work

      3. Define research objectives

  E. Parts of the paper IV: Discussion

      1. How it differs from results

      2. Possible structures (e.g., conclude–expand method)

      3. Driving home take-home messages

      4. Proper use of speculation

  F. Parts of the paper V: Literature cited

      1. Literature citation systems

      2. Citing literature in the text

      3. Listing references

  G. Parts of the paper VI: Abstract, title, and byline

      1. Critical due to exposure

      2. Title pitfalls

      3. Recommended abstract composition

  H. How to deal with editors and publishers

      1. Cover letters

      2. Responding to a review

        a. Revision (once or multiple)

        b. Response/rebuttal letter(s)

      3. Galley proofs, reprints, and offprints

III. How We Teach It

  A. In-class exercises

      1. Plagiarism group exercise

      2. Student critiques of published tables and figures

      3. Student presentations of their own tables and figures and justification for using one method over the other; followed by peer and instructor critiques

     4. Peer critiques of 1 to 2 paragraphs of materials and methods

     5. Peer critiques of 1 to 2 paragraphs of discussion

     6. Abstract preparation and peer critiques

     7. Authorship responsibility video and group exercise

B. Out-of-class exercises

     1. Getting started exercise

       a. Statement of problem

       b. Why research was important

       c. One new question that was raised by the research

       d. Quantification of authorship entitlement

     2. Take-home messages exercise

     3. Knowing your journal exercise

     4. Table and figure exercise

       a. Table from one set of own data

       b. Figure from the same one set of own data

       c. Advantages and disadvantages of each form

     5. Cover letter exercise

     6. Previously did a title, byline, index words, and abstract exercise, but now eliminated and replaced with an in-class abstract exercise and peer critique

C. Peer reviews of the manuscript

     1. Peer Review I—Results only—three peers review

     2. Peer Review II—Res., Mat./Meth. and Intro.—Two peer reviewers

     3. Peer Review III—Entire manuscript—one peer reviewer

D. Instructor reviews of the manuscript

     1. Rough draft—Three instructors completely independently

     2. Final draft—Three instructors completely independently

E. One-on-one sessions (only one of us for each session so that we limit our time commitment—state this)

     1. Session I—45 minutes—Return of getting started exercise

     2. Session II—30 minutes—Return of table and figure exercise

     3. Session III—30 minutes—Preliminary/cursory review by instructors and final questions

     4. Session IV—60 minutes—Instructor explains graded rough draft

F. Why we use these techniques

     1. The three reviewers simulate an actual submission and review process by most journals

     2. The rough and final drafts simulate an original and revised submission, and all of the work that goes with each one

G. How successful are these techniques?

     1. Rough draft evaluations

     2. Final draft evaluations

     3. Show a stacked figure of the plots of the manuscript scores for rough and final drafts for one of the years we taught. Include each of our 3 scores and a line for the mean for each of the drafts

    4. Success rate for publication of the class project—maybe we could tie this to the stacked graph (just above) to show a success rate

    5. Several students have had glowing comments about the course in their student evaluations

IV. Identification of Problems and Possible Solutions

  A. Student preparation

    1. Tried quizzes, but students considered them unnecessary

    2. Still is a problem

  B. Class interactions

    1. Tried round-table discussions, small groups, and in-class exercises—all worked well

    2. Seem to give us better group dynamics and individual participation

  C. Improving student peer reviews

    1. Went from many journals to one, so that one set of guidelines

    2. Need some mechanism to improve quality of students' peer reviews

    3. Maybe need some method to "grade" how well student does their peer review

    4. Maybe need to develop a set of guidelines/style manual for *The Journal of Publishing in Plant Science Journals*

  D. Effective information transfer—how to best get our points across

  E. Managing the volume of material to grade (all of the exercises and the manuscripts are very time-consuming to grade, but it is worthwhile in terms of what the students get out of it)

    1. Adopted page limit (12 + title page) to reduce number of pages to grade and help students focus on the topic of their article

  F. Balance between the process and the product

  G. Student evaluations

  H. As of this point, enrollment has not been a problem (probably will be if 24)

  I. Originally, we did it as a product, but each student had their journal, and we could not do peer reviews easily. Changed to one journal, and changed focus to the process rather than the product.

  J. Previously, we felt we needed more one-on-one time, so did a fourth session; now questioning whether or not it is really necessary; maybe reduce to three sessions

  K. Students whose first language is not English

V. In the future, we will be looking ahead

  A. The virtual classroom

    1. Possibly create a version all on video

    2. Possibly create a version all on CD

    3. Possibly create an Internet version

    4. Might do course reviews by using fax machine

  B. Developing a *Style Manual* for the *Journal of Publishing in Plant Science Journals*

  C. Electronic submission, review, and editing

  D. Exporting to other universities

EXERCISE 8.1 Brainstorming Your Most Straightforward Take-Home Message

Instructions

Record below Refined Take-home Message #1, or your most straightforward take-home message, and its associated supporting data you will use to prove this take-home message. This information is readily available to you as the results of Exercise 7.2 in Chapter 7. Within the framework of this take-home message in front of you, brainstorm about and record below all the associated information you will need to report about this take-home message. Exercises 8.2, 8.3, and 8.4 are associated with your other take-home messages, and if you think of information associated with those take-home messages, record that information under that respective exercise. Woodford (1968) previously offered this technique with a slightly different approach.

Refined take-home message #1: _____

_____

Data needed to support/prove message #1: _____

_____

_____

1.

2.

3.

4.

5.

6.

7.

8.

9.

10.

11.

12.

13.

14.

15.

16.

17.

18.

EXERCISE 8.2   Brainstorming your Next Most Straightforward Take-Home Message

Instructions

Now select either your second most straightforward take-home message or your Take-home Message #2 from Exercise 7.2 in Chapter 7. Again, record below the take-home message and its associated supporting data. Now brainstorm about this take-home message as you have brainstormed about your previous take-home message. Again, if thoughts about a previous or future take-home message come to your consciousness, record them under the proper take-home message.

Refined take-home message #2: _____

_____

Data needed to support/prove message #2: _____

_____

_____

1.

2.

3.

4.

5.

6.

7.

8.

9.

10.

11.

12.

13.

14.

15.

16.

17.

18.

EXERCISE 8.3   Brainstorming your Next Most Straightforward Take-Home Message

Instructions

Now select either your third most straightforward take-home message or your Take-home Message #3 from Exercise 7.2 in Chapter 7. Again, record the take-home message and its associated supporting data. Now brainstorm about this take-home message as you have brainstormed about your previous take-home messages. Again, if thoughts about a previous or future take-home message come to your consciousness, record them under the proper take-home message.

Refined take-home message #3: _____

_____

Data needed to support/prove message #3: _____

_____

_____

1.

2.

3.

4.

5.

6.

7.

8.

9.

10.

11.

12.

13.

14.

15.

16.

17.

18.

EXERCISE 8.4   Brainstorming your Next Most Straightforward Take-Home Message

Instructions

Now select either your fourth most straightforward take-home message or your Take-home Message #4 from Exercise 7.2 in Chapter 7. Again, record the take-home message and its associated supporting data. Now brainstorm about this take-home message as you have brainstormed about your previous take-home messages. Again, if thoughts about a previous or future take-home message come to your consciousness, record them under the proper take-home message.

Refined take-home message #4: _____

_____

Data needed to support/prove message #4: _____

_____

_____

1.

2.

3.

4.

5.

6.

7.

8.

9.

10.

11.

12.

13.

14.

15.

16.

17.

18.

EXERCISE 8.5   Clustering your Information About your Most Straightforward Take-Home Message

Instructions

Again, record below Refined Take-home Message #1, or your most straightforward take-home message, and the associated data you will use to support or prove this take-home message. Within the framework of this take-home message in front of you, cluster your random thoughts about this take-home message (Exercise 8.1) into the main sections of your manuscript, namely the introduction, materials and methods, results, and discussion. The essence of this exercise originated with Katz (1985). Exercises 8.6, 8.7, and 8.8 are associated with your other take-home messages, and if you think of information associated with those take-home messages, record that information under that respective exercise.

Refined take-home message #1: _____

_____

Data needed to support/prove message #1: _____

_____

_____

Introduction:

1.

2.

3.

4.

5.

Materials and Methods:

1.

2.

3.

4.

5.

Results:

1.

2.

3.

4.

5.

Discussion:

1.

2.

3.

4.

5.

EXERCISE 8.6   Clustering your Information About your Next Most Straightforward Take-Home Message

Instructions

Record below your next most straightforward take-home message, and the associated data you will use to support or prove this take-home message. Within the framework of this take-home message in front of you, cluster your random thoughts about this take-home message (Exercise 8.2) into the main sections of your manuscript, namely the introduction, materials and methods, results, and discussion.

Refined take-home message #2: _____

_____

Data needed to support/prove message #2: _____

_____

_____

Introduction:

1.

2.

3.

4.

5.

Materials and Methods:

1.

2.

3.

4.

5.

Results:

1.

2.

3.

4.

5.

Discussion:

1.

2.

3.

4.

5.

EXERCISE 8.7 Clustering your Information About your Next Most Straightforward Take-Home Message

Instructions
Record below your third most straightforward take-home message, and the associated data you will use to support or prove this take-home message. Within the framework of this take-home message in front of you, cluster your random thoughts about this take-home message (Exercise 8.3) into the main sections of your manuscript, namely the introduction, materials and methods, results, and discussion.

Refined take-home message #3: _____

_____

Data needed to support/prove message #3: _____

_____

_____

Introduction:

1.

2.

3.

4.

5.

Materials and Methods:

1.

2.

3.

4.

5.

Results:

1.

2.

3.

4.

5.

Discussion:

1.

2.

3.

4.

5.

EXERCISE 8.8   Clustering your Information About your Next Most Straightforward Take-Home Message

Instructions

Record your fourth most straightforward take-home message, and the associated data you will use to support or prove this take-home message. Within the framework of this take-home message in front of you, cluster your random thoughts about this take-home message (Exercise 8.4) into the main sections of your manuscript, namely the introduction, materials and methods, results, and discussion.

Refined take-home message #4: _____

_____

Data needed to support/prove message #4: _____

_____

_____

Introduction:

1.

2.

3.

4.

5.

Materials and Methods:

1.

2.

3.

4.

5.

Results:

1.

2.

3.

4.

5.

Discussion:

1.

2.

3.

4.

5.

REFERENCES

DAY, R.A. and GASTEL, B. 2006. *How to write and publish a scientific paper.* Sixth edition. Greenwood Press, Westport, CT.

FLOWER, L. 2000. *Problem solving strategies for writing.* Fifth edition. Harcourt Brace Jovanovich, San Diego, CA.

FOWLER, H.R. 1986. *The Little, Brown handbook.* Third edition. Little, Brown and Co., Boston, MA.

KATZ, M.J. 1985. *Elements of the scientific paper. A step-by-step guide for students and professionals.* Yale University Press, New Haven, CT.

KATZ, M.J. 2006. *From research to manuscript. A guide to scientific writing.* Springer, Dordrecht, The Netherlands.

LUNSFORD, A.A. 2003. *The St. Martin's Handbook.* Fifth edition. Bedford/St. Martin's, Boston, MA.

MATTHEWS, J.R., BOWEN, J.M. and MATTHEWS, R.W. 2000. *Successful scientific writing. A step-by-step guide for the biological and medical sciences.* Cambridge University Press, New York, NY.

MISH, F.C. 2004. *Merriam-Webster's collegiate dictionary.* Eleventh edition. Merriam-Webster, Inc., Springfield, MA.

O'CONNOR, M. 1991. *Writing successfully in science.* HarperCollinsAcademic, London, UK.

O'CONNOR, M. and WOODFORD, F.P. 1975. *Writing scientific papers in English.* American Elsevier, New York, NY.

TICHY, H.J. [with S. Fourdrinier]. 1988. *Effective writing for engineers, managers, scientists.* Second edition. John Wiley & Sons, New York, NY.

WOODFORD, F.P. (Ed.). 1968. *Scientific writing for graduate students. A manual on the teaching of scientific writing.* The Rockefeller University Press, New York, NY.

CHAPTER *9*

RESULTS I: OVERVIEW

Science is facts. Just as houses are made of stones, so is science made of facts. But a pile of stones is not a house, and a collection of facts is not necessarily science

—Jules Poincare

The fool collects facts; the wise man (woman) selects them.

—John Wesley Powell

ROLE AND NEED

By now, you should have the two to four take-home messages and the provisional title of your manuscript. These will guide you through the writing of the results section and then, subsequently, the remainder of the manuscript. The role of the results section can be summarized quite simply: it provides sufficient facts to convince the reader your take-home messages are correct and justified, and the information is new and not simply a repeat or extension of old knowledge (Peat et al., 2002). The facts are the data you will present, and sufficient means just that—enough to be convincing and no more. The norms of your peer group of scientists and of the journal to which you are submitting dictate what constitutes correctness and ample justification. The writer should gear everything toward that audience constituted by the peer group of scientists and their journal.

Preparation of the results section may seem deceptively easy. You know your take-home messages and your entire set of data better than anyone else, and all you need to do is report those facts. However, the results section can be challenging in ways you might not anticipate if you are not experienced with scientific writing. Scientists may design experiments to be more complex than necessary, and they often collect more data than are needed to justify their take-home messages. Collecting those data took time and effort and, therefore, most researchers do not want to set them aside and not report them. Often that is exactly what needs to be done, and a major pitfall of some results sections is the presentation of excessive data or extraneous information (Katz, 1985). The main job of the writer is the selection of only an appropriate subset of the entire data that gets your take-home messages across, and the determination of the form in which those data should be presented (Davis, 2005). This can be difficult, and it lends credibility to our quotation at the start of this chapter. Selection of the data should be based on a statistically summarized representation of an entire data set, and choice of only those data sets needed to support and prove the take-home messages. The writer should strive to keep presentation of the data

---

Getting Published in the Life Sciences, First Edition. By Richard J. Gladon, William R. Graves, and J. Michael Kelly
© 2011 Wiley-Blackwell. Published 2011 by John Wiley & Sons, Inc.

as simple as possible while retaining complete accuracy, precision, and honesty in reporting those data (Davis, 2005).

Journals may allow you the option of, or may require, combining the results and discussion sections into one melded unit. We encourage you to prepare stand-alone results and discussion sections whenever possible. Presentation of your results section, with no interpretation of the data, and discussion section separately makes it easier for readers to recognize and understand the new, and maybe unique, contributions you have made, relative to the contributions made previously by other researchers. Let us begin development of the results section by presenting and discussing its major components and items to avoid.

MAJOR COMPONENTS OF THE RESULTS SECTION

The two major components of the results section are (i) the presentation of the data and (ii) the written text that will be used later in the discussion section to tie the data to the take-home messages of the manuscript. Both components have a unique and specific role in establishing the importance of the results that have been obtained.

The first component of the results section is the presentation of the data. *In the construction of a manuscript, this presentation of the data must come first because one does not know what can be said as a result of the research until the data are organized into a form that is presentable.* Data may be presented in one of several modes, depending upon what is the clearest and most understandable method for relaying the necessary information from the experiments conducted. Before deciding what form the data will take, the writer must consider (i) the purpose for presenting the data and (ii) who constitutes the intended audience (Davis, 2005). The primary requirement for publication of a refereed, scientific journal article is that some portion of these data must represent new insights in your field of study (Peat et al., 2002).

The author(s) must decide what is the best format for depicting these data, or a subset of them. These data may be presented primarily as numbers in the text, numbers in the form of tables or graphs, or illustrations such as flow diagrams, photographic images, and photomicrographic images. In addition, to a lesser extent, the following are used to present data: line drawings, bar graphs, histograms, maps, pie charts, and chromatograms (Davis, 2005). This decision of what format to use for the presentation may be challenging. Fortunately for the writer, there really are only four choices for depicting the data, and the choice will depend largely on the characteristics of the data, or its subset. An exercise at the end of this chapter will help you better understand how to depict your data so they are most understandable to the reader. In addition, at the end of a subsequent chapter, another exercise will teach the writer how to decide whether to use a table or figure to report a certain set of data.

The second component is the written text, which is presented in the past tense because the experiments that generated the data were conducted previously (Day and Gastel, 2006). The text should not repeat the data within the tables or figures. Rather, the text should call attention to the main points contained within the data set. This must be done by stating the results in a meaningful way that compliments the data and forms the focus of the manuscript. This must be done, however, without the slightest hint of interpretation of what these data mean by themselves and relative to data others have discovered and published previously. These forms of interpretation belong in the discussion section of the manuscript, not the results section.

PHYSICAL ARRANGEMENT AND ORGANIZATION OF RESULTS IN THE MANUSCRIPT

The physical arrangement of the results section is unlike that of the other sections of the manuscript in that it does not at all mirror its appearance in the printed article. Most journals require that the text of the results section follows the materials and methods section. However, for some journals, the results section follows the introduction, and the materials and methods section is relegated to the end of the article, usually after the references section (e.g., *The Plant Journal* and *Plant Physiology*, among many others). In the vast majority of journals, tables, figures, and other forms of data presentation in the manuscript are not placed within the text that refers to them, but instead are placed after the references section (e.g., *The Plant Journal* and *Plant Physiology*, and many others). Before you finish your manuscript and submit it, you should check the instructions for contributors for your journal and follow the directions provided. In many cases, the tables (typically one per page) immediately follow the references section. Next comes the collective list of figure captions in numerical order on a separate page immediately after the last table. In numerical order, each successive figure then follows on its own page with no caption, but with an identification marking that links it to a figure caption. When the manuscript is converted to the published article, the various forms of data presentation are incorporated into the text to ensure a smooth flow of information as the reader proceeds through the results section.

There are many ways the writer can organize the data to be presented in the results section. The writer should start early to determine which of these organizational patterns best suits the data the author(s) want to present. The decision on the organizational pattern for presenting the data should be made before the decision is made that a particular set of data will be presented in a table, line graph, photograph, or within the text. Matthews et al. (2000) present nine patterns for organizing the data and the basis for each pattern. These patterns are chronological, geographical or spatial, functional, importance, possible solutions, specificity, complexity, pro and con, and causality. The same authors (Matthews et al., 2000) later present an excellent table that shows the writer the most effective tool (e.g., table, line graph, flow diagram) to achieve a particular goal of organizational pattern. In Chapter 11, we will address in more detail how to choose whether the data should be presented in the text, a table, or an illustration of some sort.

TENDENCIES TO AVOID

These eight tendencies to avoid might imply that what remains, the facts, will seem dry and boring. You can, perhaps should, strive to make the presentation of the data and the written text interesting. However, some level of dryness is an acceptable characteristic of reporting data in scientific studies.

- Avoid the temptation to omit negative or contradictory data (O'Connor and Woodford, 1975; Yang, 1995). As a scientist, you have a duty to be completely honest and objective in reporting your results, whether they support your theories or not. Writers of high-quality science must stay away from the pool of poor scientists "who do not let their data get in the way of their hypotheses." Many times, scientists learn what not to do in their own research by reading what fellow scientists found did not work. That said, it is important to note that it may be especially difficult to publish an article based solely upon negative data.

- Do not report the same data in more than one format within one manuscript; that is, do not report the same data in both a table and a figure. Data should be presented only once, and the format for its presentation should be decided after critically thinking about the best way to organize and depict those data as either a table or some form of illustration (Matthews et al., 2000). It is satisfactory, however, to report the same data in multiple ways within one table (e.g., as numbers, percentages, totals, means, or ratios derived from the same set of data) (Davis, 2005).

- Suppress urges to write long, scattered, or illogical sentences. All statements within the written text related to data presentation should be accurate, concise, declarative sentences.

- Your results section reports only the data *you* found in *your research*. Therefore, literature citations should never appear in the results section, because those citations acknowledge the work of others.

- Do not allow any explanations, analyses, interpretations, contrasts, or comparisons to find their way into your results section, as they belong in other parts of the manuscript, such as the discussion section.

- Likewise, do not draw any conclusions in the results section. The proper place for drawing conclusions is the discussion section of the manuscript.

- Avoid speculation.

- Do not place tables, figures, or other forms of data presentation within the text unless the formatting rules of your journal call for that.

GETTING STARTED WITH YOUR DATA

Allow plenty of time to get to know your data and to get your data into a format usable for your manuscript. You will need lots of time to critically analyze what pattern of presentation best suits your entire data set (Matthews et al., 2000). You will also need lots of time for brainstorming and critically analyzing the numerical values contained in your data set. Finally, you also will need time to devise the best, most effective manner to present those data (e.g., tables, graphs, photographs) so that the reader unequivocally understands their meaning. Do not be surprised by the fact that you may have to revise tables, figures, and other forms of data presentation several times before they are right for your manuscript. We now present several items that will help you get started with your critical thinking and analysis of your data, so you can move toward the presentation of your data in your manuscript.

- First and foremost, get comfortable with all of your data. Do not attempt to select your representative subset of information to publish until you have spent time with all of your facts, figures, and observations. If you make impulsive decisions, you might weaken your manuscript by overlooking the meaning of some subset(s) of your data.

- Begin to select the subsets of data you will present based on their pertinence to and support of your take-home messages. All data that are presented must be usable in supporting the take-home messages, or those data should not be incorporated into the manuscript.

- If you have numerical data, your first attempt at organizing and expressing them might be to make a table in a format that seems to best match what you want the reader to gain from looking at it. Do not jump immediately to presentation as a figure such as a line graph or a bar graph. Those things said, some sets of data simply lend themselves to presentation in a figure. That will come with time after you have analyzed and interpreted what the subset of data is saying about individual data points versus trends. At this point, do not worry about the format of your table or figure in relation to where you will publish it. You can work on that later, after you have determined that you are first getting the information to the reader in an appropriate manner. This may require several revisions before you feel comfortable with how you have presented the information.

- Examine tables you have made, and think critically about how they might be structured differently. Try reformatting them such that the rows (horizontal) and columns (vertical) are reversed. Brainstorm about and experiment with other ways to modify the presentation of the data in the tables. You might want to print all your versions and compare them. In most cases, data in tables can be arranged either horizontally or vertically. However, it is easiest for most readers to make comparisons when the comparisons are structured vertically rather than horizontally (Day and Gastel, 2006; Matthews et al., 2000). Unless there is a specific reason not to do so, place the independent variables in the rows and the dependent variables in columns (Matthews et al., 2000). In the end, you want an answer to the question "Which format makes it easiest for readers to make the comparisons you want to emphasize in this subset of data?"

- It is sometimes useful to transform the data used in your best table into a figure such as a line graph or a bar graph. Or, if you have started with a figure, transform it into a table. The key is to critically observe, analyze, and determine whether a table or a figure would be better for the presentation of the information you want to get across to the reader. In Chapter 11, we will give you additional information to help you select whether a table or a figure is the best mode of presentation, but we would like for you to try this at this point.

- If you have created a line graph, then try a histogram, or another type of plot, and vice versa. You may try and reverse the axes, change scales, stack similar data sets, or plot different combinations of lines as you try and determine what works best. In other words, experiment with presenting your data, but do not let anything inhibit you from looking at the data in a different form or from a different angle. You might be surprised, pleasantly we hope, by what your numbers say to you, and that may add a little serendipity to your manuscript preparation.

GENERAL GUIDELINES FOR DEVELOPMENT OF THE TEXT

After you have reworked the manner in which you will present your data, several times, if necessary, then it is time to write the text that accompanies the presentation of your data. After you have written several manuscripts, we believe you will find this section of the manuscript the easiest to write. If you really have worked the data into the proper form, then writing the text of the results section should become relatively easy. So that you may begin this process, we present here some general guidelines for writing the text of the results section. A more complete treatment of the topic of writing the text of the results is in Chapter 12.

- Because the results section is the "heart" of the manuscript, you should make sure all coauthors are also satisfied with the manner in which the data will be presented.

- If you have not done so already, prepare an outline to help guide you through the development of your text (see Chapter 8). Refer to your completed exercises on take-home messages and your provisional title in Chapter 7. *Your take-home messages should approximate the topic sentences of each paragraph of your results text, and there should be an approximate one-to-one relationship of take-home messages and paragraphs.*

- Force yourself to write in short, declarative sentences that have few adjectives, adverbs, and other forms of words that distract the reader.

- Write in a completely logical order. The order in which you present the text information should run parallel to the order in which you will describe what you did and how you did it in the materials and methods section you will write next. The order of presentation in the text should also run parallel to the order in which you present your data and information in the tables, figures, and so on. Do not skip around or move back and forth from consideration of one table/figure to another. We cannot overemphasize here the value of preparing a structured, parallel outline of each manuscript section before the writing of that particular section commences.

- As you present the information from each table in the text, write about the rows and columns in the order they appear in the table (Day and Gastel, 2006; Matthews et al., 2000). If you find this is awkward as you are trying to do it, then consider changing the structure of the tables and figures to match the text, or vice versa.

- Avoid redundancy—do not restate data that appear in tables or figures. Rather than say one mean was 10 and another was 15 (numbers that appear in a table or figure), you might say that one mean was 50% greater or 1.5 times greater than the other mean. This mode of presentation remains factual and noninterpretive, which is desired in the text of the results section. However, it also avoids exact restatement of the two means already reported in the table or figure.

- Parenthetically cite, or make reference to, tables and figures in the text. The writer should do this liberally because he or she must keep the reader oriented and in complete understanding of what is presented. A reader gets distracted easily if a statement of fact is made by the writer, but the reader cannot find the source of data that supports that statement.

- Keep it brief! The results section typically contains the shortest block of text among all manuscript sections (Yang, 1995). Although the following estimate will vary with the amount of information to be presented, a general rule of thumb is that the results section text should be one to three typed, double-spaced pages. If the text is longer rather than shorter, subheadings should be used to divide it into more manageable units (Yang, 1995). Another rough rule of thumb is that there should be about one table or figure for each 1000 words of text in the entire manuscript (Matthews et al., 2000).

- Edit your drafts critically! We have found one of the best ways to do this is to let the manuscript get "cold" for a period of time (1 to 3 days would be ideal). After this "chilling period," begin editing with a mindset that you are reviewing a manuscript from a competing scientist who is publishing an article that refutes a concept you discovered several years ago. If you can "trick" your mind into doing this, then you prepare yourself for the manner in which a peer scientist will review your manuscript. Be on the lookout for anything other than factual statements, and move all statements about methods, conclusions, analyses, interpretations, and so on to its appropriate place in the manuscript.

EXERCISE 9.1   Depiction of Data

Each writer will be assigned to one of these four take-home messages and the supporting verbal data set associated with it. Be prepared to discuss your opinions regarding the best way the data described verbally could be presented in a results section to convey the take-home message. Create a rough draft of the mode of presentation to represent your idea, and be prepared to describe it and defend it in front of your peers. Concentrate primarily on the take-home message and verbal data set assigned to you, but review the others to prepare yourself for discussion.

Take-home Message 1:

"Java King" and "Buzz" coffee differ in how position of leaves along the stem (leaf age) influences the rate of photosynthesis in the leaf.

Supporting Data Set 1:

Rates of photosynthesis for leaves of both cultivars that are of various ages (leaves at various positions along the twig, from the apex to the base).

Take-home Message 2:

Plants of "Java King" coffee can be spaced more closely in the field than plants of "Buzz" coffee without reductions in yield of beans per plant; hence, greater yields per hectare are possible with "Java King".

Supporting Data Set 2:

Yield of beans during 1996, 1997, 1998, 1999, and 2000 on a per-plant basis and a per-hectare basis from experiments in which both cultivars and various spacing treatments were randomized.

Take-home Message 3:

Expression of a newly isolated gene from "Java King" and "Buzz" coffee plants, *photoPAR*, occurs only in leaf tissue and is promoted at high rates of photosynthesis (CO_2 fixation).

Supporting Data Set 3:

Northern blot analysis shows gene transcripts in leaves but not in roots, stems, flowers, or developing fruits; quantities of transcripts increase as photosynthetically active radiation, and hence photosynthesis (foliar CO_2 fixation), increases.

Take-home Message 4:

Coffee drinkers in New York City and Ames, Iowa, differ in whether they prefer "Java King" or "Buzz" coffee; in both cities, these cultivars are preferred over the cultivars "Sunrise," "Caffeine Crazed," and "Alert."

Supporting Data Set 4:

Mean preference (taste panel) ratings (1 = bad through 10 = terrific) for (1) overall taste, (2) acidity, (3) bitter after taste, (4) appeal of aroma, and (5) induced jitters after 20 minutes indicate "Java King" is preferred in New York City, whereas "Buzz" is favored in Ames; both cultivars are uniformly preferred over the other three cultivars in both locations.

REFERENCES

DAVIS, M. 2005. *Scientific papers and presentations.* Second edition. Academic Press, San Diego, CA.

DAY, R.A. and GASTEL, B. 2006. *How to write and publish a scientific paper.* Sixth edition. Greenwood Press, Westport, CT.

KATZ, M.J. 1985. *Elements of the scientific paper. A step-by-step guide for students and professionals.* Yale University Press, New Haven, CT.

MATTHEWS, J.R., BOWEN, J.M. and MATTHEWS, R.W. 2000. *Successful scientific writing. A step-by-step guide for the biological and medical sciences.* Second edition. Cambridge University Press, New York, NY.

O'CONNOR, M. and WOODFORD, F.P. 1975. *Writing scientific papers in English.* American Elsevier, New York, NY.

PEAT, J., ELLIOTT, E., BAUR, L. and KEENA, V. 2002. *Scientific writing. Easy when you know how.* BMJ Books, London, UK.

YANG, J.T. 1995. *An outline of scientific writing.* World Scientific Publishing Co., Singapore.

CHAPTER 10

RESULTS II: PRESENTATION OF STATISTICALLY ANALYZED DATA

All models are wrong, but some are useful.

—G.E.P. Box

PLAN AHEAD

Statistical treatment of your data is an extremely important issue that should be considered as you begin to plan your research. This is not a topic that you should be pondering for the first time now, as you are preparing a manuscript for publication. If you are in the unfortunate situation of deciding during manuscript preparation how your data should be analyzed, take a vow that you will plan ahead next time. When possible, let advanced planning involve visiting a statistical consultant who can provide advice on how to design your experiment and on the best methods you can use to analyze the categories of results you anticipate. Whether we admit it or not, most scientists find themselves in a position where they know just enough statistics to be dangerous. If that is your situation, please get help as soon as you can. If a personal consultation is not feasible, you may want to review notes from classes on statistics you have taken or review your options presented in books on experimental design and data analysis. One way or another, plan ahead so that at this stage of developing your next manuscript, no uncertainties remain about the most powerful way to examine your data with the tools that the science of statistics offers.

DISCLAIMERS AND INTENT

This book focuses on manuscript preparation, so it is not intended to provide education and training in the use of statistics in research in the life sciences. Our assumption is that writers of journal manuscripts related to the life sciences have previously taken course work in which the principles of statistics, experimental design, and data analysis have been taught. By now, we hope you have formulated the take-home messages of your manuscript, those central conclusions you most want readers of your manuscript to remember. They will be the threads that hold together your manuscript. Take-home messages, however, must be based on data derived from experiments that were planned, executed, analyzed, and interpreted properly.

Getting Published in the Life Sciences, First Edition. By Richard J. Gladon, William R. Graves, and J. Michael Kelly
© 2011 Wiley-Blackwell. Published 2011 by John Wiley & Sons, Inc.

We suspect you may be similar to us, the authors of this book, in your level of statistical education and training. We learned some basics in graduate school and struggled with how to apply the information we learned as we wrote our first few manuscripts. Many years of experience leads to our acknowledgement that we benefit from guidance when it comes to experimental design, data analysis, and presentation of results. Hence, we have vowed to plan ahead and decide in advance on how an experiment should be designed and analyzed. Even with advanced planning, subsequent consultation with an expert in statistics often helps us look at our data with a fresh perspective. Seeking feedback and ideas from a statistician both before and after we have analyzed our data has proven well worth the investment in time. Give it a try!

Our experiences with consulting statisticians, manuscript reviewers, and other colleagues can be distilled into some practical advice on how, generally, statistical results are presented effectively. The intent of this chapter, therefore, is to convey our ideas on presenting your data in a manner that our experiences suggest will be viewed favorably by most reviewers and editors of refereed journals. Our advice in this chapter is deliberately simple and general. We do not want to overstep the bounds of our formal education and training in statistics. We merely want to point you in the correct direction and encourage you to seek any specific assistance you may need from statistical professionals.

DATA COLLECTION

A very simple idea that is often overlooked is the importance of recording your data neatly, on nicely organized data sheets with clearly labeled rows and columns. If you are writing data by hand, use a pencil or ink that will not fade or be destroyed by moisture or sunlight. Record dates of data collection on your data sheets, and add locations and times of day if pertinent. Leave space on your data sheets for miscellaneous notes you may need to take during data collection. For example, if you are recording the number of bacterial colonies on the surface of a medium, did you record every colony, regardless of its diameter and color? You might need to note that only colonies exceeding 2 mm in length were counted, and all colonies that were not pigmented in milky-white were ignored. Make at least one photocopy of your data sheets, or back up your files electronically if you are collecting your data via a computer. Keep copies in different places. Your data are precious and may not be replaceable if lost or destroyed by fire, water, smoke, and so on.

DATA ENTRY

You will probably be entering your data into a software package for analysis. If possible, this should be done as your data are collected. It may be, for example, that your data are generated within an electronic instrument that can interface with your computer system and statistical software directly. Or, perhaps your electronically generated data can be downloaded to a spreadsheet, which, in turn, can be transferred to your computer and statistical software package. Another possibility is that you enter your data manually, as they are generated, directly into the computer via a spreadsheet or a statistical software package. Common to each scenario above is that there is only a single episode of data entry. A two-step process of writing the data on paper and then transferring the information to a computer or software package is avoided. When entering data, it is easy to make inadvertent errors like transposing figures or placing a decimal point incorrectly. Therefore, researchers should try and minimize the opportunity for mistakes by having a direct-entry system. If a direct-entry system is not feasible, the researcher should enter data into the computer as they were entered on the data sheet so that the entered data can be proofread easily before proceeding. We suggest you ask another person who is not involved with the research to proofread for you. Objective eyes may spot errors that a scientist immersed in the numbers will miss. Remember to enter information on treatments, locations, replicate code numbers, and so on, as you transfer data from a collection sheet to an electronic form for subsequent analysis. A statistician might call these other bits of information the codes for your independent variables. They will be needed for analyses of trends and variation in the dependent variables you recorded.

KNOW YOUR JOURNAL

We are not sure whether most statisticians would agree, but we believe there may be more than one legitimate way to analyze and present most sets of data. Certainly, there is little doubt that the extent of analysis can differ. It will be important for you to weigh the various options you have for analyzing your results, ideally before the research commences. This will ensure that you are not swayed later toward one particular analysis option simply because it happens to yield statistical results that support your hypothesis or that are conveniently easy to describe in your paper (Bailar, 1986). A table of means and standard deviations may suffice sometimes, in some journals, whereas that same data set would need analysis of variance, regression analysis, or some other type of analysis if submitted to a different journal. Norms for journals vary considerably based on the audience of scientists who publish in the vast array of periodicals devoted to the life sciences. On one hand, there are life scientists who downplay the need for and value of statistics. On the other hand, some life scientists consider use of statistics essential in any experiment they do. On a broader scale, these differences are manifest in journals where these scientists publish.

You need to recognize and adhere to the norms of the journal you have chosen for your work. Indeed, perhaps the way your particular journal handles statistics will be a deciding factor in where you seek to publish your manuscript. Ahead of time, you should make a point of knowing how papers published in your journal with data sets akin to yours have treated data statistically (Fernandez, 2007; Murray, 1991; Riley, 2001). Do your best to model your manuscript after those successful publications, unless, of course, you and your consulting statistician identify better ways to convey the results to your audience. Remember to conceptualize the design and analysis of your experiment(s) before work commences; *post hoc* statistical analyses are inappropriate (Bailar, 1986).

Regardless of the norms of the journal you choose, do not lose sight of how the materials and methods section is related to the results section of your manuscript. Although it is true that the results will display the outcomes of your statistical tests, reviewers often check to ensure that the descriptions of your experimental design and data analysis in your materials and methods correspond unequivocally to what they see in your results section. Discrepancies between the two sections raise doubts about all aspects of the manuscript and must be eliminated. The next section of this chapter discusses some basics of statistics that are often evaluated by reviewers. The specific topics below are examples of issues that transcend the various sections of many manuscripts. The confidence of reviewers can be bolstered by being consistent with your handling of these issues in the materials and methods and the results sections.

BASIC STATISTICAL CONCEPTS

The Experimental Unit

A sentence in your materials and methods should state clearly what constituted an *experimental unit* in your work. These units frequently, but not always, correspond to replications used in the analysis of your data. One way to think of an experimental unit may be as an independent individual of the subject you are studying. If you are studying seedlings of a wildflower, each seedling may be an experimental unit, but only if each seedling is completely independent of the others. Independence from the others is important because it allows each unit to be assigned randomly to a treatment, and it allows for the random physical arrangement of those units in your experimental setting, such as a greenhouse, plant growth chamber, cultivated field, forest, clinic, hospital, and so on. The traditional definition of an experimental unit, the item to which a treatment has been assigned randomly, provides another way to consider the units in your work. It may be easier for you to evaluate an act of randomization you performed when establishing your treatments than it is for you to evaluate independence of the experimental units. Be certain that you know what constituted the experimental units in your work. Reviewers will want to know how many experimental units you studied. That information will first be mentioned in the materials and methods, but your results section should reflect continuity with the methods you describe. For example, if you present means of a dependent variable, say epicotyl length of seedlings, in a table in the results section, reviewers will want to see within that table that the number of experimental units measured to calculate the means is the same as what was presented in the materials and methods section. This number of units, often depicted as "*the n value*" or "$n = —$" may appear in a column or row of the table, in the table title, or in a footnote. It also may appear in the body of a figure or in a figure caption.

Samples/Observations

Experimental units can be easily confused with *samples* or *observations*. For example, a pot may contain three seedlings. Those three seedlings/plants are bound to one another in the pot and thus cannot be assigned independently to most sorts of treatments. You may take data from all three, but those data likely will be averaged together for a single value that represents the true experimental unit, which is the pot of three seedlings. Thus, the three individual seedlings are observations within one experimental unit, the pot of three seedlings. In a situation like this, your materials and methods section may state "each pot was treated as an experimental unit, with three seedlings per pot serving as observations that were combined to represent the unit." Issues with samples frequently arise in research done in growth chambers and incubators, especially when some environmental condition is the treatment being imposed. Say, for example, that you wish to determine the influence of temperature on the growth of cell cultures in incubators. You intend to use regression statistics to model how growth responds to temperatures from 15 to 40 °C, and treatments of 15, 20, 25, 30, 35, and 40 °C will be used. You have six incubators, each of which can hold 10 culture flasks. Each incubator can impose only one temperature. In such a case, the 10 flasks represent observations or samples within an experimental unit, the incubator. Replication of treatments will be necessary, but that replication can only be accomplished by *replicating over time*. Each time the treatments are replicated, it is important that the chambers be randomly reassigned to temperature treatments. Report the fact that treatments were reassigned in the materials and methods, and ensure that your results section shows that the *n* value used to compute statistics is the number of times (replications over time) the treatments were imposed. *It is extremely important that you understand the concept that an experiment cannot be replicated properly when only one unit that provides the environment is filled with test or experimental units. These test units are samples/observations, not replications or experimental units.* Depending upon the statistician with whom you work, these may be designated as subsamples, rather than samples or experimental units. Under these conditions, the subsamples will need some care in the coding and interpretation of the analyses, whereas use of the designation sample will not require the same level of care in the analysis (Philip Dixon, personal communication, 2008).

Continuous and Discrete Treatments

The next basic concept we would like to emphasize is the nature of your treatments being either *continuous* or *discrete*. This is important because it can guide your decisions about how to analyze, interpret, and present your results. Let us reconsider our example of cell cultures grown at six temperatures in incubators. Temperature is an excellent example of a *continuous* treatment or independent variable. In our example, only six treatments were imposed, but our stated objective was "to determine the influence of temperature on the growth of cell cultures." Note that our objective was not to determine the influence of six temperatures, or to assess which one promotes the most growth. By framing the objective more broadly as we have, we are saying that it is not simply the specific temperatures we chose to impose that are of interest but rather the entire range of temperatures, in our case, 15 to 40 °C. Use of regression statistics should lead to a mathematical function that will allow us to predict the growth of the cell cultures at any temperature between 15 and 40 °C. This can be much more powerful than restricting oneself to results for six temperatures only. For that reason, many journals will have a rule, often unwritten but evident in the papers they publish, that experiments involving continuous treatments like temperature be analyzed using regression. Some other examples of continuous treatments or independent variables include time, concentrations of a chemical treatment, intensity of irradiance (light), pressure, and soil moisture.

Discrete treatments or independent variables differ in that gradations between the specific treatments applied are not meaningful or possible. Comparisons may be sought between effects of two formulations of complete fertilizers on yield of cotton, among three methods of imposing a drought stress on tomatoes, or between two assays for determining the concentrations of amino acids in poultry meat. Unlike continuous treatments, one cannot imagine modeling a response over a range of such treatments in these experiments. Statistical tests other than regression are therefore appropriate here. Analysis of variance; the calculation of means, standard deviations, and mean-separation statistics; correlations; confidence intervals; and a host of other statistical tools might be used instead. The possible approaches are so numerous that it can be confusing to know how best to proceed. Our advice is as follows:

1. *Maintain your focus* on the questions you are trying to answer, the objectives you seek to meet, and most importantly, providing evidence that supports the statements of your take-home messages.

2. *Read the journals where you may submit your manuscript* with an eye for how experiments with treatments and variables similar in nature to yours are analyzed and presented.

3. *Keep your analysis and data presentation as simple and clear as possible,* yet robust enough to answer the questions you have posed.

4. *Consult with statisticians and other researchers* who have published research similar to yours.

Language and Style of Presentation

Lastly, let us briefly consider the *language and style* of manuscripts as they pertain to statistics and the reporting of statistically analyzed data. We cannot overemphasize the benefits of using recently published papers in your journal of choice as models. It is likely you will find relatively little deviation in the way that experimental design and data-analysis methods are described among papers by different authors within your niche area of science. Use these proven examples as models for statistical descriptions in your materials and methods section. Even less deviation may be evident in how statistics are presented in the results section. There are rather routine ways that researchers design a graph of their data in which a regression line is shown, along with the mathematical function of the line and related statistics. Similarly, there are standard ways that tables of means for discrete independent variables show results of mean-separation tests. The remainder of this chapter provides several examples of common approaches.

Four Examples to Clarify these Concepts

Example 10.1   Table with Means and Mean-Separation Statistics

TABLE 10.1   Concentration of Nitrate Remaining After Three Weeks in Media in Which Seedlings of *Quercus phellos* Were Grown. Values are Means of 14 Experimental Units

Medium	Nitrate $(\mu g\ NO_3^- \cdot L^{-1})$
Perlite	22 b[z]
Vermiculite	46 a
Sand	33 ab

[z]Means labeled with the same letter are not different ($\propto = 0.05$) according to the Fisher's least significant difference test.

Comments:

- The three media are discrete variables. It would not be possible to use regression analysis. Instead, a mean-separation test was used, and Fisher's test is one of many such tools. Consult with your statistician to review the differences among these tests and decide which test is best suited for your data.

- Displaying mean-separation statistics by using lower-case letters in this fashion is common. In this case, the mean for sand (33) was assigned both "a" and "b." For that reason, 33 is not different, according to Fisher's test at the alpha level shown, from either 22 (the mean for perlite) or 46 (the mean for sand). The means for perlite and sand, however, are different.

- An example of proper language to describe these results in the written text would be: "Vermiculite retained more nitrate after three weeks than did perlite, and sand retained an intermediate amount of nitrate not different from the nitrate retained in the other two media (Table 10.1)."

- In some journals, the information in the footnote could have been placed in the title of the table. Check with your journal to determine what is appropriate.

Example 10.2   Table with Means and Regression Results

TABLE 10.2   Change in Concentration of Nitrate Over Time in Perlite Used to Grow Seedlings of *Quercus phellos*

Time (weeks)	Nitrate (μg NO$_3^-$ · L^{-1})
1	47[z]
2	34
3	22
4	15
5	12
6	10
7	9
Significance[y]	
Linear	**
Quadratic	**

[z]Data are means of 14 experimental units.
[y]**Signifies significance at $P \leq 0.01$. The regression equation was nitrate = $62.7 - 17.4(\text{week}) + 1.4(\text{week}^2)$; $R^2 = 0.996$.

Comments:

- The data associated with the continuous variable time are analyzed effectively with regression. This method of presenting the statistical results highlights that both linear and quadratic terms were significant. The regression function is relegated to a footnote in this instance, but this may vary by journal.

- Compare this use of a table with the second example below of a figure (*Example 10.4*). The data are identical in both of these examples. Is one or the other, the table or the figure, more effective and informative? Why? We encourage you to make drafts of your data presented in several ways (tables vs. figures, switching rows and columns of tables, changing the type of figure [line vs. histogram, for example]) before making a final choice. Often, you will uncover a more effective way of presenting your data if you explore various approaches.

Example 10.3   Figure that Represents Results for a Discrete Variable

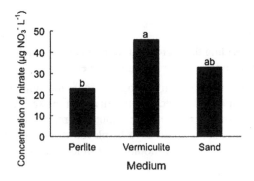

Figure 10.1   Concentration of nitrate remaining after three weeks in media in which seedlings of *Quercus phellos* were grown. Each bar represents the mean of 14 experimental units. Bars labeled with the same letter are not different ($\propto = 0.05$) according to the Fisher's least significant difference test.

Comments:

- The three media are discrete variables. It would not be sensible to use a line graph here because the portion of the line between the data points would be meaningless.

- Displaying mean-separation statistics by using lower-case letters in this fashion is common, although you should check how your journal does it. Alternate ways would include placing letters within white-boxed areas near the top of the inside of each bar or simply describing mean-separation statistics in simple situations like this in the figure caption.

- The standard deviation or standard error of each mean could be added as a vertical line rising up from the middle of each bar. In this case, the author believed any value derived from adding that information was not worth the additional clutter. Another option would be to report them in the caption.

- An example of proper language to describe these data in the text of the results section would be "Vermiculite retained more nitrate after three weeks than did perlite, and sand retained an intermediate amount of nitrate not different from the nitrate retained in the other two media (Figure 10.1)." Note that this is exactly the language used for Table 10.1 (*Example 10.1*).

Example 10.4   Figure that Represents the Results of Regression Analysis

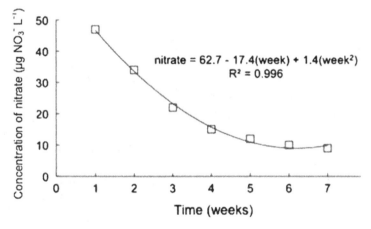

Figure 10.2   Change in concentration of nitrate over time in perlite used to grow seedlings of *Quercus phellos*. Each point represents the mean of five experimental units. The line represents the quadratic regression function that best described the change in nitrate concentration during the first seven weeks of the experiment.

Comments:

- Time is a continuous variable. Providing the regression function allows readers to predict nitrate concentration at any value of time from one to seven weeks.

- Note that the relationship between nitrate concentration and time is the emphasis here. Whether there is a statistically significant difference between any of the seven points is not the focus. The seven points are simply a means to compute the regression function that describes the relationship. The author could have displayed the standard deviations or errors about each mean. Again, authors should determine what is appropriate for their journal.

- Notice that numerical values in the equation are only shown to tenths, and the R^2 value to thousandths. Statistical software packages provide more significant digits, many of which are unnecessary or unjustified and should be removed.

- An example of proper language to describe these results in the text of a results section would be "A quadratic regression function best described the decrease in nitrate concentration in perlite during the first seven weeks of the experiment; the decrease was most pronounced during the first four weeks (Figure 10.2)."

SOME PARTING THOUGHTS

The statistical treatment of data for subsequent incorporation into the manuscript is a topic that incites fear among many first-time authors of manuscripts for refereed journals. We encourage you to take statistics and statistical considerations seriously, but you also should avoid panic or unwarranted anxiety. A wise way to remain confident about your use of statistics is to plan ahead and to seek the wisdom and expertise of a professional statistician who consults with other scientists in disciplines associated with yours. You should consider your local statistical consultant as one of the most valuable and important resources in your community of colleagues. Remember to seek consultation before you begin an experiment. After your research work has been completed, it would be wise to have your analyses and drafts of your tables and figures reviewed by a statistician. In a branch of the biomedical sciences, about one-half of experimental reports are flawed because of poor presentation of numerical results or inadequate or inappropriate statistical methodology (Murray, 1991). Gustavii (2003) also presents an excellent review of common statistical errors in the biomedical sciences, and similar information is available for the plant sciences (Dyke, 1997; Fernandez, 2007).

As you write, remember to adhere to the norms for use of statistics within your discipline and by your disciplinary colleagues and your chosen journal (Fernandez, 2007; Murray, 1991; Riley, 2001). This is an essential ingredient in the recipe for success of your manuscript. For most journals, the norms include a clear definition of an experimental unit in your work, specification of the experimental designs used, and consistency between the statistical descriptions in the materials and methods and the presentation of statistics in the results section. This also extends to your analysis and interpretation of the data in your discussion section. Present your data and statistics as clearly and simply as possible (Dyke, 1997).

Although there are no guarantees about the fate of manuscripts submitted to refereed journals, following the basic ideas presented in this chapter should increase your probability of success. Keep in mind that just as all the best writing in the world will not guarantee that a work of research will be published if it were done inappropriately, all the best statistical analyses in the world will not guarantee that an inappropriately completed research project will be published.

... AND WHY THIS IS SO IMPORTANT

Mish (2004) defines inference two ways, and both of them are applicable here. Inference is (i) "the act of passing from one proposition, statement, or judgment considered as true to another whose truth is believed to follow from that of the former" or (ii) "the act of passing from statistical sample data to generalizations (as of the value of population parameters) usually with calculated degrees of certainty." Virtually all of our new knowledge is generated, and science progresses, via inferences made by a scientist about his or her data that were generated through experimentation. In nearly all instances of research, the data published in a journal article are a subset of the entire set of data generated via conduct of the research. Therefore, almost all publication avenues rely upon data reduction that leads to inferences (O'Connor and Woodford, 1975).

In a powerful essay about science and its progress, Platt (1964) discussed why certain scientific disciplines, such as molecular biology, have progressed rapidly, whereas other branches of science have moved at a snail's pace. Platt's essay discusses what he calls "strong inference," a technique portrayed as key to the rapid expansion of knowledge in certain disciplines. Thus, if the scientist/writer does not have her or his inferences based on appropriately designed experiments and data analysis, the scientist/writer could deceive herself or himself into believing a set of data leads to a certain conclusion, when it really does not. If caught early enough this only harms the scientist. If not caught early, other scientists read the results, use them as a basis for their own experiments, and lead a group of other scientists down a faulty trail of endeavor.

A common joke among scientists revolves around the idea that we can draw any conclusions we want from our data as long as we analyze them statistically in a deliberate way that fulfills our wishes. "If the analysis conflicts with your preconceived notions, analyze a different way," so the story goes. (Please see our quotation at the start of Chapter 11.) Such anecdotes can be funny, perhaps because they are based in truth. The use of statistics, even among well trained and informed scientists, is subject to human judgment and the "dark side" of human nature that may tempt us to be deceitful with our data. It is difficult to deny that many of us enter into experiments wishing for a certain outcome. That can lead to inappropriate selection of data that prove the tenet of the scientist. It can also lead to inappropriately selective choices from among the multitude of statistical treatments available to make meaning from our results. The correct choice of statistical analyses will permit you to become bias-free in your presentation and analysis and interpretation of your data (Peat et al., 2002). We urge you to question yourself, subject yourself to voluntary review by respected peers, and to be on the lookout for bias in yourself and others, particularly about the use of statistical procedures.

Bailar (1986) wrote an excellent treatment of the potential ethical dangers inherent in the use of statistics in medical research. We encourage you to read this short article, as it has many excellent points and lists eight practices that can distort scientific inference. For now, we close this chapter by summarizing four of the most important practices that Bailar cites as potentially "distorting scientific inferences":

- Post hoc *hypotheses*: Remember to restrict your statistical analyses to tests that you planned *before you started* the research.

- *Fragmentation of reports*: In this world of publish or perish, scientists sometimes divide the results of one to several experiments into several manuscripts with the hope of increasing the number of publications (Broad, 1981). This gives the reader *an incomplete view of how the research was conducted*, and it *often is done with a* post hoc *analysis*, both of which are ethically questionable.

- *Using a statistical test or tests that have low power*: It is possible to deceive a reader into thinking that a lack of a statistically significant test necessarily means that the effect was not present. Bailar (1986) encourages us to *report on the potential power* of the statistical tests we use.

- *Selective reporting of findings*: Avoid the temptation to report only the results that conform to the story you would like to tell; *strive to be unbiased* and to reveal the complete truth to your readers.

REFERENCES

BAILAR, J.C., III. 1986. Science, statistics, and deception. Ann. Int. Med. 104:259–260.

BROAD, W.J. 1981. The publishing game: Getting more for less. Science 211:1137–1139.

DYKE, G. 1997. How to avoid bad statistics. Field Crops Research 51:165–187.

FERNANDEZ, G.C.J. 2007. Design and analysis of commonly used comparative horticultural experiments. HortScience 42:1052–1069.

GUSTAVII, B. 2003. *How to write and illustrate a scientific paper.* Cambridge University Press, New York, NY.

MISH, F.C. 2004. *Merriam-Webster's collegiate dictionary.* Eleventh edition. Merriam-Webster, Inc., Springfield, MA.

MURRAY, G.D. 1991. Statistical aspects of research methodology. Brit. J. Surg. 78:777–781.

O'CONNOR, M. and WOODFORD, F.P. 1975. *Writing scientific papers in English.* American Elsevier, New York, NY.

PEAT, J., ELLIOTT, E., BAUR, L. and KEENA, V. 2002. *Scientific writing. Easy when you know how.* BMJ Books, London, UK.

PLATT, J.R. 1964. Strong inference. Science 146:347–353.

RILEY, J. 2001. Presentation of statistical analyses. Exptl. Agric. 37:115–123.

CHAPTER 11

RESULTS III: OPTIONS FOR PRESENTING RESULTS

Oh, yeah, him ... yeah, he never lets his data get in the way of his hypotheses.

—Grady Chism

REFRESHING THE DEFINITION

Let us start this chapter by restating that the goal of the results section is to provide sufficient facts to convince the reader that your take-home messages are correct and justified. The facts (your data) are provided to the reader by placing them in the text, in tables, or in figures in the results section, and then subsequently bringing out the important issues in the text of your results. In this chapter, you will learn more about choosing among these three options for presenting your data, and you will be advised about table and figure preparation. This chapter is about converting your measurements and other forms of data into results that allow you to convey your take-home messages in a story that is factual and justified.

Tables and figures are costly to produce relative to the costs associated with the production of the text of an article (Day and Gastel, 2006). Thus, use them wisely. Most reviewers of journal article manuscripts are asked if any tables or figures could be removed without affecting the overall quality of the manuscript. You do not want to be in a position where one or two of your tables or figures needs to be removed as you revise and resubmit your manuscript. Do not report data just because you have those data and worked hard to obtain them. Please review the quotations at the beginning of Chapter 9.

MOST DATA SHOULD APPEAR IN TABLES OR FIGURES

Your data are the heart of your article, and all other sections of your manuscript are dependent upon the data you choose to present. The tables and figures should stand alone, which means they are comprehensible without reference to the title, introduction, materials and methods, maybe even the results, and discussion sections of the manuscript (Katz, 1986; Woodford, 1968). All other sections of your manuscript should be prepared only after you are completely satisfied with the rough draft of your results. You should not waste time preparing the introduction first if, in the end, you elect to make a change in data presentation that requires the tone of the introduction to change from what you anticipated originally.

Tables and figures are set apart physically from the text of the manuscript, but later they will become distinct, but integrated, parts of the entire results section of the published

Getting Published in the Life Sciences, First Edition. By Richard J. Gladon, William R. Graves, and J. Michael Kelly
© 2011 Wiley-Blackwell. Published 2011 by John Wiley & Sons, Inc.

article (Day and Gastel, 2006). Attention is drawn to tables and figures, which is good, as long as the data in the tables and figures merit the focus. For manuscripts for most journals in the life sciences, tables will appear one per page starting immediately after the end of the literature cited/references section. Tables are ordered in the same sequence they are referenced in the text. After all tables have been added, the next page starts the figure caption(s). These figure caption(s) are placed sequentially on this page and subsequent pages if there are several captions. Next, each line graph, line drawing, bar chart, histogram, photograph, photomicrograph, and so on, will appear in the same order as they are first mentioned in the text. The captions preceding the figures should also be in this sequence. Some journals will have authors make a marginal mark in the body of the manuscript where each table or figure should be located in the printed article, although this practice is increasingly rare. More recently, some journals are requiring the authors to embed the tables and figures into the results section so that there is less chance of a mix-up of which table or figure is which, and where it should be placed (Day and Gastel, 2006). Check the instructions to authors document, the style manual, or recent issues of the journal for such details.

Tables and figures provide you with an opportunity to organize your data in logical and structured ways. This enhances the ability of your readers to comprehend your information, and, again, this is good. It is easier for the reader to make rapid assessments and interpretations of a series of means within one column in a table, for instance, than it would be for the reader to digest the same information listed within several sentences of text. In addition, complex relationships between two or more factors can be depicted visually in figures. The message from such figures could not easily be conveyed in text, or even tables. Photographs, micrographs, and other visual images are also important types of figures, and some journals refer to them as plates. Normally, formatting rules for plates and figures are similar or identical. However, authors should check the instructions to contributors in each journal to determine whether plates can be used, and if so, how they are to be formatted within the manuscript.

Several resources provide writers of journal articles with steps for synthesizing, or recipes for constructing, tables and figures. Works by Davis (2005), Day and Gastel (2006), Woodford (1968), and Yang (1995) are valuable references for inexperienced writers. However, probably the greatest help for a inexperienced writer is to read a lot of published journal articles, especially in the journals where they envision they will publish their work. Very careful scrutiny of these published journal articles often gives direction that no amount of preparatory reading and completion of exercises can replace.

MAKING YOUR CHOICE

Before we get into specifics, let us first take a look at some overarching factors that will help the author determine whether the data should be placed in a table, a figure, or the text. Often, it is best to make both a table and a figure from the same data set. Then, perhaps, after rearranging and working with those data several times, choose the best way to present the information. We have provided an exercise in the back of this chapter in which you will do exactly that, and at the same time, learn how to discern which option is best for your set of data.

You might favor placing your data in a *table*, rather than a figure or in the text, if you:

- want readers to see the exact numerical values;
- wish to present your same set of data in multiple formats (e.g., as absolute numbers, percentages, totals, means, ratios) within one unit of presentation;
- have more than about five treatments, or some other form of repetitive data, to compare.

You might favor placing data in a *figure*, rather than in a table or the text, if you:

- wish to show the relationship(s) between two or more factors;
- have information in the form of a photograph or other visual image; or
- believe that any trends your data show are more important to convey than the exact numerical values.

You might favor placing data in the *text*, rather than in a table or figure, if you:

- have only a few numbers to report; or
- do not mind the idea that many readers may overlook these data (because those readers may focus only on data in tables and figures).

CONSTRUCTING A TABLE

The requirements for table construction for each journal may be different, but many journals have similar or identical rules for how to prepare tables. Study the example that follows these paragraphs, because it exemplifies many of the common rules you may have to follow in constructing your tables. Your recognition of the categories of issues to consider as you prepare tables is more important than memorizing the particular rules for a particular journal. Look at the table again, but this time concentrate on the subtle details associated with its construction. Then proceed to the next page, where you will learn about some of the categories of issues that need to be considered when constructing a table.

Yang (1995) offers an excellent review of the components of a table and the design considerations necessary for you to produce a good table. We suggest you think about and incorporate these components and design considerations into your scheme for approaching the synthesis of your tables.

These are the components of a table:

- *The table number*. These are listed in order of appearance in the manuscript. This arches over all sections of the manuscript as tables can appear in the introduction, materials and methods, and discussion, as well as the results. Usually, the number is Arabic, but some journals will use a Roman numeral.
- *The table title*. This is a brief description of the contents of the table. Some suggest that this title be no more than two phrases, clauses, or sentences long (Day and Gastel, 2006).
- *Column headings*. These headings indicate and describe the content of each of the columns in the table.
- *The horizontal rules (lines)*. Normally, there will be three rules. The first rule separates the table title from the column headings, and this rule may be a single line or a double line, depending upon the journal. The second rule appears between the column headings and the set of data, and the third rule appears at the end of the data. Below this third rule there may be nothing, or there may be footnotes that explain items in the table. Sometimes there may be a fourth rule, and this occurs when subheadings are used in construction of the table. Normally, a table will have no vertical rules.
- *The stub*. This is the left-most column of the table, and it lists the categories or subjects in the other columns. It indicates the content of each horizontal row in the table. The stub only requires a heading if its message is unclear without one.
- *The body of the table*. This is the set of numerical data you will be presenting. In almost all cases, these data should be aligned by decimal point, unless there are no

decimal points. A missing datum or data can be shown as designated by the journal. Most journals will use one, two, or three em-dashes or ellipses (. . .).

- *The footnotes.* Each new footnote should start on a new line, and it should be flush left. The designator for each footnote will be mandated by the journal, but they are usually a, b, c, or z, y, x, or superscripted 1, 2, or 3, or some list of symbols that are presented in a hierarchy determined by the journal.

There are several design considerations that authors should follow. However, most journals reset, or typeset, each table. Nonetheless, you should follow the structure presented in published articles in the journal, as this gives you a chance for the most seamless conversion from the manuscript to the printed article.

- Design the table to fit the column format (how many columns of text appear on a page) used by the journal where you will publish the manuscript. Some journals use a one-column format, but most use a two-column or three-column format.

- Keep the design of the table and the presentation of the data simple and easy to follow.

- Do not have a lot of wasted (white) space in the middle of the table. If this is the case, consider reformatting the table.

- Round the data entries to the nearest significant digits.

- Finally, ask yourself, is this table really necessary?

Table 11.1 is an example of a table you might construct.

TABLE 11.1   Root-Zone Oxygen (O_2), Leaf Nitrogen (N) Accumulation, Number of Root Nodules per Plant, and Shoot Dry Weight of *Alnus maritima* subsp. *maritima* Grown in a Greenhouse for Eight Weeks in Pots with Four Root-Zone Moisture Treatments. Germinated Seedlings were Watered Daily, and N Fertilizer was Provided Three Times Weekly for Four Months Before Treatment. Treatments Began 5 June 2002

Root-zone moisture treatment[z]	Root-zone O_2 (kPa)[y]	Total leaf [N] ($\mu g \cdot g^{-1}$)	No. of nodules/ plant	Shoot dry weight (g)
Watered daily, drained	17.3 a[x]	2.53 a	4.8 b	2.55 a
Partial flood	13.1 b	2.21 b	8.3 a	2.70 a
Total flood	1.2 c	1.41 c	3.9 bc	1.40 b
Total flood with argon	0.9 c	1.34 c	1.3 c	1.33 b

[z]n = 11 seedlings per treatment.

[y]Values represent the average of two measurements per plant, after four and eight weeks of treatment.

[x]Means within columns followed by the same letter are not different at $P \leq 0.05$ according to Fisher's least significant difference test.

ADDITIONAL IMPORTANT ISSUES TO CONSIDER DURING CONSTRUCTION OF TABLES

- Everything in the table is double-spaced—*everything*—unless yours is the rare journal that specifies otherwise.

- The table title is the basic description of the contents of the table. Each table title is placed on the same page as the remainder of that table. Each table begins by having its number specified (e.g., Table 11.1) at the top of the table, followed by its table title. The table title must include enough information to allow the finished table to stand independently of the remainder of the text. Table titles should be in complete sentences or complete phrases, so that they are intelligible units of information (Katz, 1986; Yang, 1995).

- Taken together, information in the table title, the set of data, and the footnote(s) of the table should allow the reader to have a clear picture of how the work was conducted without consulting the text. Our example table (Table 11.1) tells the reader plants were grown in pots in a greenhouse, not in fields outdoors or in test tubes in an incubator. We also know when treatments began, the duration of the treatments, the number of plants in each treatment, and so on. Generally, it is better to provide too much information in the table title and footnote(s) rather than to provide too little. Let a reviewer or editor tell you to delete what they consider excessive detail rather than have reviewers or editors confused about what you did.

- Often, there are quite specific rules dictating which letters in the headings/ subheadings for the columns and rows should be capitalized. Each journal will have its set of rules that should be explained in its instructions to contributors or style manual. In our example Table 11.1, note that only the first letter of the first word in each heading/subheading is capitalized.

- Words in a column should be left-justified, unless you have been instructed otherwise. Data within columns may need to be left-justified, centered, right-justified, or tabbed in a certain way, depending upon your journal. Our example shows decimal tabs, i.e., all decimal points are lined up vertically.

- If your experimental procedures contained control treatments, be sure to place control values so they are the first item the reader sees. This may call for them being across the top or down the left side of the table. The control values should receive "equal billing" with your treatment data. The comprehensibility of the table will be increased greatly by the reader first understanding what were the control values, and then being able to compare those values with the values for the treatments (Gustavii, 2003).

- All nonstandard abbreviations need to be defined, even if they were defined previously in the text. In our example table, note how O_2 and N are defined in the table title so they can be used in abbreviated form in the column heads. Few things create more havoc for the reader than to have abbreviations that do not make sense or are not defined appropriately. We recommend that you do not use abbreviations unless they are necessary.

- Avoid using exponents in table column headings, as the placement of the exponent often gets lost through repeated electronic submissions, especially when different software programs are used (Day and Gastel, 2006). Appropriate use of SI prefixes should remedy this problem.

- There may be rules (written or unwritten but apparent from published articles) for how the results of mean-separation tests should be displayed. Conform to the required or typically used format.

- Avoid presenting detailed results for data sets where there are no statistical differences. However, occasionally, there may be cases where such presentations are justified because they help you develop your take-home messages.

- Be sure you report your data to the appropriate number of significant figures. Most scientists cannot use a hand-held measuring instrument (a ruler) that reads to the nearest 0.0001 millimeters.

- Footnote formats vary considerably among journals. You should look in the instructions to authors of your journal to determine what is required. In our example table, note that the footnotes are double-spaced, end in periods, and have superscripted letter labels in order beginning at the end of the alphabet and proceeding backwards. Other journals may use letters that start at the beginning of the alphabet and proceed forward, and others may use numbers, usually superscripted, starting with the number 1. Some may also use symbols such as daggers, double daggers, asterisks, and so on.

- A table may be too large to fit on one manuscript page. That is fine, because you can continue tables on subsequent pages, provided you identify the succeeding pages as a continuation of the first page of the table (e.g., Table 11.1, continued). This labeling for the continuation should be separate from the pagination of the manuscript. If your table overflows to an additional page by only one or two lines, you may experiment with ways to reduce it slightly. You could try decreasing the font by 1 point, changing the line spacing slightly for a portion of the table (e.g., line spacing in the footnotes might be 1.9 instead of 2), etc. Do not go beyond slight modifications such as these, however.

- If it is necessary to conserve space, tables may be arranged in landscape format for certain tables, even if the remainder of the tables in the manuscript are in portrait format. Most journals do not require all tables to be in one format or another.

You should make notes here regarding any other formatting details you notice in our example table. You should also note any questions you should ask editors of your targeted journal about how to prepare tables. Construction of tables and figures may be very specific for each journal, and this is why you should choose your journal before you get too far into writing your manuscript.

CONSTRUCTING A FIGURE

Figures may contain data plots, photographs, photomicrographs, or other illustrations such as flow charts and line drawings.

Each figure or illustration will consist of two distinct parts that have distinct functions. The first part is the figure caption. It consists of a title for the figure, which orients the reader, and may be followed by an explanation of the symbols and images within the figure. The title portion of the caption provides details that allow the reader to understand the presented data. The second part of the figure is either the observed, visual image or the data that have been organized in a way that permits the reader to have better comprehension of the data (e.g., a graph).

Most figures, especially graphs, serve two purposes. First, the figure summarizes the set of data into a spatial form that permits the reader to understand the meaning of the data. Second, a figure helps us scientists/readers to recognize patterns that are inherent in the data (Katz, 1986).

Woodford (1968), Davis (2005), and Matthews et al. (2000) have presented some general, overarching issues to address before work commences on the construction of a figure. These general issues, mostly in the form of questions to the writer, are as follows:

- Before you start work on your figure(s), reread the instructions to contributors for your journal, as you will find that many of your later questions will be answered quickly if only you had read the instructions before starting.

- What is the purpose of the figure? Is it for evidence, efficiency, or emphasis? What point are you trying to make by using the figure? In most cases, figures are used to promote understanding of numerical data in a way that the numbers themselves cannot relay properly. Figures should be used when the author wishes to communicate that a trend or relationship is more important than the presentation of the original numbers.

- Is the figure both independent of the other items used to present data and indispensible to your argument?
- How will your audience of readers respond to the figure? Will they be able to interpret it correctly?
- What form will the data presentation assume? Will you use plotted lines in a graph (for continuous variables), or will you use a bar graph or histogram (for discrete or discontinuous variables)? A photograph?
- Pay attention to size and scale. Design the figure/illustration to fit the columns in the journal where you will publish your manuscript. This will reduce the amount of reduction or enlargement of your figure that almost always causes loss of clarity and definition. A good way to see what will happen is to use a photocopy machine that can enlarge/reduce the object, and this will show you what the image will look like when it is printed.
- If a plotted line is used, can you ethically connect the series of points, or must you present a best-fit line through the data points (i.e., regression analysis), or must you place the individual data points on the plot without any form of connection of the data points?
- Is the figure completely clear, yet attractive? Keep your figures simple—they should be comprehended at a glance, rather than only after intense scrutiny. Is there "junk" or "glitz" in the figure that causes it to not be as succinct as it could be?
- Are the data presented in a scientifically truthful manner, or are they massaged into a more eye-appealing manner that is not 100% truthful?
- If your figure does not relay a clear, truthful appropriate message, you should probably consider abandoning it and searching for some other way to relay your information.
- In most instances, figures associated with techniques are found in the materials and methods section, figures of data are found in the results section, and figures of synthesis of data, abstractions, and models are found in the discussion section.
- If all else fails, there are several good resources that can help with technical illustrations in journal articles (Briscoe, 1996; CBE, 1988; Cleveland, 1994; Hodges, 1989; MacDonald-Ross, 1977; MacGregor, 1979; Tufte, 1983).

An example figure caption (Figure 11.1), and its associated figure, are shown on the next two pages. Inspect the figure caption and its figure carefully. Then review some of the categories of issues that need to be considered when preparing a figure and placing it appropriately within your manuscript. *NOTE: A caption like the one on the next page would be placed on a separate page, usually after the references section or the table(s), if there were a table or tables, that precedes the page on which the figure is printed. In the example figure below, note that our example is the first figure from its manuscript. When there is more than one figure, multiple captions should be placed in numerical order on the same page, or consecutive pages, but each figure is printed on a separate page, and its identity is shown in a conspicuous location.*

Figure Caption

Fig. 11.1. Net photosynthetic rate of *Acer rubrum*, *Betula nigra*, *Magnolia virginiana*, *Prunus serotina*, and *Syringa meyeri* irrigated with solutions of tap and ocean water with salinity of 0, 3, 6, and 12 g kg^{-1}. One fully expanded leaf of each plant was sampled after 7 and 21 d of treatment. No difference existed between the two dates, so the means of the two measures for each plant were used for analysis and presentation. The least significant difference (LSD) value is for testing any two means at $\alpha = 0.05$.

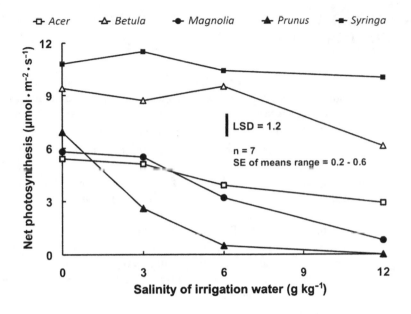

(Fig. 11.1)

ADDITIONAL IMPORTANT ISSUES TO CONSIDER DURING CONSTRUCTION OF FIGURES

- Yes, it is really true. A figure caption does not appear on the same page as the corresponding figure. The figure caption or captions are placed in numerical order on a separate page, or successive pages, immediately before any of the figures appear, and each figure, on its own page, is placed immediately after the last page of figure captions.

- Also note, as shown for the example, Figure 11.1, that the identifying figure number is placed in a conspicuous location on the same page as the figure. In this instance, it is near the bottom, right-hand corner of the page. This identifying figure number may be placed anywhere on the page, but it *must* be unequivocal, conspicuous, and in the same position on each figure. This prevents confusion among reviewers and editors as to which figure it is. There have been untold instances where a reviewer or editor opened the package containing the manuscript and could not decipher which figure corresponded to each figure caption. In this type of situation, the editor or reviewers will more than likely send the manuscript back to the author to clarify the situation before they will complete their review.

- Plots should be constructed as simply as possible, so that you afford the reader an easier time interpreting its meaning (Davis, 2005). Resist the urge to embellish plots with such gimmicks as unnecessarily using three-dimensional plots, unusual data-point symbols, or different line patterns (dashed lines, dotted lines, and so on) (Yang, 1995).

- It is good practice to make all lines (axes and data lines) should be presented only as solid black lines. The identification of different treatments or times data were taken occurs through the use of different symbols, not different patterns of the data lines.

- If your journal requires it, or permits it and you would like to do so, you may want to make the axis lines slightly thinner so that the data lines appear bolder and draw more attention. It is acceptable, and perhaps even preferred, to have data lines a bit bolder and/or broader than the axis lines. In any case, do not make the axis lines bolder than the data lines, because you do not want to emphasize the axes. You want to emphasize your data.

- Do not plot an excessive amount of information in one panel. In our example, five lines are depicted, and that is approaching the maximum amount of information to include in one panel. For most journals, reviewers, and editors, the upper limit is usually six lines in one panel (Davis, 2005; Day and Gastel, 2006). Generally, eight to ten bars are the maximum for a bar chart or histogram (Davis, 2005). If your graph has more than six to eight lines, you should consider dividing it into two graphs.

- Position tic (and subtic) marks inward from the axis and toward the data lines and data points. In no instances should tic marks protrude outward from the axis lines. Tic marks are remnants from the lines of an imaginary sheet of graph paper used to construct your plot. Avoid the clutter caused by too many tics, and try several choices of number of tics per unit of axis until the plot looks uncluttered.

- Axis labels should be simple and contain as few numbers and letters as possible. For example, convert labels such as 1000, 2000, 3000 g into 1, 2, 3 kg.

- Axis labels should be large, and in some cases, they may look disproportionately large. Axis labels should be completely clear, and they should show the units of measure. The size of most figures is reduced, sometimes by 50 or 75%, when printed in journals. Starting with larger axis labels will mean the axis labels will remain legible after reduction.

- Use only common symbols. Open and closed circles, squares, and triangles are the preferred symbols. These six symbols are all you should need for almost all figures, if you have followed the above guideline of no more than six lines in a panel (Davis, 2005; Day and Gastel, 2006; Yang, 1995). Pairs of these can be used sometimes for comparisons within the data set. For instance, the data associated with a closed circle may be compared with the data associated with an open circle, and the data associated with an open square may be compared with the data associated with a closed square—all in the same graph. Do not use cutesy symbols, such as a | , *, @, daggers, and so on, as they often confuse rather than clarify the presentation, and are often reserved for specific applications.

- Are your data expressed along a continuous variable like time or concentration? If so, line plots are probably ideal. If your data represent comparisons of discrete treatments like soil types or extraction methods, then bar charts, histograms, or other plot styles may be preferred.

- It is easier for the reader to interpret the meaning of a figure when the continuous variable is plotted on the x-axis, rather than the y-axis. The y-axis then should contain the dependent variable such as the response over time (say, growth of the cells or plant), or it should contain the response to the applied concentration of the treatment.

- Without regard to plot style, the width of the plot (length of the x-axis) should usually exceed the height of the plot (length of the y-axis). Typically, the ratio of the width to the height should be about 1.3 to 1, or about 4 to 3. In many cases, this ratio may extend out to 1.6 to 1, or about 5 to 3. In cases where this ratio is inverted, that is, the height (length of the y-axis) is greater than the width (length of the x-axis), the plot may look ridiculous, and the meaning may not lend itself at all to interpretation. (You may want to plot an "inverted" set of your data just to prove this to yourself.)

- If you find you have two or more figures that have identical x- or y-axes, then strongly consider stacking them together to save space. See the example figure that follows. Normally, stacking can best be done when the x-axes are identical, and the individual graphs are stacked one on top of the other.

- Always show the 0 (zero) value (origin of either axis) along numerical axis scales. Use an axis break, or several breaks if necessary, if the values for your plot area begin far from zero, or there are major gaps between groups of data. See the example figure that follows.

- Select numerical scales for your axes so that your data fill the internal area of the figure bounded by the axes. Reduce the amount of wasted space or white space to a minimum.

- The printed figure submitted should be no smaller than the actual size anticipated in the published article. It is also best if the figure in the manuscript is not much more than 1.25 to 1.5 times as large as the published figure will be. Even if the editorial office enlarges the plot, which they almost never would do, the quality of the figure would leave much to be desired.

- Units of measure should always be in parentheses.

- Consider carefully whether a legend that identifies each treatment is helpful or creates unnecessary clutter within the figure. If needed, place it within the field of the figure, and do not place a box around it (Davis, 2005; Day and Gastel, 2006). Depending upon the journal, a legend may be required within the figure, or it may be required as part of the verbiage in the caption for that figure. Regardless, readers must be able to interpret the plot plus the caption without reliance on the text of the paper. The legend should define all symbols, curves, and all abbreviations not identified as a part of a curve.

- Never allow a label or legend to increase the dimensions of the figure beyond those required for the presentation of the data. In all cases, your figure axes should not extend farther than your data permit, and you should not extrapolate your data or your interpretations beyond the limits of the data.

- If two y-axes are used, the reader must be able to match the proper data set with each axis. See the example figure that follows. Try and refrain from using two, three, or more axes on one plot because matching the data set to its axis can be very confusing to the reader. Often, this problem can be alleviated by the use of stacked graphs, because there will be a common x-axis. Many times three-dimensional plots look great, but they are extremely difficult to interpret unless the reader is highly skilled in that area.

- Situations where multiple colors in plots of data are justified are extremely rare.

- Plots may be boxed, so that the data set is completely enclosed, or they can remain unboxed and bounded by the axes. See your intended journal to determine whether one way or the other is preferred, or required. The example figures that follow include a figure that is boxed.

- Dependent variables are typically associated with the y-axis, whereas independent variables are usually found in association with the x-axis. You can help yourself remember this by use of the saying "y is a function of (depends upon) x."

- If several figures are constructed, be sure to apply your dimensions, lines, symbols, and so on, consistently across all figures, so that the reader is not confused as she or he moves from one figure to another.

- Never allow computer software to limit the quality of your work. If the software package you are using does not have the features you need to construct a figure properly, then change software.

- Check the instructions to contributors regarding how photographs or other visual images should be submitted. Whether they are submitted as figures or plates, or are handled electronically is an issue. How they should be printed (paper type, image-quality standards, and so on) is another important factor.

- Color may be permitted in visual images within figures or plates, although some journals may not allow it. Be aware, however, that journals often charge high fees for color images. Also be aware that the charges for color are in addition to the charges you will already pay for page charges.

THE SPECIAL CASES FOR PHOTOGRAPHS

Photographs present issues that may not be associated with the production of bar charts, histograms, line graphs, and so on. They need additional care because the writer cannot go to the computer software package that created the graph and make adjustments in line widths, symbol size, axis labels, and so on. A photograph, photomicrograph, etc., must be handled carefully because the entire experiment may have to be conducted again if the artwork is not suitable for publication. Day and Gastel (2006) presented excellent information about how to handle photographic materials that are part of the manuscript. Here we summarize pertinent information from that resource.

- As with tables and figures, be sure you really need the visual image to tell your story. Sometimes a picture is worth a thousand words, but other times one or two sentences can relay the information contained in the visual image.

- Be sure the image is of the highest quality because its quality will get reduced as the item moves through the publication process. In addition, if the visual image is of extreme importance to your manuscript, then choose a journal that specializes in high-quality transmission of visual images. Not all journals will care about or have the equipment necessary for maintaining a high-quality image. The acceptance of one or more of your take-home messages may depend upon transmission of a high-quality image.

- Although most journals now request or require use of electronic images, some do not have the equipment necessary for their use. Be aware of this as you decide which journal you will use.

- Be cognizant of the column format of your journal, and make your visual image to fit the journal in one, two, or however many columns the journal uses. The highest quality resolution will occur when there is neither reduction nor expansion of the image. The usual case is that the image will need to be reduced, and if this is the case, strive to produce an image that requires no more than a 50% reduction in size.

- For most original photographs, the entire image may not be needed nor wanted, so you can use cropping to get to the correct size to fit the needs of the journal. If you cannot crop the figure yourself, provide the editorial staff an original photograph that contains the area of focus delineated by arrows, letters, or some other means so that the journal may crop the photograph properly.

- For photomicrographs, place a measure of distance in the photograph so that the length of the image is unequivocal to the reader. This often allows the writer to skip a statement of the magnification in a legend or the figure caption, and it will also be clearer to the reader.

- Submit color images only if your journal allows it and only if the color helps readers understand your related take-home message.

- If a line drawing is necessary, then use a professional illustrator as the quality of their work often pays for itself quickly. Use of professional illustrators is quite common in the medical professions.

PLACING DATA IN THE RESULTS TEXT

Very often, information presented in a bar graph or a histogram can be converted into in-text statements, especially if the data are discrete treatments and there are only a few of them. In most cases, if only a maximum of six to eight numbers must be presented, it may be best to write those numbers as part of a declarative sentence and place it directly in the text of the results section. Peat et al. (2002) present an excellent discussion on the rules for reporting numbers, especially when they are reported in the text, and here are a few key issues. When presenting numbers in the text, be sure you do not start a sentence with a number or symbol. For instance, use thirteen rather than 13 to start the sentence. Also, do not present a number less than one within a sentence by starting with the decimal point. Always place a zero to the left of the decimal point, and use 0.05 rather than .05.

Example of a figure in which the minimum value along the x- or y-axis is not zero

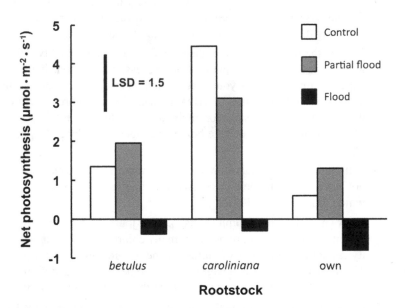

Example of a boxed figure with two y axes

Use of both a left y-axis and a right y-axis can facilitate showing similarities or differences between two dependent variables. The right y-axis in this example has been improved by having 0 (zero) represented at its upper end and an axis break included. Boxing plots, regardless of whether they have a right y-axis, is optional, and dependent upon your journal.

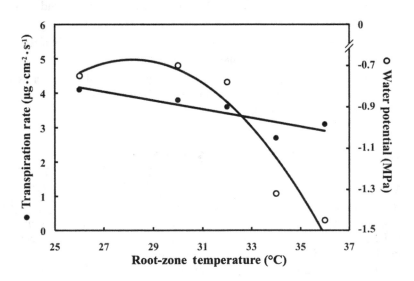

Example of a stacked figure with broken axes

Stacking these plots saves space, avoids unnecessary repetition of the x-axis labels, and allows readers to compare the data in the upper and lower panels easily. Breaking axes between 0 and the first tic mark prevents what otherwise would be a large unused white space within the plots.

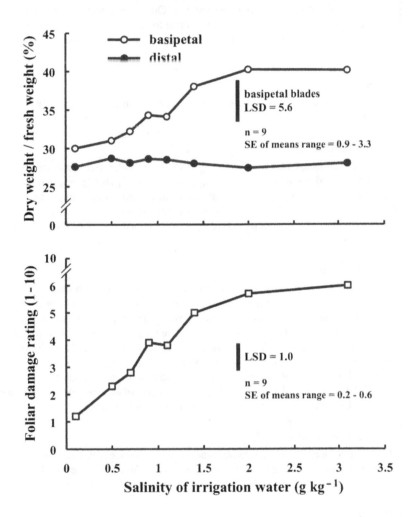

EXERCISE 11.1   Constructing a Table and Figure from One Set of Your Data

Background

This exercise has been designed to fulfill three purposes that are very essential in beginning to build your manuscript. Our first purpose is to have you refine further the take-home message you feel has the most straightforward set of associated data. Our second purpose is to have you use those data to practice constructing a table and figure suitable for use in your manuscript. Our third purpose is to cause you to think critically about the advantages and disadvantages of presenting this set of your data in either tabular or graphic formats.

General Instructions

Please answer each part on a separate piece of paper. Complete this exercise by following each set of specific instructions for that part.

Part 1 Instructions

Write several paragraphs in which you provide a *general critique* (not specific to your data) of the advantages and disadvantages of using tables versus figures (plots, graphs, etc.) to depict data. You should be able to provide at least two or three each of advantages and disadvantages for the use of both tables and figures. If you cannot recall several of these advantages and disadvantages, return to your reading assignments to refresh your memory about them before you complete this exercise.

Part 2 Instructions

Further refine one of your take-home messages from the research data on which your manuscript will be based. This take-home message should be associated with your most straightforward and unequivocal set of data related to one of your take-home messages. It also would be good to choose a take-home message that only required one set of data to prove/provide evidence for this take-home message. Your refined take-home message should be limited to *one clear and direct statement or sentence* that represents an appropriate balance of conciseness and detail. Below this refined take-home message, list the dependent variables (variable names, not the actual data) that you plan to include in your manuscript as experimental evidence that supports or defends the take-home message you chose to use in this exercise.

Part 3 Instructions

Choose one category of quantitative, experimental evidence (data) you cited above. Create one table and one figure by using the *same set of data* as your source of evidence. Be sure to follow style and formatting guidelines for how a table and a figure should be constructed.

Part 4 Instructions

Choose one of these formats as *the best one* for presenting that set of data. Defend your choice of format as *the best one* for the data you used in Part 3 by addressing these questions in your response:

A. Which format did you choose for presentation of your data?

B. Why do you prefer that format to the other format for your set of data?

C. How did you decide what type of plot, graph, etc., to use for your figure?

D. Would it be preferable to place some/all these data in the text rather than a table or figure? Why? How would placing the data in the text affect writing your manuscript?

EXERCISE 11.2   Constructing Additional Tables and Figures

General Instructions

Repeat Exercise 11.1 for each of your other take-home messages. You do not need to repeat Part 1, but you should complete all of Parts 2, 3, and 4 of Exercise 11.1. In Exercise 11.1, we asked you to use a straightforward set of data to conduct that exercise. Things may change at this point. The data that you may be using for these additional take-home messages may not be as simple and straightforward as the data used for Exercise 11.1. You may also need more than one table and/or figure to provide satisfactory evidence for proving these additional take-home messages. You should do what you need to do to prove these additional take-home messages. When Exercises 11.1 and 11.2 have been completed, you will be in excellent position to begin to write the text of your results section. As you become more skilled at fulfilling our three purposes for these exercises, it will take you less time. Ideally, as you really become experienced with publishing, you may get to a point where you mentally conduct Exercise 11.1 *before you start your research*. We hope you get to this point during your professional career. Exercises 11.1 and 11.2 may be the most critical exercises you will complete in this book.

REFERENCES

Briscoe, M.H. 1996. *Preparing scientific illustrations: A guide to better posters, presentations, and publications.* Second edition. Springer, New York, NY.

[CBE] Council of Biology Editors. 1988. *CBE style manual.* Fifth edition. CBE, Bethesda, MD.

Cleveland, W.S. 1994. *The elements of graphing data.* Second edition. CRC Press, Boca Raton, FL.

Davis, M. 2005. *Scientific papers and presentations.* Second edition. Academic Press, San Diego, CA.

Day, R.A. and Gastel, B. 2006. *How to write and publish a scientific paper.* Sixth edition. Greenwood Press, Westport, CT.

Gustavii, B. 2003. *How to write and illustrate a scientific paper.* Cambridge University Press, New York, NY.

Hodges, E.R.S. 1989. *The guild handbook of scientific illustration.* Van Nostrand Reinhold, New York, NY.

Katz, M.J. 1986. *Elements of the scientific paper. A step-by-step guide for students and professionals.* Yale University Press, New Haven, CT.

MacDonald-Ross, M. 1977. How numbers are shown. A. V. Communic. Rev. 25:359–409.

MacGregor, A.J. 1979. *Graphics simplified.* University of Toronto Press, Toronto, Canada.

Matthews, J.R., Bowen, J.M. and Matthews, R.W. 2000. *Successful scientific writing. A step-by-step guide for the biological and medical sciences.* Second edition. Cambridge University Press, New York, NY.

Peat, J., Elliott, E., Baur, L. and Keena, V. 2002. *Scientific writing. Easy when you know how.* BMJ Books, London, UK.

Tufte, E.R. 1983. *The visual display of quantitative information.* Graphics Press, Cheshire, CT.

Woodford, F.P. (Ed.). 1968. *Scientific writing for graduate students. A manual on the teaching of scientific writing.* The Rockefeller University Press, New York, NY.

Yang, J.T. 1995. *An outline of scientific writing.* World Scientific Publishing Co., Singapore.

CHAPTER *12*

RESULTS IV: PREPARING THE WRITTEN TEXT OF THE RESULTS SECTION

> It is better to remain silent, and be thought a fool, than to speak and remove all doubt.
>
> —Mark Twain (Samuel Clemens)

NEED FOR THE WRITTEN TEXT

The written text of your results section serves several functions. First and foremost, the written text functions as the engine that transforms your data, which are usually dry facts and measurements, into vibrant results that set the basis for the story your research results will tell. Second, it is one of the three places where your data may be reported. Third, all other sections of the manuscript are subservient to the results, because they provide the background needed for the reader to understand why and how the research was conducted, and what the research means. Recall that most manuscripts present the bulk of the data in either of the two other places available to you, the tables and figures. It is difficult for readers to comprehend long lists of numerical data in a series of sentences in the results text, and if this is the case with your data, then those data should be relegated to the tables or figures. Likewise, a picture may convey a thousand words. What might take paragraphs to describe in the text can be shown more elegantly and honestly with a table, figure, photograph, or other visual image.

CHARACTERISTICS OF THE RESULTS SECTION TEXT

If tables and figures contain the bulk of your data, what is to be written in the text of the results section? How should this text look? What are the typical characteristics of the text of the results? In answering these questions we can formulate the following characteristics.

- This block of text is usually the shortest section in the manuscript. Often, the text of the results section will comprise just one to two typed, double-spaced pages, although in some instances, there may be a third or fourth page. When the results text is justifiably this long, the writer should consider dividing the results text into units, each with a subheading.
- For many writers, the text of the results section is often the easiest to prepare. In many cases, all you need to do is "walk" readers through your findings. Introductory

Getting Published in the Life Sciences, First Edition. By Richard J. Gladon, William R. Graves, and J. Michael Kelly
© 2011 Wiley-Blackwell. Published 2011 by John Wiley & Sons, Inc.

material, methodology, and interpretation do not belong here—only the straight-forward facts that constitute your new contribution to knowledge.

- Emphasize only the most important information, focused on the questions asked and developed in the rationale for doing the work and the objectives that you wished to complete in the work.

- In all cases, your data should provide the basis for your take-home messages. Often, if you have done a good job of developing your take-home messages, a simple rewording of those take-home messages will suffice as the topic sentences for each paragraph in the results text.

- Sentence structure should be particularly simple. Short, declarative sentences, all in the past tense, should predominate. Consequently, the text of the results may seem mechanical or dry to some readers. However, reviewers, editors, and other scientists expect this and value it. Such a terse style probably means you are reporting only the facts and are reporting those facts accurately.

- No repetition should exist within your results text paragraph(s). Once you have finished reporting a category of data, you are done considering that information within this section of the manuscript.

- Some repetition will probably exist between your results text and the titles of tables or captions of figures. Remember, readers should be able to assess the way you conducted your work when looking at each table or figure—they should stand by themselves. Readers should not need to refer to the text as they try to comprehend the content of tables and figures.

- Parenthetical references to all specific tables and figures are both expected and required. Unless it creates ambiguity, place the location that is the source of the data parenthetically at the end of the sentence; for example, "(Table 1)". If there is a problem of ambiguity, then place the location in a manner that removes the ambiguity. In particular, this becomes an issue when several tables or figures are referenced in order to make a point in the written text.

- Consider deleting any table or figure you do not feel compelled to cite in the results text, unless the table or figure is specifically in the manuscript to augment information contained within either the introduction or the materials and methods sections. In a case such as that, the table or figure should be contained within the introduction or the materials and methods sections, rather than the results section.

- Few, if any, data shown in tables and figures should be presented or relisted in the same way within the text. You should refer readers to tables or figures to see the details, not repeat those details verbatim in the text. For example, you might say, "The putative new hormone increased root development by 34% after four weeks (Table 1)." In such an instance, the table would probably contain two means, one of which is 34% greater than the other. It would not contain "34%" per se. As writers gain experience, they often find creative ways to report the data in tables and figures to emphasize an important point (the putative hormone increased root development) without repeating the exact numbers reported in the table or figure. This is a very useful technique to master.

- Describe your results in a logical order, usually chronologically, and cite tables and figures in numerical order (i.e., do not refer to Table 3 before you refer to Table 2). In addition, describe the data within tables and figures in the order in which they appear, in left to right, top to bottom fashion.

- It is a good idea to order the information within a table in the same sequence in subsequent tables and to use the same sequence when presenting the information in the text of the manuscript. For example, if you have data from several locations, you should follow this guideline. Construct the tables so that the locations occur in the same sequence in all tables.

- If the data you are presenting should/must be analyzed statistically, then complete those analyses before you write the manuscript.

CHECKLIST OF WHAT SHOULD NOT BE PRESENT IN THE RESULTS TEXT

As you write the text of your results section, avoid introducing any of the following characteristics into the text:

- inconsistencies between what is stated in the text and what is stated within the data in the tables and figures (this problem is avoidable, if one takes care and devotes attention to detail in preparing the take-home messages, the tables and figures, and the text);

- interpretations;

- conclusions;

- redundancy between the text, tables, and figures;

- comparisons to other work (yours or those of other scientists);

- materials and methods;

- more data than are necessary to convince readers of your conclusions;

- deceptions (e.g., omission of data that contradict your take-home messages; do not refuse to present information and address its value simply because it does not agree with and support your original hypotheses developed during the early stages of the research project; please see quotation at the beginning of Chapter 11);

- the word "significant" (if you say two means were "different," and your materials and methods text contains your method of statistical mean comparisons and separations, then readers will know the two means you describe as "different" differed significantly);

- references;

- excuses;

- opinions;

- speculation(s);

- explanations;

- biases.

When you have completed a draft, review it carefully and edit out any of these you find.

EDITING EXAMPLES OF POORLY WRITTEN RESULTS TEXT

Use information from the previous pages and chapters as you examine these examples of problematic results text. Edit these two results sections rigorously and be prepared to discuss your reactions and editorial comments.

EXERCISE 12.1 Poorly Written Results I

The water deficit treatment resulted in a pre-dawn stem water potential of -1.9 Mpa (Table 1), which has been defined as severe (Hsiao, 1973). The weeklong water deficit treatment did not impair the ability of the seedlings to recover from drought stress, as re-watered seedlings had water potentials commensurate with the control seedlings 1 week after watering (Table 1). Application of ABA to well-watered seedlings did not significantly affect the water potentials of the treated seedlings compared to the buffer-control or normally watered seedlings.

The probe used in the current study hybridizes to a 1.8-kb transcript that accumulates in peach bark during cold acclimation (Artlip et al., 1997). A transcript of the same size was observed in all tissues in this work suggesting that the same gene is being expressed in these tissues (Fig. 1). Minimal expression of _ppdhn1_ transcript was seen in bark collected from seedlings prior to potting (PP) or prior to treatments (FL). The 1.8-kb transcript, however, was clearly induced in response to water deficit treatment (Fig. 1). The transcript was still present 1 week after recovery from the water deficit but at a lower level. The ABA application also induced transcription in bark tissues, at a higher level than the ABA control (Fig. 2). The likelihood of the differential transcript accumulation being due to isogenes is slight. DNA blot analysis utilizing this probe indicated that, under stringent hybridization and washing conditions used by Artlip et al. (1997) and this research, the probe hybridizes to a single DNA fragment, suggesting a single gene (Artlip et al., 1997).

EXERCISE 12.2 Poorly Written Results II

To test the effect of reduced catalase activity on chilling stress tolerance, T_2 seedlings from transgenic 11–12 plants and from control plants (untransformed wild-type tomato and azygous 11–16 plant) were grown under a normal greenhouse conditions and transferred to a dark, 4 °C cold room for 4 days. The chilling treatment was performed in the dark to avoid the complication of the effect from photooxidative stress. We have previously reported on the diurnal variation in the chilling sensitivity of tomato seedlings (Kerdnaimongkol et al., 1997). In these experiments, chilling sensitivity was highest in seedlings chilled at the end of the dark period, and these seedlings became more resistant to chilling injury on exposure to the light. Thus in this study, seedlings were chilled at the end of the light period when plants were at the most critical point to overcome the sensitivity of seedlings to low temperature. Transgenic and control tomato seedlings exhibited no visible injury during the 96 h dark chilling. However, when the plants were removed from chilling and placed under greenhouse conditions for 48 h, all of the transgenic ASTOMCAT1 line 11–12 were severely damaged and subsequently died. In contrast, all of control seedlings survived the 96 h dark chilling. These results indicate that antisense suppression of pTOMCAT1 and catalase activity in transgenic tomato seedlings leads to enhanced sensitivity to hydrogen peroxide treatment and chilling temperature.

CHAPTER 13

REVISING AND EDITING

You cannot solve a problem with the same thoughts that created it.

—Albert Einstein

We are the products of editing, rather than authorship.

—George Wald

DEFINITIONS

Several definitions of key words are in order (Mish, 2004). It is important to understand the differences in the meaning and usage of these often-misused words.

- Revising is to look over again to correct or improve (e.g., a manuscript), or, alternately, to make a new, amended, improved, or up-to-date version.
- Reviewing is to give a critical evaluation of, or, alternately, to go over or examine critically or deliberately.
- Editing is preparation of writing for publication, or, alternately, altering, adapting, or refining to bring about conformity to a standard or to suit a particular purpose. Thus, editing encompasses both functions of revising and reviewing.
- Proofreading is to read and mark corrections in a document.

This chapter is focused mainly on the function of revising. However, we will address reviewing as it pertains to an author seeking either an internal or external review of the manuscript during preparation. Peer review, which will be discussed in Chapter 24, occurs after the manuscript has been developed completely and submitted for consideration for publication. Although we have included the definition of proofreading in this chapter, we will not address it here. Proofreading is done only on a galley proof or page proof of the manuscript, immediately before publication. Actions associated with proofreading will be introduced and discussed within Chapter 26.

NEED FOR AND PRINCIPLES OF EDITING

You have finished preparation of your tables, figures, and the associated text that constitutes the results section of your manuscript. Now is the first time for you to revise your work, in part by seeking evaluations from colleagues. We have previously defined editing as the act(s) of preparing a piece for publication or, alternately, altering, adapting, or refining to bring about conformity to a standard or to suit a particular purpose. In its most basic sense, editing

Getting Published in the Life Sciences, First Edition. By Richard J. Gladon, William R. Graves, and J. Michael Kelly
© 2011 Wiley-Blackwell. Published 2011 by John Wiley & Sons, Inc.

comprises the latter parts of this definition. Revisions are completed by the author with the express purpose of editing the manuscript so that it (i) meets (conforms to) the standards required by the journal where the author wishes to publish the manuscript or (ii) suits the particular purpose of improving the manuscript. Both revising and reviewing are editing functions, because they refine the manuscript to meet the standard of quality demanded by the author(s), reader(s), and journal. Thus, editing encompasses both revising and reviewing, and it will not be addressed separately in this book.

Word choice and interpretation are the keys to the transfer of written information from a writer to a reader (Matthews et al., 2000). First, the writer must choose specific words that express thoughts and ideas unequivocally—that is, the writer must transfer thoughts into words. Second, the reader must convert those words into thoughts and ideas. Thus, the thought processes of the writer are transferred to the reader via the writer's choice and use of words. The inherent difficulty is ensuring that your thoughts as a writer unequivocally become the reader's concept of the written message. Here we can better understand the quotation from one of America's foremost writers:

> The greatest possible merit of style is, of course, to make the words absolutely disappear into the thought.
>
> —Nathaniel Hawthorne

Orators use timing and inflections to develop ideas, but these are not available to the writer, who must rely upon good word choice and usage. Often, a writer's oral description of the work becomes the first draft of the manuscript. There are many differences between spoken and written English, and effective revision points out these differences and allows the writer to structure the information as written, rather than translated spoken work. Properly completed revisions fine-tune this transfer process (Matthews et al., 2000).

Most writers revise their work, or its component parts, several times before anyone else sees it. When the writer has reached some level of acceptance of their written work, she or he will then ask a colleague or two to review it to make it better. It is at this point that our quotation from Dr. Einstein at the start of the chapter is especially true. The best thing you can do as a writer is to get the manuscript to a certain point of acceptance by yourself (i.e., by revising it), and then follow that with review by an informed outsider.

We believe there are two major ways to improve your writing via revisions. The first way is to learn how to critique the writing of others, because detection of faults in the writing of another scientist will help you detect faults in your own writing. Successful revisers are good readers (Alley, 1996). Ultimately, you will learn what to do, and what not to do, in your own writing. To help you achieve this, we have put together Exercise 13.1, and you should complete this exercise before moving to the next chapter of this book. The second way to improve your writing is to learn how to criticize your own writing (Barrass, 2002). The best writers invest time revising, criticizing, and reviewing their own work. The writer must learn how to become detached from their writing so that they can judge it objectively. The reviser must edit stringently, and in detail, lest a reviewer will, creating an uncomfortable situation for the writer.

The point of editing, revising, and reviewing is to make the manuscript meet a standard of quality that permits its publication in a refereed journal. However, remember that the published article will bear your name, not the names of the reviewers and editors, and be sure that you concur with the changes they suggest to your manuscript. Be aware that others' suggestions and editorial changes may give a different focus or meaning that was not your original intention (Davis, 2005). It is fine for you to accept constructive criticism, but you must keep in mind that it must also not change the intent of what you wished to say. Of course, sometimes we find that what we wrote was not true, and then we really become thankful for the work of reviewers.

Most authors go through several revisions of all or part of their manuscript, and it will often be reviewed several more times before it is ready to be published. It has been suggested that all writers are of relatively equal capabilities, but writers who can revise well become good or great writers, and good revising begets good writing (Day and Gastel, 2006; Matthews et al., 2000; McMillan, 2006). Many authors find that reading the revised manuscript aloud permits them to identify errors they could not catch when they simply read the manuscript. If you are in a mode where you are trying to find the best methods that will help you to prepare a manuscript, you may want to try reading it aloud to yourself. Often, reading printed copy helps revisers, reviewers, and editors find errors more quickly than reading the same document on a computer screen.

PRINCIPLES ASSOCIATED WITH REVISING

After a draft of the entire manuscript has been completed, or at least a component part, we recommend you allow your writing to get "cold" for a few days before attempting to revise it. Most writers who do this say they cannot believe they wrote all those mistakes into the rough draft, when they tried to be so careful. Truth be told, probably the main reason written work improves tremendously upon revision is that the author has become more objective in their view of their work after a cooling-off period (Tichy, 1988). Most major problems with a manuscript fall into one or more of these three categories: (i) problems with the specific observations (data), (ii) problems with the logic statements (analysis of the data), and (iii) the organization (presentation) of the information, in both written and numerical form (Katz, 1985).

It is now best to resign one-self to the realization that more than one draft will be required. After the period during which the draft gets "cold," it is time to begin revising the manuscript. Leaving the manuscript to get "cold" also allows the writer to shift into a different mode. Writing and revising are completely distinct activities, with completely distinct functions. Many authors erroneously allow themselves to procrastinate on their writing by inserting periods for revision into their time for writing, and this should not be done (Tichy, 1988). Authors should write the entire first draft without revisions. Many authors place too much importance on generation of the perfect first draft, and this leads to difficulties trying to fix problems with spelling, grammar, and punctuation. The first draft should be used to develop the ideas and organization of the manuscript, along with sensitivity to writing for the intended audience (McMillan, 2006). We finish with an apropos quotation:

> Much of the unpleasantness associated with writing is due to postponing writing too long and then trying to accomplish the impossible—a polished paper in the first draft.
> —Tichy (1988)

It has been suggested that two important events must occur in order for a writer to complete a successful revision of a manuscript (Alley, 1996). First, a good job of revising requires the reviser to separate herself or himself from ownership of the piece so they can conduct the revision objectively. The second event is that the reviser must be a good reader. No two people write the same, nor do any two people read the same. If you are, or have coached yourself into becoming, a good reader, you will understand that some people focus on structure when they read something, whereas others will focus on grammar or word usage. A good reviser will understand that there are different approaches to the revision of a manuscript, and they will revise using those approaches during each revision.

Usually, multiple revisions must be made, and some writers find the best way to revise is to focus each revision on a specific task. For instance, a writer may make one revision that focuses on clarity only, followed by a revision that only focuses on brevity, and so on, until all aspects of cleaning the manuscript have been addressed (Alley, 1996; Davis, 2005; Luellen, 2001). Another suggestion is to first focus on the necessary content of the manuscript, then its structure, and, finally, its style (Matthews et al., 2000; O'Connor, 1991). Tichy (1988) suggests there are five focal points in revising, and if each revision is assigned a specific focal point, cleaning the manuscript can be done bit by bit, rather than parts of each focal point being done during each revision. These five focal points are, in alphabetical order, increasing *brevity*, increasing *clarity*, better defining *content*, improving *style*, and meeting appropriate *standards of correctness* (Tichy, 1988). Tichy (1988) recommends the following flow of focal points during revisions: content → clarity → standards of correctness → brevity → improving style. We suggest each writer develop their own flow of events for each revision of their manuscript.

Matthews and colleagues (2000) approach this in a slightly different, but more thorough, manner, and they have an excellent treatment of the subject. If you are having trouble getting the revision process started, or completed, please see their work. They propose the series of tasks become successively smaller as each revision is approached: basic organization and logic → clarity, word usage, and style → brevity and the fine points of grammar and punctuation.

They also address sentence and paragraph length. They suggest a sentence length of 15 to 20 words, with 12 words being too few, and more than 40 words too many. They also suggest an average of about 150 words per paragraph. Thus, an average paragraph should contain seven to ten sentences. In most cases, it is not acceptable to have a one-sentence paragraph, and that sentence should be incorporated into the previous paragraph. You may consider these suggestions as you revise, but keep in mind there is no perfect formula for sentence and paragraph structure. Short sentences might be preferred when presenting concepts that are particularly complex. Longer, more complicated sentences might be appropriate when the ideas they contain are straightforward.

The general organization of the manuscript for most scientific journals comprises an abstract, and sections that present the introduction, materials and methods, results, discussion, and references. However, writers must be cognizant of the location of each of these sections for the journal of choice, as the order varies. Regardless of the order in which the sections are presented, the first subdivision of a section is the paragraph, and there are several items to consider in the revising process. Be sure that each paragraph starts with a topic sentence that orients the reader to the content of the paragraph. Next, make sure it is developed appropriately so that it moves to a conclusion of the information presented. During revision, be sure to add missing relevant information and delete irrelevant information in each paragraph. The paragraph should flow from one sentence to the next one, and sentence length should be varied by addition of shorter, longer, or maybe even compound sentences throughout the paragraph. Finally, be sure to use good word choice in building and revising the sentences (see Chapter 20).

PRINCIPLES ASSOCIATED WITH REVIEWING

Overview

The process of reviewing a manuscript may occur in two ways. You may be a reviewer of someone else's manuscript. In this case, you have a responsibility to the writer to spend the time and energy required to conduct a thorough review of the manuscript. Subsequently, the original author should address all the suggestions made by the reviewer.

Conversely, this same person may have written a manuscript and sought a colleague (you) to provide a review of the manuscript. In both cases, the reviewer must be thorough, honest, critical, and constructive in his or her appraisal of the manuscript. This book has been designed to help you through the process of producing a manuscript about your research. Thus, we will not focus on how you would conduct a review of someone else's manuscript. In this chapter, we will focus our effort on what you can expect from a reviewer, and how you might respond to a critical review.

Your Manuscript is Being Reviewed

Peer review, both in-house and after submission, is one of the most important foundations of science. If someone has accepted your invitation to review your manuscript, you have every right to expect they will conscientiously accept the responsibilities associated with that review. A reviewer of your manuscript should accept this responsibility only if he or she (i) feels comfortable that the content of the manuscript is in his or her area of expertise; (ii) believes the review process will not create a problem with conflict of interest; and (iii) can meet the deadline(s) associated with the review of that manuscript (McCabe and McCabe, 2000).

The writer should enter the review process with a positive attitude that the reviewer(s) will help make the manuscript better. If you enter this stage with a positive attitude, then the rewards of the review process are there for you. If the writer does not think positively about the review process, then the entire act of publishing one's work will be fraught with anger and fear. As no manuscript is perfect, each writer must be prepared to perform another revision of the manuscript.

Diplomacy in review criticisms is encouraged so that authors will be as receptive as possible to the results of this expenditure of time and effort by the reviewer. Good reviews should be clear, straightforward, and objective in their criticisms without putting the author in a defensive posture. *An excessively critical review can devastate an inexperienced author completing her or his first few manuscripts. Sarcasm and acrimony are unprofessional, and they have no place in a review. The reviewer should consider herself or himself a colleague of the author(s), helping the author(s) to get the manuscript into shape for publication. All criticisms should be helpful, constructive, and explicit.*

Upon receipt of reviews and suggestions for making the manuscript better, the author must look at the reviews objectively. There will be many suggestions for improving the manuscript, but there may be several suggestions the author(s) do not want to use. If you as the writer feel justified that the suggestion(s) given by one or more reviewers should not be made, then do not incorporate those comments (Alley, 1996; Davis, 2005). The writer must be very sure the suggestions are either incorrect in the context of the manuscript or will add confusion, rather than clarification, to the understanding of the content of the manuscript. Also keep in mind that the reviewer suggesting changes should supply a reasonable justification for why these should be done. A review that states certain things should be done, but without an explanation of why they need to be done, often leads to miscommunication between the author(s) and the reviewers.

All reviewers need to keep in mind that the content of a manuscript is the property of the author(s) until the manuscript has been released to copyright of the publisher. At that time, the information becomes part of our base of published research results accessible within the public domain. Thus, the manuscript must be treated as a confidential communication during the entire review process. Reviewers should not discuss the content of the manuscript under review with colleagues, except those who act as an additional confidential reviewer, called upon to clarify an issue for the reviewer. This occurs when a reviewer solicits help judging the science in the manuscript when they cannot do an adequate job on a particular body of information in the manuscript. If the reviewer needs to solicit too much help from an additional reviewer or two, then maybe they should consider returning the manuscript, rather than performing a less-than-adequate job on the review.

The main responsibilities of reviewers are to determine the verifiability of the results and to evaluate the merit or quality of the work to be published. In addition, reviewers should attempt to detect the rare instances of fraud or plagiarism by the author(s). The issue of a conflict of interest is a difficult one. On one hand, it is good for science to have competitors review one another's manuscripts before the work is accepted for publication. However, competitors who do not act ethically (e.g., hold up the review process of the competitor's manuscript so that their paper can get accepted first) create many problems in the publication process.

To communicate unequivocally, reviewers should use the set of proofreader's marks shown later in this chapter. Additional editing and proofreading marks can be used, but the list we provide includes the marks reviewers need most commonly.

CHARACTERISTICS AND QUESTIONS TO ADDRESS IN REVISING, REVIEWING, AND EDITING

As you complete sections of your manuscript, you may want to have others review those sections, in addition to your own revisions. As you go through the revisions you wish to make in your manuscript, at some point you, as the author, must step back from writing the manuscript and address issues that are of central importance to producing a high quality manuscript. You will want to address these issues related to your work before it goes any further, because poor answers to several of these questions and issues may affect the manuscript adversely. The instructions you will use here will focus on making your manuscript easier to review and edit for subsequent reviewers and editors.

Please keep in mind that each manuscript must first pass muster for good English and grammar, followed secondly by good content, which, thirdly, can then be adjudicated for scientific merit, the last stage in the acceptance of your manuscript. You should focus your energy on providing a critique of your own writing by addressing these items as the basis for your revisions. Subsequently, your peer reviewers will look for some or all of these characteristics when they review your manuscript. These items were the basis for the characteristics of good scientific writing introduced in Chapter 6. Each item may be rearranged to form a question, and both revisers and reviewers should reformulate these into questions and address them as they conduct their respective revision or review.

- *Accuracy and precision* must be evident in everything written.
- *Proper organization* leads the reader through structured, understandable text.
- A *logical flow* of words, sentences, and then paragraphs, moves the writing from point to point with understanding.
- *Clarity* is the basis for understanding, so your reviewer should question everything that could be misunderstood. Write so that you cannot be misunderstood.
- *Economy of words* eliminates words that do not have a function or are not concise. However, never sacrifice understanding for economy.
- *Informative words* are more powerful and tell you more than indicative words.
- *Proper choice of words* leads to understanding, because many words have multiple meanings or can be interpreted in several ways.

- *Write objectively* so that readers do not suspect bias.
- *Consistency* in your writing allows your work to flow from word to word, sentence to sentence, paragraph to paragraph, and section to section. If the writing is not consistent throughout the manuscript, then the words and structure become a distraction to the reader.
- *Interesting words* keep the reader attentive and focused on the messages you wish to communicate through your manuscript. However, be sure to use simple words, not complex words or jargon, in keeping the interest of your reader.

In addition, we also suggest revisers and reviewers look for problems with and seek answers for the following questions (Day and Gastel, 2006).

- Have the authors complied with the entire set of instructions for preparing the manuscript?
- Is everything organized logically—word to word, sentence to sentence, paragraph to paragraph, and manuscript section to manuscript section?
- Are the words in the text clear and unequivocal?
- Have all forms of information been presented concisely with neither too little nor too much verbiage?
- Have appropriate grammar, spelling, syntax, punctuation, and choice of words been used throughout the manuscript?
- Are all tables structured in a manner that makes the data easy to understand by the reader? Are the figures easily understood by the reader at first look?
- Is the presentation of the results (the tables, figures, photographs, etc.) lucid and unequivocal?
- Is the information presented consistently as the reader moves from section to section as she or he reads the article?
- Does the manuscript lack information it should contain? Does it contain information (either text or data) that should not be included?
- Is this manuscript addressing a topic relevant to the discipline?
- Is the content of this manuscript original, or have the author(s) or other scientists published the base of information previously?
- Is the manuscript or its parts too long or short in relation to the importance of the content or the amount of data provided?
- Are there errors of fact, interpretation, or calculation; that is, are the text and data accurately presented?
- Is the presentation of the information biased in such a way that fact cannot be distinguished from assumptions and suppositions?

REFERENCES

ALLEY, M. 1996. *The craft of scientific writing.* Third edition. Springer-Verlag New York, Inc., New York, NY.

BARRASS, R. 2002. *Scientists must write. A guide to better writing for scientists, engineers and students.* Second edition. Routledge, New York, NY.

DAVIS, M. 2005. *Scientific papers and presentations.* Second edition. Academic Press, San Diego, CA.

DAY, R.A. and GASTEL, B. 2006. *How to write and publish a scientific paper.* Sixth edition. Greenwood Press, Westport, CT.

KATZ, M.J. 1985. *Elements of the scientific paper. A step-by-step guide for students and professionals.* Yale University Press, New Haven, CT.

LUELLEN, W.R. 2001. *Fine-tuning your writing.* Wise Owl Publishing Co., Madison, WI.

MATTHEWS, J.R., BOWEN, J.M. and MATTHEWS, R.W. 2000. *Successful scientific writing. A step-by-step guide for the biological and medical sciences.* Second edition. Cambridge University Press, New York, NY.

McCABE, L.L. and McCABE, E.R.B. 2000. *How to succeed in academics.* Academic Press, San Diego, CA.

McMILLAN, V.E. 2006. *Writing papers in the biological sciences.* Fourth edition. Bedford/St. Martin's, Boston, MA.

MISH, F.C. 2004. *Merriam-Webster's collegiate dictionary.* Eleventh edition. Merriam-Webster, Inc., Springfield, MA.

O'CONNOR, M. 1991. *Writing successfully in science.* HarperCollinsAcademic, London, UK.

TICHY, H.J. [with S. Fourdrinier] 1988. *Effective writing for engineers, managers, scientists.* Second edition. John Wiley & Sons, New York, NY.

EXERCISE 13.1   Revising, Editing, and Proofreading

Instructions

We have created one hundred (100) errors in this short manuscript, and we would like you to sharpen your revising, editing, and proofreading skills by finding as many errors as you can. We have not assigned a value for each error, although there are errors of varying magnitude. This exercise will contain spelling errors, mistaken punctuation marks, more serious errors such as maybe forgetting to include a major part of the manuscript, problems with the in-text use of references, references themselves, and several other forms of minor and major errors. We purposefully will not tell you the exact number of errors of any one type or anything about the types of errors in the manuscript because you may focus on finding a certain number or type of errors rather than learning how to revise, edit, and proofread.

Also, this manuscript purposefully does not contain a cover letter of submission and/or a submission form, so do not allocate an error to the lack of either of these documents.

We wish you "happy hunting." Please keep this cover page with your completed exercise, as it contains your name.

Name: _____

Score: _____ /**100**

Housing Developments Adjacent to Prairie Remnants Cause Increased Gall Formation on Goldenrod

Lars A. Brudvig

Ecological Restoration Laboratory, Savannah River Site, Aiken, S.C. 29803

Received for publication _____. This research was supported in part by grants from Iowa State University, The State of Iowa Department of Natural Resources, and The National Science Foundation under Grant No. 305-24595832. The author acknowledges and expresses appreciation to Max and Delaney Brudvig for help with counting and recording formation of galls and wasp populations.

Environmental Stress

Housing Adjacent to Prairie Remnants Cause
Increased Gall Formation on Goldenrod

ADDITIONAL INDEX WORDS encroachment, wasp, ecology, ecological balance, ecosystem, ecosystem disturbance.

Abstract. Encroachment of housing developments into remnants of prairies upsets ecological balances by removing one or another of interacting species of plant and animal. My objective was to determine whether a reduction in the number of wasps caused by housing-development encroachment causes an increased incidence of galls on goldenrod growing in the prairie renmant. I located nine housing developments in Iowa that encroached on remnants of native prairies. I recorded the number of wasps present and the incidence of gall formation on goldenrod plants inhabiting those prairie remnants. Seven of the nine housing developments exhibited increased gall formation on goldenrods in adjacent prairies, and in two developments, there was no increase in the number of galls. Wasp populations decreased in the first 100 m of the undeveloped prairie. I also found a concomitant increase in the incidences of gall formation on goldenrod plants within the same first 100 meters of the prairie. I conclude that natural prarie renmants are affected adversely by encroachment of housing developments. State Departments of Natural Resources should take action to increase wasp populations in the first 100 m of native prairie remnants adjacent to recently constructed housing developments.

Introduction

Introduction and Review of Literature

Goldenrod gall fly (*Eurosta solidaginis* Fitch) is a parasite of a variety of tallgrass prairie goldenrod species, including late goldenrod (Solidago altissima L.). Female *E. solidaginis* lay eggs inside stems of *S. altissima* by using a specialized ovipositor. Eggs hatch within 10 d and begin feeding on the inside of the stem of *S. altissima*. At the same time, they produce a chemical that induces abnormal growth in the area of feeding. This growth eventually forms a gal which is about 35 mm in diameter and that protects the *E. solidaginis* larva from freezing during winter. In spring, larvae pupate and then exit the gall as a mature, adult *E. solidaginis*. Gallflies, in turn, are preyed upon by a variety of animals, such as downy woodpeckers (*Picoides pubescens* L.) (Confer and Paicos, 1985) and the parasitic wasps *Eurytoma gigantea* Walsh and *Eurytoma obtusiventris* Gahan (). These wasps have a specialized ovipositor that is used to inject eggs into goldenrod galls. These larva then pray upon *E. solidaginis* larvae and exit the gall when mature (Milne and Milne, 1980).

Because recent research suggests that ecosystem disturbance can produce trophic cascades, that can manifest in losses of top predators (Peace et al., 1999), I set out to measure goldenrod/fly/wasp dynamics as an indicator of ecosystem health. A basic definition of a trophic cascade as a chain-reaction where changes to one trophic level (or part of the food web) results in modifications to another trophic level, through predator/prey interactions.

This hypothetical data set explores goldenrod/fly/wasp interactions in 9 tallgrass prairie remnants adjacent to recently constructed housing developments in Iowa. The prairie remnants are fairly sizeable, high-quality remnants ($\sim$15 ha; $\sim$400$\times$400 m) adjacent to advancing housing developments. In each case, the western edge of the prairie remnant borders on the housing development, but in all cases, the remainder of the prairie is surrounded by field, row-crop agricultural production. Conservation groups have expressed concerns about the proximity of these housing developments to the adjacent prairie remnants. In this work, I will determine what, if any, affects the recently constructed housing development has on the prairie remnant and its wasp populations and gall formation on goldenrod.

Materials & Methods

Nine recently constructed housing developments adjacent to prairie remnants were located in Iowa (Table 1). In each case, the western edge of the prairie was also the eastern edge of the housing development. One housing development was located in each of these nine counties: Polk, Scott, Woodbury, Linn, Johnson, Pottawattamie, Blackhawk, Dubuque, Cerro Gordo, and Story. 1-m × 2-m quadrats (sampling rectangles) were established every 20 m, starting at the western edge of the prairie, on west-east linear transects. In total, 5 transects were established, each with 10 quadrats. Previous research has shown that *S. altissima* stems are distributed uniformly throughout each of the prairie remnants (Brudvig, unpublished data). Within each quadrat, all *S. altissima* stems were harvested and returned to the laboratory, with quadrat identity intact. In the laboratory, stems were sorted by presence of galls. Galls were opened and noted for presence or absence of *Eurytoma* wasps. Effects of edge distance on gall formation and wasp presence were tested with one-way analysis of variance. The proportion of goldenrod stems containing galls and the proportion of galls containing wasps were treated as dependent variables, and the distance from the edge of the development was treated as a fixed factor (Quinn and Keough 2002).

Result

The proportions of stems containing galls and the proportions of galls containing wasps showed significant edge effects (Tables 2,3). The greatest proportion of *S. altissima* stems infected with galls was on the western prarie remnant edge, with 60% of stems containing galls. The proportion of infected stems decreased between 0 and 100 m from the western edge, at which point the proportion of infection stabilized at 20% for the duration of the sampling transects (Figure One). Conversely, the proportion of galls that contained wosps was lowest at the western prairie edge, at 10%, and the proportion of galls increased with movement eastward. Wasp values also stabilized at 100 m, and they did not differ between 100 and 180 m from the western edge (Fig. 2).

Discussion and Conclusions

We conclude that housing developments are adversely affecting the prairie remnant for up to 100 m from the edge. The presence of the housing development increases gall formation through decreased wasp abundance. I speculate that some aspect of the housing developments, potentially wasp predators, has reduced the abundance of wasps, which has resulted in an increased number of galls forming gadflies. Although, at this point, *S. altissima* is equally abundant throughout the prairie remnant, I fear that if *E. solidaginis* populations continue to remain high, this could eventually have a negative impact on the number of *S. altissima*. As the impact of the housing developments seems to persist up to 100 m into the prairie remnant, and the prairie remnant is approximately square with sides of 400 m, a development that surrounded the prairie remnant would effectively impact 75% of the prairie remnant. I note that I tested only one trophic interaction in this prairie renmant; however, many other interactions occur, such as plant/pollenator, and may be in jeopardy. I recommend that housing developers abandon plans to continue building these housing developments, and that the remainder of the prairie remnants continues to be surrounded by fields used for the production of horticultural and agronomic crops such as tomatoes, potatoes, corn, and soybeans.

Literature Cited

Brudvig, L.A. and J.M. Brudvig. 2010. Personnel communication.

Confer, J.L. and P. Paicos. 1985. Downey woodpecker predation at goldenrod galls. Journal Field Ornithology 56:64–56.

Milne, L. and M. Milne. 1980. National Audubon Society Field Guide to North American Insects and Spiders. Alfred A. Knopf, New York.

Pace M.L., J.J. Cole, S.R. Carpenter and JF Kitchell. 1999. Tropic cascades revealed in diverse ecosystems. Trends Ecol. Evol. 14:488–483.

Quinn, G.P., M.J. Keough. 2020. Experimental design and data analysis for biologists. Cambridge Univ. Press, NY.

TABLE One.   Attributes of nine Iowa prairies used to study the affects of housing development on goldenrod gall formation and gad fly parasitism

Sight	County	Latitude (°N)	Longitude (W)	Size (ha)	Year subdivision established
Prairie 1	Polk	42.027866	-93.644704	15.1	2008
Prairie 2	Scott	41.609795	-90.589142	27.3	2006
Prairie 3	Woodbury	42.483618	-96.347694	17.2	2007
Prairie 4	Linn	41.95298	-91.719275	15.9	2007
Prairie 5	Johnson	41.663968	-91.554673	12.5	2004
Prairie 6	Pottawattamie	41.273807	-95.800438	21.7	2008
Prairie 7	Blackhawk	42.551247	-92.390728	20.0	2006
Prairie 8	Dubuque	42.509122	-90.69068	18.6	2005
Prairie 9	Cerro Gordo	43.126296	-93.399582	19.1	2007

Figures

Figure One. The porportion of golden rod (*Solidago*) stems containing gall declines with distance from housing development edge. Data was collected in nine Iowa Prairies along West/east oriented transects, originating at western prairie edges, which were adjacent to housing developments. The mean proportions were separated as shown by using Fisher's least significant difference test.

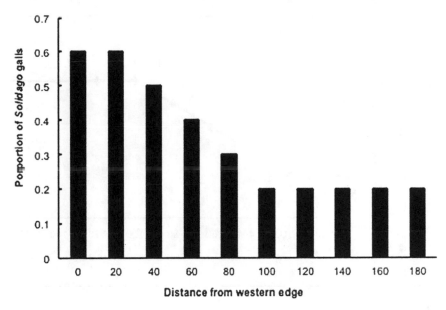

(Fig. 1)

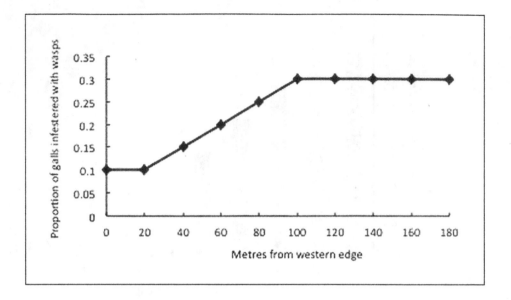

TABLE 2. Analysis of variance table for goldenrod gall data.

Source	Degrees of freedoms	Sum of squares	Mean square	F	P
Distance	9	2.34	0.26	775.91	< 0.0001
Error	80	0.027	0.00034		
Corrected total	89	2.37			

TABLE 3.   Analysis of variance table for wasp data.

Sources	Degrees of freedom	Sum of squares	Mean square	F	P
Distance	9	1.75	0.19	2.22	<0.029
Error	80	7.00	0.087		
Corected total	89	8.75			

CHAPTER 14

REFERENCES

Knowledge is not knowing all the answers. Knowledge is knowing how or where to find the answers.

—Dale Van Der Haar

DEFINITIONS AND NEED

The references section is the compilation of bibliographic descriptions of published information (references such as journal articles, books, and so on) used by the author(s) (CSE, 2006). In different journals it may be variously called "Literature," "Citations," "Literature Cited," "List of References," "References Cited," or any one of several other monikers. Use the moniker specified by your journal. Do not use a different one, and do not make one up. The Council of Science Editors (CSE, 2006) suggests that scientists use "References" rather than "Literature Cited" because "References" is more unequivocal than calling these citations, as the word citation has several meanings. Also, CSE (2006) distinguishes between references and a bibliography. References have been published and are cited in the course of the work. Items in a bibliography may or may not have been published, and they may or may not have been cited as the manuscript was produced, but are included because the author feels it is important for the reader to know about them. Davis (2005) provides an alternate set of definitions.

In this chapter, we will cover in-text references, the components of the systems used for those citations within the text, and the components of the compilation of the references (bibliographic reference information) at the end of the manuscript, or in another place if the information is not at the end of the manuscript. As we will see later, the system used for in-text references controls the format of the list of references at the end of the manuscript. The CSE Manual (Scientific Style and Format: The CSE Manual for Authors, Editors, and Publishers) (CSE, 2006) is considered the standard for all manuscripts written about the natural sciences. Virtually all journals defer to this reference when controversy arises.

PRINCIPLES

Three over-arching rules must be followed when a manuscript author works with previously published literature (Day and Gastel, 2006). First, authors should use, and cite, only references that are significant or critical to a facet of the manuscript. Cite the science, not the authors, lest one be accused of name-dropping (Peat et al., 2002). Second, all items that define the reference must be checked against the original reference, not a citation of it by another author. This second rule covers both the content of the reference and its defining characteristics such as the journal name, the volume, inclusive pages, and so on. These must be verified before the manuscript is submitted for publication, and this verification is the responsibility of the author(s), not the journal, the editor, or the copyeditor (Peat et al., 2002). Third, the author(s) must ensure that all references cited in the text also appear in the list of references, and every reference that appears in the list of references also is cited at least once, somewhere in the manuscript text. An additional rule is to cite only references that are readily available to others in the scientific field (Katz, 1985).

In-text reference citations may occur in the written text of all parts of the manuscript except the results section, and in many cases, the abstract. These sections respectively are a description of what was found and an overall summary of the content of the manuscript, and there is no need to accredit the work(s), idea(s), or word(s) of others in these parts of the manuscript. Other parts of the manuscript, such as the discussion, introduction, materials and methods, table titles, and figure captions, may contain in-text reference citations. This is especially important in the introduction section, because an author can only summarize the observations of others, and the discussion, where the author needs to present, or link together, the context of both previous observations and conclusions and those of the present manuscript (Katz, 1985).

There are several reasons why a reference section is a major component of a scientific manuscript. First, references establish your credibility as an author with the reader. Second, the references allow the reader to gain a better understanding of and give credit to previous work in the discipline. It is from these sources that you gathered much of your information for the conduct of your research and in the preparation of your manuscript. Third, references give the reader a clear understanding of how to locate the information.

At times, especially for inexperienced writers, the question of how many cited references should be used can be problematic. Writers want to make sure all bases are covered when they present background information, but often this leads to a long list of references. At some point, the writer must look earnestly at the number of references used and whether expression of that information can be condensed, especially in terms of the number of references. A complete set of references is important because it allows the reader to understand the scope and depth of the research reported (Katz, 1985). Then, the difficult part comes in removing some references so the presentation of the information can be condensed. There are no hard and fast rules that govern the number of references appropriate for a given manuscript. However, do not in any way "pad" the number of references cited, because you may lose all credibility with this foolish trick:

> Manuscripts containing innumerable references are more likely a sign of insecurity than a mark of scholarship.
>
> —William C. Roberts

Most journals do not limit the number of references that can be cited in their published articles, but some do. Perusal of the instructions to authors or the style manual for the journal often gives writers some guidance in this area. By their nature, it is likely that review articles will not have a restriction on the number of references. However, for journals that do impose restrictions, the limits are often 20 to 40 references for a regular research report and 10 references for a short communication, with some journals allowing fewer (Mathews et al., 2000; Peat et al., 2002). As is the case with the presentation of your data, it is more important that you select the correct references than that you are all-inclusive in reporting what has been done (Peat et al., 2002).

During manuscript preparation, and certainly before submission, the author(s) must confirm that every in-text reference citation is matched with its reference in the references section. Likewise, every entry in the list of references must be matched to at least one citation in the manuscript. Often, several key references will be cited several times in the text. The presence of these items must be checked "both ways," because in-text reference citations and references intended for use can be deleted during editing and revision without the author(s) remembering to remove them in all places in the manuscript. Errors in this category are almost as prevalent as errors in the factual content of the entries of the references section, but can be avoided easily if authors simply acknowledge the presence of each reference somewhere in the manuscript.

To accomplish this, the author(s) should conduct an electronic search for author names or other words unique to each citation. Or, try producing a hard copy of the references section and placing it adjacent to the manuscript while the manuscript is being revised. You can place a check mark next to the entry in the references section of the manuscript each time you come across an in-text reference citation of that reference. This method allows an author to ensure that all references have been used at least once in the manuscript, and it also tells the author(s) how many times each reference was cited. This last piece of information can become useful if a reviewer or editor mandates that some percentage of the references should be eliminated. Exercise 14.1 at the end of this chapter will give authors practice with this technique.

All parts of all in-text reference citations and the references list presented in the end section must be checked for accuracy by matching each reference with the location of the original work. Any journal editor will tell you that the greatest number of mistakes in an article are found in the references section or the in-text reference citations. Authors should check their original work when they finish the final draft of the manuscript, and they might consider doing it again at the galley or page-proof stage, just in case references have been added or deleted. Check thoroughly all parts of every reference [author(s), year, title, journal, volume, maybe issue number (if used), inclusive pages, etc.] against the original. Some authors make a photocopy of the references section of the manuscript then go to the library to confirm that each reference has been cited correctly and completely. Some journals require the corresponding author to provide the first page of all references cited within the manuscript so that these may be checked for accuracy before the manuscript is sent out to reviewers (Gustavii, 2003). Exercise 14.2 at the end of this chapter gives you practice finding these errors.

In the future, when you conduct a literature search to start work on a research project, deposit *all citation information* in a computer document that is mobile and readily accessible for changes; that is, create a master list of your references associated with that project. Later, when you are preparing the manuscript, you will be able to use this document as the basis for the references section (Day and Gastel, 2006; Gustavii, 2003). This should be done without considering the system or format used by the journal where you will first submit the manuscript, because the manuscript may be rejected by one journal but accepted by a journal with a completely different set of requirements. Accumulate this list of references in alphabetical order, as all journals that do not use the citation-sequence system will call for an alphabetized references section. It is also easier to add and delete references when the list is prepared alphabetically. If you have all this information in a computer document, then you can easily adjust it to match the required style and format of each journal, as there are software programs available to do this (Day and Gastel, 2006; Gustavii, 2003). Never use the master list directly, but instead make an electronic copy of it for the manuscript and edit it to match the requirements of the journal. You should deposit in your document everything that possibly could be used by a journal to cite a particular reference. It is easier to have more information and not use some of it than to have to go back to the library and look up titles or the inclusive pages for all of the references in the manuscript.

Probably the most important thing for a first-time author of a manuscript to remember is that there are numerous styles and formats within any one of the referencing systems, and "one size does not fit all." As you look at different journals and make your choice of where you will submit your manuscript (see Chapter 5 and exercises contained in it), please be aware that each journal will have its own style and format requirements within the system that it uses for referencing the literature used by the author(s), and you must follow them to the letter (and the punctuation mark). Gustavii (2003) provides an excellent comparison of the two main systems used for references in scientific publications. Although we are providing a lot of background information to you in this chapter, the best thing you can do, and the sooner the better, is to obtain and study in detail the instructions to authors document that every journal produces. In addition, peruse the articles that have been published recently. Between these two sources, you should be able to determine which system your journal is using and the exact formatting characteristics of your journal.

It is of the utmost importance that you format your in-text references and your references section exactly in the manner required by your journal. If you do not, there is a high probability that your manuscript will be returned to you before it is even sent out for review. In addition, incorrect formatting raises two significant "red flags" you really do not want to raise upon submission. One, you did not follow directions in writing the manuscript – did you follow them in conducting the research? Two, was this manuscript sent somewhere else first, rejected, and then submitted here without changing the format to match that of the new journal?

Sometimes, references are published in a foreign language. If this is the case with one of your references, then see the CSE Manual (2006). This source gives complete details on how to handle references in a foreign language, and that is beyond the scope of this book.

REFERENCE CITATION SYSTEMS

Overview of Citing Literature in the Text

There are three main in-text reference citation systems (CSE, 2006). Your journal's choice of system for in-text referencing dictates a particular system for presenting the references in the references section of the manuscript. However, there are numerous "adjustments" within each of these systems, and each journal may choose which "adjustment" to its particular system it would like to use. In a survey of 52 journals, more than 33 distinctly different formats were observed (O'Connor, 1978). More recently, 250 different formats were identified in scientific literature (Garfield, 1986).

It is important that you look in the style manual or instructions to authors document of your journal, as well as in recent issues of the journal, to determine the exact details of the system used by your journal. The following are some common examples of variations: complete title versus no title for the reference; complete, partial, or no punctuation in the entry; variable placement of the year of publication of the reference; portions of the entry in boldface type, and so on.

Below, we present the three main systems used today, and nearly all this information may be found in the CSE Manual (CSE, 2006). The first one, the name–year (or Harvard) system, is most common. The second system is the citation–name (or Vancouver) system. It is less popular than the Harvard system, but it is used more frequently than a third system, the citation–sequence system. The major common characteristic of the name–year (Harvard) and citation–year (Vancouver) systems is that the references section is ordered by alphabetizing the references by the last name of the senior author of each reference. Within each system, we will present the advantages and disadvantages of using each one.

Three additional methods for citing certain sources are not covered by these systems. These methods should be used infrequently because the information has not been scrutinized by peer review. In all three cases, the information is not recorded in the references section of the manuscript, and it appears only within the text of the manuscript. The first method covers information given by a presenter at a scientific meeting. You would like to accredit this individual, but the work has not been published yet. The best way you can cite the work is to paraphrase, and after that statement, parenthetically insert the words (Jane Doe, personal communication, date of communication). This same method of accreditation is used when you gained the information via other means (e.g., a telephone call, an e-mail message, a letter to you). You must obtain the person's approval to use this type of citation, as his or her interpretation may not be exactly the same as yours.

The second method occurs when you have found pertinent information in a report or some other written form of communication that has not been published and made generally available to the public. Again, the citation would occur only in the text, and it would be shown parenthetically as (Richard Roe, unpublished data) or (Richard Roe, unpublished results).

The third method is citation of an abstract, either published (in the program of a scientific conference, for example) or unpublished. A major reason abstracts should not be cited in the published references is that only about one-third of published abstracts become part of a peer-reviewed journal article (Liu, 1996). For this reason, some journals do not allow citations of abstracts in their published articles (Gustavii, 2003).

The Name–Year (or Harvard) System

This system was "invented" in 1881 by the Harvard zoologist Edward Mark (Chernin, 1988). Owing to confusion with reference works within the legal literature, which was called "Harvard," it is best to call this the name–year system (CSE, 2006).

The name–year system is a very friendly system; it has several advantages for authors because the citations are not numbered, so citations can be added or deleted without retyping the entire references section. It also helps readers because it identifies the author(s) and the year of publication in the text, adjacent to the information. This helps readers place information into its proper context more easily. Because the references are alphabetized in the references section, a reader can see quickly what a particular author has done because the alphabetized naming will group the articles by the person. This same thing can be said about locating the works of a specific person. This system is also very convenient when writing the manuscript. *Many resources that present information about how to write a journal article suggest that this system be used for writing all drafts of the manuscript. Then, at the point when it is finished, the writer switches to the system required by the journal where the manuscript will be submitted.*

The name–year system also has several disadvantages. The most important is that there are many rules that must be followed in using the in-text reference citations. This is especially the case for multiauthor papers that are referenced, and especially when several of these multiauthor papers have been published in the same year. Many of these rules conflict with recommended, standard bibliographic formats, and this can become a major issue of frustration for the writer. In addition, long strings of author names and years of publication can distract readers. Use of this system takes a lot of space in the text of the manuscript, which increases publication costs.

In the manuscript text, the in-text reference is displayed as (author surname, year) or (surnames of authors, year). Sometimes, a comma does not separate the name(s) and year of the citation, only a space. When multiple citations from the same author(s), but in different years, are cited, a comma separates the years. When different authors, or the same author(s) in different years, are cited at one place, a semicolon separates the citations. When there are multiple citations from *exactly the same author(s) in exactly the same order in exactly the same year*, the year designation is followed by a lower case a, b, c,. . . to separate the citations. When a comma is used to separate the author(s) and date of a citation, a semicolon is used to separate citations, e.g., (Smith, 2004; Jones, 2006). Most possibilities are covered in the examples below, and if one is not, then consult the CSE Manual (CSE, 2006).

In most cases, the "et al. rule" for these citations is that, in the text, you list the author if there is only one author, you list both of the authors if there are two authors, separated by an "and", and if there are three or more authors, the first author is followed by et al.

Either of the in-text citation types below may be used, depending upon the syntax of the sentence in which the citation was used. Normally, we suggest you strive for uniformity in writing style, but it is satisfactory to move from one type to the other type as the word flow makes one type preferred over the other. The reference compilation in the references section should always be alphabetical. When multiple citations from exactly the same author(s) occur in the same year, those citations are arranged by the lower-case letter following the (same) year designation. These lower case letters must remain with each citation, as that is how they are differentiated. When multiple, different sets of authors with the same senior author occur, the citations are arranged alphabetically by the second, third, etc., author, within that same year.

The Latin words *et alia*, which mean "and others," have now been anglicized, and they are generally not italicized when used for in-text reference citations in most journals. However, refer to the instructions for authors of your journal of choice to determine whether or not you should italicize it. There is no period after et, as the complete word is used, but there is a period after the abbreviation for alia, and therefore, we use et al. to designate more than two coauthors.

There are several additional rules, shown below, for in-text reference citations for additional cases in which there are unusual combinations of the number of authors and years. When a type of citation cannot be found below, please see the CSE Manual (2006).

Examples of In-Text Reference Citations, Including Both Types, Using the Name–Year (Harvard) Reference Citation System

One author. . . Jones (1995) *or* (Jones, 1995)

Two authors. . . James and Smith (1994) *or* (James and Smith, 1994)

Three or more authors. . . Jones et al. (1994) *or* (Jones et al., 1994)

Two or more citations. . . James and Smith (1994), Jones (1995) *or* (James and Smith, 1994), (Jones, 1995) *or* (James and Smith, 1994; Jones, 1995)

Two citations by one author in one year. . . Jones (1995a), Jones (1995b) *or* (Jones, 1995a, 1995b) *or* (Jones, 1995a; Jones, 1995b)

Multiple-year citations of one author. . . Jones (1989, 1993, 1995) *or* (Jones, 1989; Jones, 1993; Jones, 1995)

Multiple-year citations by different authors. . . James and Smith (1989), Smith (1991), Jones et al. (1993) *or* (James and Smith, 1989; Smith, 1991; Jones et al., 1993)

The Citation–Name (or Vancouver) System

The second system is the citation–name, or Vancouver, system. In this system, as the author is writing, he or she will link a citation number within the text to a reference. After the manuscript has been written, all references are alphabetized and listed by the first letter in the surname of the senior author. The numbers applied previously are now changed to the order once the list of references has been alphabetized, and new numbers are applied to each reference and the in-text reference citations are changed to their corresponding new number. This alphabetic ordering is carried on to additional authors of the reference, if necessary. When all authors are exactly the same on two separate publications, this is carried on to the year of publication. If the author(s) and their sequence are exactly the same and the year is exactly the same, then the writer distinguishes between references by placing a lower case a, b, c, and so on, immediately after the year. Subsequently, each reference is numbered in this alphabetic sequence, and this forms the references section. In the text, the writer places the number, or numbers if more than one reference is cited, in parentheses in the line of the sentence where the information was used. For some journals, the in-text reference number is placed as a superscript, without use of parentheses. In either of these cases, when multiple in-text reference citations are used, they are presented in numerical order, and not alphabetical order of the senior author. In addition, a comma, with or without a space between, separates the multiple in-text reference citations whether the list of numbers appears within parentheses or as a set of superscripts. In cases where the ideas of several sets of authors are cited, the manuscript writer must ensure that the reader unequivocally understands which in-text reference citation is tied to which idea of an author. In this case, it is more suitable to have citations at appropriate places within the sentence, rather than at the end only. If your journal uses in-line, in-text reference citations, then be sure other numbers appearing in the sentence within parentheses are followed by an appropriate designator term, unit, or symbol that keeps those data unequivocally separated from the in-text reference citation numbers (CSE, 2006). Matthews et al. (2000) describe 35 examples of how to reference many types of resources the researcher may want to use in his or her manuscript. It is an excellent resource for writers using the citation–name system. The CSE Manual (2006) also has examples for nearly every possible scenario.

This system has several advantages. The main one is that the reader is distracted minimally by the in-text reference citations. This is especially true when the writer makes a long string of consecutive in-text reference citations, because they can be denoted as (3,7,10,11,23), rather than by a combination of many names and many years. This system also saves space and reduces production costs for journals. It can be a very good system for short articles, such as notes, and in particular, review articles where many references are cited throughout the text of the article. Another advantage is that there are relatively few formatting rules to follow with this system compared with the name–year system.

There also are several distinct disadvantages of this system. In longer manuscripts, such as review articles and book chapters, the reader continually must go back and forth between the text and the list of citations to determine who did what they just read. In addition, because there is not a year with the in-text reference citation, the reader does not learn immediately when the work in the reference was completed. These disadvantages may be overcome by carefully crafting the sentences containing the references so that either the name(s) of the researcher(s) and/or the dates are incorporated into the sentence. Difficulties also arise when the reader cannot link together articles by the same author(s) to understand what direction an author or group of authors is taking in their work. The third disadvantage is that the visibility of authors is challenged, because a number and not their name references their work. A fourth disadvantage is that the list of references does not stand well by itself. The fact that the same author has produced several papers does not readily appear because the works may not be referenced sequentially in the text. It may be very difficult to see that one author has made many contributions because references to his or her work are scattered throughout the references section.

Probably the most serious shortcoming of this system is that revisions and reviews often lead to additions and/or deletions from the list of references, and every time a reference is added or deleted, all numbered references after that one need to be changed. This becomes a confusing mess that may only get fixed at the end of the review process, and sometimes does not get fixed then. The Harvard system has largely replaced this system, because it is so much easier to add and delete citations at a later time, that is, after each revision or review of the manuscript. If the Vancouver system must be used, then it is best to use author–date citations in the text until the final draft of the manuscript has been developed. Then, the author(s) can change to the Vancouver system and designate each citation with its appropriate number. Computer software is available to make this conversion at the appropriate time.

The Citation–Sequence System

In the citation-sequence system, the author numbers the in-text reference citations in order from first to last as they appear in the manuscript, without regard to alphabetical order or section of the manuscript. The reference section is built in this numerical order based upon the use of each reference. In-text reference citations may be included in the text, table titles, and figure captions. Depending upon the journal, the in-text reference citations will be most often presented in one of two ways. Other systems exist, but these two ways are the most common. In the first way, the first citation in the text is given number 1 in parentheses, without punctuation, immediately after the presentation of the information or at the end of that sentence. In the second way, the number 1 is used as a superscript without punctuation at the point where it was used to accredit some form of information.

The information about the reference cited may occur in either of two places. First, it may be placed as number 1, with punctuation, in the list of references that becomes the references section at or near the end of the manuscript. Second, it can be used without punctuation as a footnote on the bottom of the page where the citation occurred. Each subsequent citation is given the next greater number until all literature references have received a number in order as they appeared in the manuscript. If authors refer to a given citation a second, third, or more times, the original number used for the citation is used each additional time they refer to that work.

The advantages and disadvantages of this system are similar to those for the Vancouver system. Advantages are that there are fewer distractions to the reader, and the references are arranged using a standard bibliographic format.

Disadvantages are that the reader cannot associate references with a specific person, persons, or year the work was done because there is only a number associated with the citation in the text. In addition, every time revisions or reviews cause one or more additions or deletions, the references must be renumbered for all references that follow the addition or deletion. This system is excellent for shorter manuscripts, which usually have fewer references, and it is common for biomedical journals (ICMJE, 2002).

LISTING REFERENCES IN THE REFERENCES SECTION OF THE MANUSCRIPT

New references you encounter should be logged into your literature database. *Never place a reference in the references section of your manuscript until you have seen it and read it. The problem here is that the words do not change when someone else reads the article, but the interpretation of those words may vary from person to person.*

- The listing for each reference should contain a complete list of all authors, in the correct order for the journal. Be sure you have not changed the order of the authors, especially the senior author of the citation. List all authors unless your journal explicitly instructs you to limit the references section entries to fewer (three, six, etc.) authors.

- The year the work was published should be designated.

- The listing should contain the complete title of the work cited, with no abbreviations that are not in the original title. If you have cited the title incorrectly, this will probably cause fewer problems relative to other types of mistakes. However, you should strive for no errors. Place the title in your entries unless your journal explicitly tells you not to.

- List the resource that holds the published work, whether it is a journal, a book, a proceedings, and so on. It is easy to give the wrong name, or especially the incorrect abbreviation, for the journal. Care must be exercised to avoid falling into a trap of something "looking like" it should be abbreviated in a certain way that is not correct. A table of abbreviations of journal title words may be found later in this chapter. *Do not make up your own abbreviations.*

- Contain all pertinent information about the book, if that is the original source.

- Give the exact descriptors of the location within the resource (volume, issue number, inclusive page numbers, or first page only). Be careful because it is easy to transpose volume or page numbers, especially the first page number of the resource. This makes retrieval exceptionally difficult for a reader. Also be sure you know whether you must include an issue number of a volume when it is needed to find the citation, as certain journals start each issue within a volume with page 1. Enter the inclusive pages of each reference unless the journal explicitly tells you to give only the first page of the reference.

- Use the in-text reference citation system of the journal where you will submit the manuscript, if that is known during manuscript preparation. When this is not known, we would suggest that the default system should be the name–year, or Harvard, system.

- Use the correct format for listing the references in the section where the references are compiled, if that is known during manuscript preparation.

- Make sure the punctuation format used for each in-text reference citation and in the list of references matches the system required by the journal. See the examples below of how the systems vary.

- Place references to unpublished work and verbal communications in the text in parentheses, and not in the references section. They are not published items and therefore do not belong in the references section of the manuscript.

- Place in-text reference citations in the sentence in the text where they describe who did the work. Do not necessarily try to fit all in-text reference citations into the end of the sentence only. However, it is best to place as many in-text reference citations as possible at the end of a given sentence because it causes the reader less distraction as they are reading the manuscript.

GENERIC PROCEDURE FOR LISTING REFERENCES

For all three systems, references are listed in the references section in essentially the same manner. The information on how to list the references in the references section is presented generically, and *the author must determine the exact system and format used by the journal where the manuscript will be submitted.*

As an example, the entry will start with the surname of the senior author, followed by a comma, then his or her first name or initial, followed by the middle name or initial, depending upon the journal. Each initial may be terminated by a period. Immediately after this period, if an initial was used, or the complete middle name, there will be a comma that ends the name of the senior author. After one space, there will begin the name(s) of the junior author(s) in the exact same sequence that appears in the by-line of the reference. The name(s) of the junior author(s) will be listed sequentially with the person's first or given name, or initial, followed by a period. Next will be the initial of the middle name followed by a period, or the complete middle name, either of which is then followed by the entire surname followed by a comma. Some journals use this format, and some use a format in which the name(s) of the junior author(s) is reported in this order: first name (not followed by a period) or initial followed by a period, middle name (not followed by a period) or middle initial followed by a period, followed by the surname (last name). This sequence is repeated until the last author of the sequence is completed, and that surname or given name or initial is followed by a period. If there are two coauthors, then a comma may or may not be placed before the "and," and an "and" may or may not be placed between the two coauthors. If there are more than two coauthors, a terminal comma may or may not precede the "and," and the word "and" may or may not be placed immediately before the last coauthor is listed. *All authors must be cited here, and they must be presented in the exact order as listed in the byline of the original reference!*

The name(s) will be followed by the year the reference was published. A lower case letter may accompany the year if it is needed to clarify same-author, same-year publications, followed by a period. Next will come the complete title of the work, and no abbreviations may be used. Not all journals require listing the title of the published article. The remainder of the entry will vary with the type of reference and format of the journal. Journal articles are the most common reference in scientific writing, and those entries will contain, in this order, the name of the journal, followed by a period in some cases, the volume number, the issue number in parentheses (if necessary), either of which will be followed by a colon, and the inclusive pages, without abbreviations and linked by a hyphen. Some journals do not require the inclusive pages, but rather only the first page of the published article. This system varies with other, nonjournal references, and below we have provided a somewhat generic sample of how the reference should appear.

Virtually every journal, or other publishing unit, will abbreviate the names of the journals listed in the references in the references section. This saves much precious space in the published journal. Below, we have listed the abbreviations used for these journals ([ANSI-NISO] American National Standards Institute, Inc., 1977, 1985), and you should use these abbreviations for all references. *Again, do not make up your own abbreviations.* Elsewhere, we have stated that many different formats for the presentation of these reference descriptors have been identified, and you should determine exactly which format your journal will follow when you start to write your manuscript. Some journals have a format that looks similar to that of another journal, but sometimes there is no punctuation, certain descriptors are in boldface font, or one of the descriptors is in an unusual place in the entry. Attention to detail is of utmost concern.

EXAMPLE FORMATS FROM REFERENCE SECTIONS OF KEY JOURNALS IN THE BIOLOGICAL SCIENCES

The references presented below have been taken verbatim from the journal name immediately before the reference. Notice differences in hanging indents, use of boldface font, italics, periods, commas, colons, capitalization, parentheses, et al., placement of entries (year, volume number), ampersands, small capitals, spacing, and so on. Perusal of this set of references should make you understand why a generic format cannot be used.

This portion uses a different reference from each journal:

American Journal of Science

Goldhammer, R. K., Dunn, P. A., and Hardie, L. A., 1987, High frequency glacio-eustatic sealevel oscillations with Milankovitch characteristics recorded in Middle Triassic platform carbonates in northern Italy: American Journal of Science, v. 287, p. 853–892.

Science

T. Ito et al., *Proc. Natl. Acad. Sci. U.S.A.* **97**, 1143 (2000).

Nature

Bennett, R. J., Keck, J. L. & Wang, J. C. Binding specificity determines polarity of DNA unwinding by the Sgs1 protein of *S. cerevisiae. J. Mol. Biol.* **289**, 235–248 (1999).

Plant Physiology

Kim J, Harter K, Theologis A (1997) Protein–protein interactions among the Aux/IAA proteins. Proc Natl Acad Sci USA **94:** 11786–11791

American Journal of Botany

McKenna, D. D., and B. D. Farrell. 2006. Tropical forests are both evolutionary cradles and museums of leaf beetle diversity. *Proceedings of the National Academy of Sciences, USA* 103: 10947–10951.

Journal of Cell Biology

Fuchs, E., and D.W. Cleveland. 1998. A structural scaffolding of intermediate filaments in health and disease. *Science.* 279:514–519.

Cell

Caton, M.L., Smith-Raska, M.R., and Reizis, B. (2007). Notch-RBP-J signaling controls the homeostasis of CD8- dendritic cells in the spleen. J. Exp. Med. *204*, 1653–1664.

Genetics

Koshland, D., J. C. Kent and L. H. Hartwell, 1985 Genetic analysis of the mitotic transmission of minichromosomes. Cell **40:** 393–403.

Journal of Dairy Science and Journal of Animal Science (same system)

McSweeney, P. L. H., P. F. Fox, and N. F. Olson. 1995. Proteolysis of bovine caseins by Cathepsin-D – Preliminary observations and comparison with chymosin. Int. Dairy J. 5:321–336.

Microbiology

Yan, B., Methe, B. A., Lovley, D. R. & Krushkal, J. (2004). Computational prediction of conserved opers and phylogenetic foot printing of transcription regulatory elements in the metal-reducing bacterial family *Geobacteraceae. J Theor Biol* **230,** 133–144.

American Journal of Physiology

Erlebacher JA, Danner RL, Stelzer PE. Hypotension with ventricular pacing: an atria vasodepressor reflex in human beings. *J Am Coll Cardiol* 4: 550–555. 1984.

Journal of Biological Chemistry

Humphry, M., Consonni, C., and Panstruga, R. (2006) *Mol. Plant Pathol.* **7,** 605–610

An, S. J., and Almers, W. (2004) *Science* **306,** 1042–1046

Journal of Food Science

Cameron AC, Boylan-pett W, Lee J. 1989. Design of modified atmosphere packaging systems: modeling oxygen concentrations within sealed packages of tomato fruits. J Food Sci 54:1414–6.

Ecology

Lloyd, M., G. Kritsky, and C. Simon. 1983. A simple Mendelian model for 13- and 17-year life cycles of periodical cicadas, with historical evidence of hybridization between them. Evolution 37:1162–1180.

The American Naturalist

Nagylaki, T., and M. Moody. 1980. Diffusion model for genotype-dependent migration. Proceedings of the National Academy of Sciences of the USA 77:4842–4846.

Proceedings of the National Academy of Sciences of the United States of America

Celie PH, et al. (2004) Nicotine and carbamylcholine binding to nicotinic acetylcholine receptors as studied in AChBP crystal structures. *Neuron* 41:907–914.

Mazor Y, Van Blarcom T, Mabry R, Iverson BL, Georgiou G (2007) Isolation of engineered, full-length antibodies from libraries expressed in Escherichia coli. *Nat Biotechnol* 25:563–565.

This portion uses the same reference cited in these different journals:

American Journal of Science

Bennett, R. J., Keck, J. L., and Wang, J. C., 1999, Binding specificity determines polarity of DNA unwinding by the Sgs1 protein of *S. cerevisiae*: Journal of Molecular Biology, v. 289, p 235–248.

Science

R. J. Bennett et al., *J. Mol. Biol.* **289,** 235 (1999).

Nature

Bennett, R. J., Keck, J. L. & Wang, J. C. Binding specificity determines polarity of DNA unwinding by the Sgs1 protein of *S. cerevisiae. J. Mol. Biol.* **289,** 235–248 (1999).

Plant Physiology

Bennett R, Keck J, Wang J (1999) Binding specificity determines polarity of DNA unwinding by the Sgs1 protein of *S. cerevisiae*. J Mol Biol **289**: 235–248

American Journal of Botany

BENNETT, R. J., J. L. KECK, and J. C. WANG. 1999. Binding specificity determines polarity of DNA unwinding by the Sgs1 protein of *S. cerevisiae*. *Journal of Molecular Biology* 289: 235–248.

Journal of Cell Biology

Bennett, R. J., J. L. Keck, and J. C. Wang. 1999. Binding specificity determines polarity of DNA unwinding by the Sgs1 protein of *S. cerevisiae*. *J. Mol. Biol.* 289:235–248.

Cell

Bennett, R.J., Keck, J.L., and Wang, J.C. (1999). Binding specificity determines polarity of DNA unwinding by the Sgs1 protein of *S. cerevisiae*. J. Mol. Biol. *289*, 235–248.

Genetics

BENNETT, R. J., J. L. KECK and J. C. WANG, 1999 Binding specificity determines polarity of DNA unwinding by the Sgs1 protein of *S. cerevisiae*. J. Mol. Biol. **289**: 235–248.

Journal of Dairy Science and Journal of Animal Science (same system)

Bennett, R. J., J. L. Keck, and J. C. Wang. 1999. Binding specificity determines polarity of DNA unwinding by the Sgs1 protein of *S. cerevisiae*. J. Mol. Biol. 289:235–248.

Microbiology

Bennett, R. J., Keck, J. L. & Wang, J. C. (1999). Binding specificity determines polarity of DNA unwinding by the Sgs1 protein of *S. cerevisiae*. *J Mol Biol* **289**, 235–248.

American Journal of Physiology

Bennett RJ, Keck JL, Wang JC. Binding specificity determines polarity of DNA unwinding by the Sgs1 protein of *S. cerevisiae*. *J Mol Biol* 289: 235–248. 1999.

Journal of Biological Chemistry

Bennett, R. J., Keck, J. L., and Wang, J. C. (1999) *J. Mol. Biol.* **289**, 235–248

Journal of Food Science

Bennett RJ, Keck JL, Wang JC. 1999. Binding specificity determines polarity of DNA unwinding by the Sgs1 protein of *S. cerevisiae*. J Mol Biol 289:235–8. (Actually, 289:235–248)

Ecology

Bennett, R., J. Keck, and J. Wang. 1999. Binding specificity determines polarity of DNA unwinding by the Sgs1 protein of *S. cerevisiae*. J. Mol. Biol. **289**:235–248.

The American Naturalist

Bennett, R. J., J. L. Keck, and J. C. Wang. 1999. Binding specificity determines polarity of DNA unwinding by the Sgs1 protein of *S. cerevisiae*. Journal of Molecular Biology 289:235–248.

Proceedings of the National Academy of Sciences of the United States of America

Bennett RJ, et al. (1999) Binding specificity determines polarity of DNA unwinding by the Sgs1 protein of *S. cerevisiae*. *J. Mol. Biol.* 289:235–248.

ABBREVIATIONS OF WORDS USED IN THE TITLE OF A JOURNAL

Before we list the words for journal title names and their corresponding abbreviations, we present several important items. For a journal title word not on the list below, consult Index Medicus (www.nlm.nih.gov), and, if that fails, write out each word of the title in its entirety, because no abbreviation is better than a bad or poor abbreviation (Gustavii, 2003).

- Words to be abbreviated must be part of the journal name in its fullest form. The name of the sponsoring organization of the journal is used only when it is tied to the proper name of the journal (e.g., *Journal of the American Medical Association*).

- Do not abbreviate single-word titles (e.g., *Science, Nature*). Normally, there is no punctuation after a single-word title.

- Do not include articles, conjunctions, and prepositions (low-impact words) in the description of the reference. The only exceptions are when the article, conjunction, or preposition is (i) part of a personal name; (ii) part of a place name; (iii) in a scientific term; (iv) in a technical term; or (v) part of a standard phrase.

- Abbreviation should be done by truncation whenever possible, and two or more letters must be truncated in order for it to be an abbreviation.

- Words ending in -ical, -otic, -logical, -ology, and so on, are usually abbreviated by truncating the al, ic, ogical, and ogy off the end of the word, respectively. Similar truncations should be used for corresponding words.

- Both cognates (from another language) and variants of a root word with the same stem will use the same abbreviation.

- Words may be abbreviated by contraction, provided the resulting contraction is unequivocal. Normally, formation of this type of contraction does not use an apostrophe, nor does it use a period.

- One-syllable words and words that contain five or fewer letters are normally not abbreviated. There are some exceptions, but they are rare.

- The first letter in each abbreviated word is capitalized.

- Some journals do not use any punctuation, including apostrophes, in the journal title, and each word, abbreviated or not, is followed only by a space. Other journals will use all forms of punctuation, including a period at the end of every abbreviated word. You must determine the format your journal uses and follow that format.

- All diacritical marks, ampersands (&), and dashes are omitted.

- All hyphenated words are treated as two words, if each can stand by itself.

STANDARD ABBREVIATIONS OF JOURNAL WORD TITLES

Below are some standard abbreviations of journal title words used frequently in the biological sciences ([ANSI-NISO] American National Standards Institute, Inc., 1977, 1985). A more complete list is contained in Index Medicus (www.nlm.nih.gov). In many cases, abbreviations are followed by a period. However, some journals use no periods, and you must determine this usage of periods before you submit your manuscript.

Journal Title Word	Abbreviation	Journal Title Word	Abbreviation
Academy	Acad.	General	Gen.
Acta	No abbrev.	Genetic(s)	Genet.
Advances	Adv.	International	Int.
Agricultural	Agric.	Journal	J.
Agriculture	Agric.	Methods	No abbrev.
Agronomy	Agron.	Microbiological	Microbiol.
American	Am.	Microbiology	Microbiol.
Annals	Ann.	Molecular	Mol.
Annual	Annu.	National	Natl.
Applied	Appl.	Official	Off.
Association	Assoc.	Pathology	Pathol.
Bacteriological	Bacteriol.	Physiology	Physiol.
Bacteriology	Bacteriol.	Proceedings	Proc.
Biochemical	Biochem.	Progress	Prog.
Biochemistry	Biochem.	Quarterly	Q.
Biological	Biol.	Report	Rep.
Botanical	Bot.	Research	Res.
Botany	Bot.	Review	Rev.
Bulletin	Bull.	Science	Sci.
Canadian	Can.	Scientific	Sci.
Cell	No abbrev.	Society	Soc.
Cellular	Cell.	Study	Stud.
Current	Curr.	Surgery	Surg.
Development(al)	Dev.	Survey	Surv.
Disease(s)	Dis.	Symposia	Symp.
Ecology	Ecol.	Symposium	Symp.
Edition	Ed.	Technical	Tech.
Environment	Environ.	Technology	Technol.
Environmental	Environ.	Therapeutics	Ther.
European	Eur.	Transactions	Trans.
Experimental	Exp.	United States	U.S.
Federal	Fed.	University	Univ.
Food	No abbrev.	Zoological	Zool.
Forest	For.	Zoology	Zool.

AVOID OR WATCH OUT FOR THESE THINGS

- Be sure all in-text reference citations are presented as references in the references section. Every item in the references section must be cited at least once somewhere in the manuscript. This is of paramount importance in publishing, as the greatest number of errors in a manuscript occur by the loss of a citation either in the text or the references section. Please complete the exercise at the end of this chapter. It is designed to give you practice developing a system that minimizes these errors.

- Cite only significant, published work that is easily and readily accessible by all who might read your article. Be careful about citing works not yet published, such as verbal or written personal communications, unpublished data, abstracts, theses, dissertations, and other secondary references. If you must cite these types of references, be sure you know how to cite them in the journal publishing the manuscript.

- Do not clutter the references section with references to theses, dissertations, abstracts, and so on, unless they are absolutely necessary. These references have not stood the rigors of peer review and, therefore, are not as fundamentally sound as a peer-reviewed journal article. Remember, one-third of these never get published (Liu, 1996).

- Try and not use information from the World Wide Web, because the source, quality, and accessibility to the information may change too quickly.

DEPENDING UPON THE MANUSCRIPT, AND WHERE IT WILL BE PUBLISHED, THESE SUGGESTIONS MAY BE EITHER ESSENTIAL OR BENEFICIAL

- Journals vary in their system and format for the references section and citation of works in the text, and you must follow the guidelines of your journal. Many journals use an established system, such as the Harvard system, but they make their own "adjustments" to the system. For example, the journal may not use any punctuation, except for a terminal period at the end of the reference. You must follow the requirements of your journal. You cannot use just any system you wish, and you cannot make your own "adjustments" to the format mandated by your journal.

- If you absolutely must cite a personal communication or reference to an unpublished work, include it in the text in parentheses within the statement of reference to that source. Other secondary sources of information should be cited as a standard reference in the text and listed in the references section.

- If a manuscript submitted for publication has been accepted *by the editor that makes the final decision about acceptance*, then you may cite that article as "Accepted for publication," and it may be listed in the references section. Until this final authority has accepted that manuscript, you must either not refer to it or refer to it as a personal communication of unpublished work. If the galley proofs or page proofs (see Chapter 26) have been corrected and returned to the editor, then you may refer to the manuscript as "In press," and it may appear in the references section. Some journals use these terms interchangeably, so you should consult the instructions to authors document, the style manual, or a recent issue of the journal to determine which format your journal uses.

REFERENCES

[ANSI-NISO] American National Standards Institute, Inc., National Information Standards Organization. 1977. *American national standard for bibliographic references.* ANSI/NISO Z39.29-1977. NISO Press, Bethesda, MD. [Available from NISO Press Fulfillment, P.O. Box 338, Oxon Hill, MD 20750-0338, USA. FAX: 301-567-9553.]

[ANSI-NISO] American National Standards Institute, Inc., National Information Standards Organization. 1985. *Abbreviations of titles of publications* ANSI/NISO Z39.5-1985. NISO Press. Bethesda, MD. [Available from NISO Press Fulfillment, P.O. Box 338, Oxon Hill, MD 20750-0338, USA. FAX: 301-567-9553.]

CHERNIN, E. 1988. *The "Harvard system": A mystery dispelled.* Brit. Med. J. 297:1062–1063.

[CSE] Council of Science Editors, Style Manual Committee. 2006. *Scientific style and format: the CSE manual for authors, editors, and publishers.* Seventh edition. The Council, Reston, VA.

DAVIS, M. 2005. *Scientific papers and presentations.* Second edition. Academic Press, San Diego, CA.

DAY, R.A. and GASTEL, B. 2006. *How to write and publish a scientific paper.* Sixth edition. Greenwood Press, Westport, CT.

GARFIELD, E. 1986. The integrated Sci-Mate software system. Part 2. The editor slashes the Gordian knot of conflicting reference styles. Current Contents 17(11):81–88.

GUSTAVII, B. 2003. *How to write and illustrate a scientific paper.* Cambridge University Press, New York, NY.

[ICMJE] International Committee of Medical Journal Editors. 2002. Uniform requirements for manuscripts submitted to biomedical journals [Updated October, 2001; cited 20 Jan 2002; http://www.icmje.org].

KATZ, M.J. 1985. *Elements of the scientific paper. A step-by-step guide for students and professionals.* Yale University Press, New Haven, CT.

LIU, L. 1996. Fate of conference abstracts. Nature 383: 20.

MATTHEWS, J.R., BOWEN, J.M. and MATTHEWS, R.W. 2000. *Successful scientific writing. A step-by-step guide for the biological and medical sciences.* Cambridge University Press, New York, NY.

O'CONNOR, M. 1978. Standardisation of bibliographical reference systems. Br. Med. J. 1(6104):31–32.

PEAT, J., ELLIOTT, E., BAUR, L. and KEENA, V. 2002. *Scientific writing. Easy when you know how.* BMJ Books, London, UK.

EXERCISE 14.1 Treasure Hunt for Errors—Missing Citations and References

This passage from the introduction in a Master of Science thesis has been changed for this exercise with the permission of the author. There are errors in the text, in the transition of the information from the text to the references section, and in the references section. There are 50 errors in this exercise. Please practice your editing skills by finding as many errors as you can. This exercise has been double-spaced so that you may make your corrections in the text and the references section as you find the errors.

Introduction

The plant hormone ethylene (C_2H_4) is involved in many physiological processes such as fruit ripening, abscission, senescence, leaf epinasty, and seed germination, and it is physiologically active at extremely small concentrations (<0.1 $\mu L \cdot L^{-1}$) (Abeles et al., 1990; 1992). Horticultural commodities are classified as climacteric or nonclimacteric based on their capacity to produce C_2H_4. Climacteric fruits undergo a large increase in carbon dioxide production from respiration, and changes in C_2H_4 production rates are accompanied by changes in color, composition, and texture of the commodity (Abeles et al., 1993). Ethylene biosynthesis occurs via the conversion of methionine to S-adenosyl methionine, which then is converted to 1-aminocyclopropane-1-carboxylic acid and C_2H_4. These three reactions are catalyzed by the enzymes S-adenosylmethionine synthetase, 1-aminocyclopropane-1-carboxylic acid synthase (ACS), and 1-aminocyclopropane-1-carboxylic acid oxidase (ACO), respectively (Adams and Yang, 1978; Kendy, 1993; Saltveit, 1989; 1999).

The effects of C_2H_4 can be both desirable and undesirable. Loss of chlorophyll and development of carotenoids due to C_2H_4 in fruits such as tomatoes, apricots, peaches, and citrus are considered desirable changes. Other beneficial effects of C_2H_4 include promotion of thinning of apples and cherries, flower induction in pineapples, and degreening of nonclimacteric fruits such as citrus (Abeles and Saltveit, 1992). Undesirable effects of C_2H_4 include loss of chlorophyll and yellowing of broccoli (Ku et al., 1999; Wills, 2000), russet spotting on lettuce (Rude, 1956; Rood and Nasty, 1956), sprouting of potatoes (Alam, 1994), toughening of asparagus (Saltveit et al., 1999), and loss of flowers and/or leaves from abscission-sensitive plants (Abeles, 1992). The most damaging effect of C_2H_4 is its effect on the postharvest shelf life of horticultural commodities. Even trace amounts of C_2H_4 reduce the shelf life of horticultural commodities throughout the marketing chain (Wills, 1999; 2001).

Over the years, several strategies to reduce the damaging effects of C_2H_4 have been studied. Maintaining lower concentrations of C_2H_4 and preventing its accumulation around commodities by controlling the surrounding atmosphere has successfully reduced C_2H_4-related damage. Several compounds such as aminoethoxyvinyl glycine and (aminooxy)acetic acid (Armhein and Winker,

1997), which inhibit C_2H_4 biosynthesis, have been identified (Amrhein and Winker, 1979). However, these compounds do not offer protection against exogenous C_2H_4 (Amrhein and Wenker, 1997). Another approach is the use of compounds that inhibit C_2H_4 action. These include compounds such as silver thiosulfate (Beyer, 1976), 2,5-norbornadiene, and certain organic molecules such as diazocyclopentadiene and 1-methylcyclopropene (1-MCP) (Sisler and Serek, 1997a).

1-Methylcyclopropene (1-MCP) is an inhibitor of C_2H_4 action (Willis, 2002). It acts by binding to C_2H_4 receptors, and it thus blocks subsequent C_2H_4-regulated processes (Sisler and Serek, 1997; 1997b). 1-MCP has proved to be more beneficial than other C_2H_4 action inhibitors in several ways. It is active at lower concentrations ($nL \cdot L^{-1}$), it is more stable than other organic molecules, and it is nontoxic (Sisler and Serek, 1979; Serek and Sisler, 1997). Another advantage to both humans and plant materials is its efficacy on a wide range of horticultural commodities such as apple (Fan, 1999; 2001), banana (Jiang, 1998; Jiang et al., 1999), broccoli (Wills and Ku, 1999), tomato (Mir et al., 2002; 2004) and ornamentals (Sisler and Serek, 1994; Serek et al., 1994; 1997). 1-MCP also may decrease or delay C_2H_4 production depending upon the commodity. The Food and Drug Administration has approved the use of 1-MCP on ornamental plants, and more recently, apples.

1-MCP effectively inhibits C_2H_4 action in several stages of tomato ripening. However, in tomatoes at less mature stages such as breaker and turning (USDA, 1975) the effect is too strong, and ripening is delayed to the point where 1-MCP cannot be used commercially (Rohwer, 2001; Rohwer, 2002; Ku and Wills, 2002). In commercial wholesale operations, the emphasis is on handling the produce for a predictable, short period of time so that more produce is moved to market without incurring losses due to fluctuating market conditions. Thus, the overall objective of this research was to determine whether or not 1-MCP could be used to control, but not inhibit completely, the ripening rate of less-mature tomatoes. The first specific objective of this research was to determine whether ripening could be reinitiated by subsequent treatment with C_2H_4 in tomatoes treated previously with 1-MCP. The second specific objective was to determine the concentrations of 1-MCP that would be most appropriate for commercial use on less mature stages of ripeness of tomatoes.

REFERENCES

ABELES, F.B. 1992. *Ethylene in plant biology.* Third edition. Academic Press, San Diego, CA.

ABELES, F.B. and M.E. SALTVEIT, JR. 1992. *Ethylene in plant biology.* Second edition. Academic Press, San Diego, CA.

ABELES, F.B., P.W. MORGAN, and M.E. SALTVEIT, JR. 1992. *Ethylene in plant biology.* Second edition. Academic Press, San Diego, CA.

ADAMS, D.O. and S.F. YANG. 1977. Methionine metabolism in apple tissue. Plant Physiol. 60:892–896.

ADAMS, D.O. and S.F. YANG. 1979. Ethylene biosynthesis: Identification of 1-aminocyclopropane-1-carboxylic acid as an intermediate in the conversion of methionine to ethylene. Proc. Natl. Acad. Sci. USA. 76:170–174.

ALAM, S.M.M., D.P. MURR, and L. KRISTOF. 1994. The effect of ethylene and of nucleic acid syntheses on dormancy break and subsequent sprout growth. Potato Res. 37:25–33.

AMRHEIN, N. and D. WENKER. 1979. Novel inhibitors of ethylene production in higher plants. Plant Cell Physiol. 20:1635–1642.

AMRHEIN, N. and D. WINKER. 1997. Novel inhibitors of ethylene production in higher plants. Plant Cell Physiol. 20:1635–1642.

BEYER, JR., E.M. 1976a. Silver ion: A potent antiethylene agent in cucumber and tomato. HortScience 11:195–196.

BEYER, JR., E.M. 1976b. A potent inhibitor of ethylene action in plants. Plant Physiol. 58:268–271.

FAN, X., S.M. BLANKENSHIP, and J.P. MATTHEIS. 1999. 1-Methylcyclopropene inhibits apple ripening. J. Amer. Soc. Hort. Sci. 124:690–695.

JIANG, Y., D.C. JOYCE, and A.J. MACNISH. 1999. Responses of banana fruit to treatment with 1-methylcyclopropene. Plant Growth Regul. 28:77–82.

KENDE, H. 1993. Ethylene biosynthesis. Annu. Rev. Plant Physiol. 44:283–307.

KIM, W.T. and S.F. YANG. 1992. Turnover of 1-aminocyclopropane-1-carboxylic acid synthase protein in wounded tomato fruit tissue. Plant Physiol. 100:1126–1131.

KU, V.V.V. and R.B.H. WILLS. 1999. Effect of 1-methylcyclopropene on storage life of broccoli. Postharvest Biol. Technol. 17:127–132.

MIR, N., M. CANOLES, R. BEAUDRY, E. BALDWIN, and C. MEHLA. 2004. Inhibiting tomato ripening with 1-methylcyclopropene. J. Amer. Soc. Hort. Sci. 129:112–120.

OLSON, D.C. et al. 1991. Differential expression of two genes for 1-aminocyclopropane-1-carboxylic acid synthase in tomato fruits. Proc. Natl. Acad. Sci. USA. 88:5340–5344.

OLSON, D.C., J.A. WHITE, L. EDELMAN, R.N. HARKINS, and H. KENDE. 1991. Differential expression of two genes for 1-aminocyclopropane-1-carboxylic acid synthase in tomato fruits. Proc. Natl. Acad. Sci. USA. 88:5340–5344.

ROHWER, C.L. and R.J. GLADON. 2001. 1-Methylcyclopropene delays ripening of pink and light red tomatoes. HortScience 36: 466 (Abstr.).

ROOD, P. 1956. Relation of ethylene and post-harvest temperature to brown spot of lettuce. Proc. Amer. Soc. Hort. Sci. 68:296–303.

SALTVEIT, M.E. 1999. Effect of ethylene on quality of fresh fruits and vegetables. Postharvest Biol. Tech. 15:279–292.

SEREK, M., E.C. SISLER, and M.S. REID. 1994. Novel gaseous ethylene binding inhibitor prevents ethylene effects in potted flowering plants. J. Amer. Soc. Hort. Sci. 119:1230–1233.

SEREK, M., E.C. SISLER, and M.S. REID. 1995. Effects of 1-MCP on the vase life and ethylene response of cut flowers. Plant Growth Regul. 16:93–97.

SISLER, E.C. and M. SEREK. 1997. Inhibitors of ethylene responses in plants at the receptor level: Recent developments. Physiol. Plant. 100:577–582.

USDA. 1975. Color classification requirements in tomatoes. USDA visual aid TM-L-1. U.S. Dept. of Ag. Agric. Marketing Ser., Fruit and Vegetable Div.

WILLS, R.B.H. 2002. 1-MCP extends ripening of green tomatoes and postharvest life of ripe tomatoes. Postharvest Technol. 22:75–91.

WILLS, R.B.H. and V.V.V. KU. 2002. Use of 1-MCP to extend the time to ripen of green tomatoes and postharvest life of ripe tomatoes. Postharvest Biol. Technol. 26:85–90.

WILLS, R.B.H., M.A. WARTON, and V.V.V. KU. 2000. Ethylene levels associated with fruit and vegetables during marketing. Aust. J. Exp. Agric. 40:465–470.

EXERCISE 14.2 Treasure Hunt for Typographical Errors in In-Text Citations and the List of References

This passage from a Master of Science thesis has been modified from the original and is used with the permission of the author. There are 50 errors in this exercise. Twenty-five of them occur in the in-text citations or in the transition of the information from the in-text citation to the References section. There also are 25 errors in the References section.

Introduction

The plant hormone C_2H_4 is involved in many physiological processes such as fruit ripening, abscission, senescence, leaf epinasty, and seed germination, and it is physiologically active at extremely small concentrations (<0.1 $\mu L \cdot L^{-1}$) (Abeles, 1992). Horticultural commodities are classified as climacteric or nonclimacteric based on their capacity to produce C_2H_4. Climacteric fruits undergo a large increase in carbon dioxide production from respiration, and changes in C_2H_4 production rates are accompanied by changes in color, composition, and texture of the commodity (Abeles al., 1929). Ethylene biosynthesis occurs via the conversion of methionine to S-adenosyl methionine, then 1-aminocyclopropane-1-carboxylic acid and C_2H_4 in three reactions catalyzed by the enzymes S-adenosylmethionine synthetase, 1-aminocyclopropane-1-carboxylic acid synthase (ACS), and 1-aminocyclopropane-1-carboxylic acid oxidase (ACO), respectively (Saltviet, 1999).

The effects of C_2H_4 can be both desirable and undesirable. Loss of chlorophyll and development of carotenoids due to C_2H_4 in fruits such as tomatoes, apricots, peaches, and citrus are considered desirable changes. Other beneficial effects of C_2H_4 include promotion of thinning of apples and cherries, flower induction in pineapples, and degreening of nonclimacteric fruits such as citrus (Abeles et., 1992). Undesirable effects of C_2H_4 include loss of chlorophyll and yellowing of broccoli (Ku-Wills, 1999), russet spotting on lettuce (Rood, 1965), sprouting of potatoes (Alma et al., 1994), toughening of asparagus (Slatveit, 1999), and loss of flowers and/or leaves from abscission-sensitive plants (Abels et al., 1992). The most damaging effect of C_2H_4 is its effect on the postharvest shelf life of horticultural commodities. Even trace amounts of C_2H_4 reduce the shelf life of horticultural commodities throughout the marketing chain (Wills, 2000).

Over the years, several strategies to reduce the damaging effects of C_2H_4 have been studied. Maintaining lower concentrations of C_2H_4 and preventing its accumulation around commodities by controlling the surrounding atmosphere has successfully reduced C_2H_4-related damage. Several compounds such as aminoethoxyvinyl glycine and (aminooxy)acetic acid, which inhibit C_2H_4 biosynthesis, have been identified (Anrhein and Wenker, 1997). However, these compounds do not offer protection against exogenous C_2H_4. Another approach is the use of compounds that inhibit C_2H_4 action. These include compounds such as silver thiosulfate, 2,5-norbornadiene, and certain organic molecules such as diazocyclopentadiene and 1-methylcyclopropene (1-MCP) (Serek and and Sisler, 1997).

1-Methylcyclopropene is an inhibitor of C_2H_4 action. It acts by binding to C_2H_4 receptors, and it thus blocks subsequent C_2H_4-regulated processes (Sisler and Serek, 1979). 1-MCP has proved to be more beneficial than other C_2H_4 action inhibitors in several ways. It is active at lower concentrations ($nL \cdot L^{-1}$), it is more stable than other organic molecules, and it is nontoxic (Sisler et al., 1997). Another advantage to both humans and plant materials is its efficacy on a wide range of horticultural commodities such as apple (Fan and Blankenship, 1999), banana (Jing et al., 1999), broccoli (Ku, 1999), tomato (Mire et al., 2040) and ornamentals (Serek and Sisler, 1994). 1-MCP also may decrease or delay C_2H_4 production depending upon the commodity.

The Food and Drug Administration has approved the use of 1-MCP on ornamental plants, and more recently, apples.

1-MCP effectively inhibits C_2H_4 action in several stages of tomato ripening. However, in tomatoes at less mature stages such as breaker and turning (UDSA, 1975) the effect is too strong, and ripening is delayed to the point where 1-MCP cannot be used commercially (Rohwer and Gladen, 2001; Wills and Ku, 2000). In commercial wholesale operations, the emphasis is on handling the produce for a predictable, short period of time so that more produce is moved to market without incurring losses due to fluctuating market conditions. Thus, the overall objective of this research was to determine whether or not 1-MCP could be used to control, but not inhibit completely, the ripening rate of less-mature tomatoes. The first specific objective of this research was to determine whether ripening could be reinitiated by subsequent treatment with C_2H_4 in tomatoes treated previously with 1-MCP. The second specific objective was to determine the concentrations of 1-MCP that would be most appropriate for commercial use on less mature stages of ripeness of tomatoes.

REFERENCES

ABELES, F.B., P.W. MORGAN, M.E. SALTVEIT, JR. 1992. *Ethylene in plant biology.* First edition. Academic Press, San Diego, CA.

ALAM, S.M.M., D.P. MURR and L. KRISTOF. 1994. The effect of ethylene and of nucleic acid syntheses on dormancy break and subsequent sprout growth. Potato Res. 37:25–33.

AMRHEIN and D. WINKER. 1979. Novel inhibitors of ethylene production in higher plants. Plant Cell Physiol. 20:1635–6142.

FAN, X. and J.P. MATTHEIS. 1999. 1-Methylcyclopropene inhibits apple ripening. Journal of the American Society for Horticultural Science 124:690–695.

JIANG, Y., D.C. JOYCE, and A.J. MACNISH. 1999. Responses of banana fruit to treatment with 1-methylcyclopropene. Plant Growth Regul. 28:82–77.

KU, V.V.V. and WILLS. 1999. Effect of 1-methylcyclopropene on storage life of broccoli. Postharvest Biol. Technol. 17:127–132.

MIR, N., M. CANOLES, R. BEAUDRY, E. BALDWIN, and C. MEHLA. 2004. Inhibiting tomato ripening with 1-methylcyclopropene. 129:112–120.

ROWER, C.L. and R.J. GLADON. 2001. 1-Methylcyclopropene delays ripening of pink and light red tomatoes. HortScience. 36: 466 (abstr.).

ROOD, P. 1956. Relation of ethylene and post-harvest temperature to brown spot of lettuce. Proc. Amer. Soc. Hort. Sci. 68–296: 303.

SALTVEIT. 1999. Effect of ethylene on quality of fresh fruits and vegetables. Postharvest Biol. Tech. 15:297–292.

SEREK, M., E.C. SISLER, and M.S. REID. 1949. Novel gaseous ethylene binding inhibitor prevents ethylene effects in potted flowering plants. J. Amer. Soc. Hort. Sci. 119:1233–1233.

SISLER, E.C. and M. SEREK. 1997. Inhibitors of ethylene responses in plants at the receptor level: Recent developments. Physiol. Plant. 00:577–582.

USDA. 1957. Color classification requirements in tomatoes. USDA visual aid TM-L-1. U.S. Dept. of Ag. Agric. Marketing Ser., Fruit and Vegetable Div.

WILLS, R.B.H., M.A. WARTON, and V.V.V. KU. 1000. Ethylene levels associated with fruit and vegetables during marketing. Aust. J. Exp. Agric. 04:465–470.

WILLS, R.B.H. and V.V.V. KU. 2020. Use of 1-MCP to extend the time to ripen of green tomatoes and postharvest life of ripe tomatoes. Postharvest Biol. Technol. 26:85–0.

CHAPTER *15*

MATERIALS AND METHODS

There is a certain method in their madness.

—Horace

DEFINITION AND NEED

We define the Materials and Methods section as a condensed description of how the research work was conducted, including the experimental design and statistical analysis. Depending upon the journal, there may be other monikers for this section of the manuscript, including Experimental, Experimental Procedures, Methods, Procedures, and Protocols (Matthews et al., 2000; McCabe and McCabe, 2000; O'Connor, 1991).

The materials and methods section sets forth exactly the materials used to conduct the research and the methods, or procedures, used to generate the data contained within the manuscript (Day and Gastel, 2006; Katz, 1985; Matthews et al., 2000). Content pertaining to the materials section should include the biological materials, chemical substances, other substances, supplies, tools, instruments, appliances, apparatus, and so on, used to complete the research. Methods-related information includes techniques, procedures, protocols, formulations, recipes, solutions, formulae, transactions, and algorithms/computer programs.

The results of your research will be accepted only if their procurement is based on methods that are acceptable within the discipline where the work will be published. Therefore, this section defines, or sets limits on, the acceptability of the data, and reliable data only come from a well defined system (Katz, 1985, 2006; Peat et al., 2002). If the materials and methods section has been written correctly, reviewers will rarely suggest major edits to it (Katz, 1985).

The materials and methods section provides answers to the two following questions (O'Connor, 1991). What materials did you use? How did you use them to obtain the results reported? At its most basic level, this section is needed because it supplies a *competent, informed reader* with enough detail that he or she will (i) have confidence in how the research was conducted and (ii) be able to repeat it as needed (O'Connor, 1991). It is important to understand that this section of the manuscript is written for a competent scientist. This permits the writer to use more technical words than in other sections of the manuscript. Jargon should still be avoided, and if specialized terms or abbreviations are used, define them properly.

We certainly cannot speak for all manuscript reviewers and editors from all journals. However, we suggest that most reviewers will ask, either consciously or subconsciously, at

Getting Published in the Life Sciences, First Edition. By Richard J. Gladon, William R. Graves, and J. Michael Kelly
© 2011 Wiley-Blackwell. Published 2011 by John Wiley & Sons, Inc.

least one of the following questions as they complete their review of the manuscript (Barrass, 2002):

- Are the technical and experimental methods appropriate for the study, adequately described, and replicable?
- Are the controls adequate and described properly?
- Are the treatments adequate and described properly?
- Are the statistical design and analysis of the treatment effects appropriate for the experiment?

It will be in your best interest to be sure that these questions have been answered or at least addressed when you think you have completed the materials and methods section.

PRINCIPLES

The materials and methods section may be easy to write, because it is simply a compilation of the actions of the researcher(s). It should be written in a specific, factual, and straightforward manner, with no colorful or verbose statements that may confuse the reader. Nothing presented within this section should be analyzed or interpreted at this point.

At least the first draft of this section should be written as the work has been completed or soon afterward, as the details will slip from your mind quickly unless you take meticulous notes. This may be the one section of the manuscript that can be written in draft form before the study has been initiated. Subsequently, the author may modify the draft to reflect any changes in the plan of work. We often suggest that a student or inexperienced writer who becomes stalled in writing his or her results section should move to writing or cleaning up the materials and methods section, then return to the results section.

Because the materials and methods section is written for the *competent, informed reader*, minor details may be eliminated, and the writer should strive to provide just the right amount of information, at an appropriate level of detail (Day and Gastel, 2006). Limit the presentation of information to only what is necessary for an informed reader to understand how the presented results, and the resultant take-home messages evolving from those results, were generated. It must be logical, almost to a fault. Loose, scattered, illogical, or colorful language or organizational style has no place in this section. For many manuscripts, this section will be one to five typed, double-spaced pages, and it may contain references, tables, figures, or photographs, as necessary. The writer must use Système Internationale for every descriptor that carries a dimension.

You may have developed a new procedure or new apparatus, and reporting this may be an important part of the manuscript you are now writing. However, it may warrant its own publication if the importance of the development is relevant to many other scientists. Or, it may warrant its own subsection in the materials and methods section, so the writer may emphasize this discovery as a major part of the entire body of work.

The materials and methods section must be able to "stand alone," without reference to other parts of the manuscript. When another researcher reads this section, he or she should have no questions about how this research was conducted, the experimental design, data analysis, what new procedures and apparatus were developed, and how reproducible and consistent will be the results from this research when compared with those of previous research (Day and Gastel, 2006).

TWO IMPORTANT GUIDELINES

Readers of your manuscript, especially reviewers, must get past the grammar, syntax, punctuation, and so on, to get to the point where they can analyze, interpret, and evaluate the science in your manuscript. This is particularly true about the materials and methods (Day and Gastel, 2006).

For some authors, especially inexperienced authors, deciding how much information to incorporate into this section is very difficult. You must "walk" the reader through what you have done sequentially, or he or she will get lost. You must also decide what level of description to use and how to describe any modifications from standard protocols done during the research. Please remember that reader understanding and ability to repeat by the reader is critical (Katz, 1985), and these two guidelines will help you prepare the materials and methods section.

Give a *competent, informed reader* enough information to

- duplicate the study if he or she had the necessary equipment, supplies, instruments, and so on, at hand, and

- judge whether or not the procedures were done correctly (i.e., is the science satisfactory and appropriate inferences about the data are permissible).

CONTENTS OF THE "USUAL" SUBSECTIONS

Subsections are often, but not always, used in the materials and methods to partition information into logical units that are easy for readers to reference. You may have the option of not using subsections, and this may be the best approach when the materials and methods section is brief and straightforward. For more complex situations, the seven potential subsections that follow provide a good example of subsection categories used in publishing in the life sciences. Recent articles in your journal of choice will give you other ideas to consider and may be especially helpful. As you peruse journals, you may see materials and methods sections printed in a font smaller than that used in the other sections of text. You also may see materials and methods sections that do not contain several of these subsections, and other journals may require some of these, along with other subsections not included here. Nonetheless, submit the manuscript to the journal with all sections in the proper order with the same font and font size.

Source and Handling of Biological Material

Include the genus and specific epithet of species you name. Also include authorities who named the species, if that is the norm for your journal and if authorities have not been reported earlier in the manuscript. List descriptors such as cultivar name, breeding line, and so on, of all biological materials used. If the biological material is not routinely available, you may want, or be required, to list its source by company name and location. Italicize the genera, and specific epithets, because they are Latin. You should describe exactly how you produced (propagated, bred, generated) the organism, and you should present all information about any substances used during its growth. If human subjects or other animals were used, be sure to state that a committee oversaw the conduct of the work and that it was done in an ethical manner (Gustavii, 2003; Matthews et al., 2000).

A complete description of the environmental conditions under which the organism was produced is mandatory. A synoptic table that summarizes the various environmental conditions used for organism growth has been published (Langhans and Tibbitts, 1997). This subsection should include information on the root medium used to grow the plant or the agar-based medium used to grow the cells, tissue, fungus, or bacterium. It is common knowledge that many classical experiments in the biological sciences cannot be reproduced properly because the entire set of environmental conditions were neither monitored nor reported. In some cases, it is appropriate to explain briefly why a particular taxon, chemical, or procedure was used. The use of a chemical, its purity, or why a particular piece of equipment or apparatus had to be used might also be justified in this section.

Details of Treatments and How They Were Applied

If one of your objectives was to impose a treatment (e.g., chemical, temperature) on your organism, then describe the treatment in detail, including when and how it was applied. You should also include the stage of growth when the treatment was made, the concentration of the applied chemical, and so on. In some instances, an environmental effect (e.g., temperature or light) may be synergistic with the chemical treatment, and the details associated with those interactions must also be presented. Applied chemicals should be described with correct chemical names by using International Union of Pure and Applied Chemistry nomenclature. The International Union of Pure and Applied Chemistry name for your chemical(s) can be found either on the container in which it was received, the material safety data sheet (MSDS), or by using Chemical Abstracts Service indices. It may be suitable to define the chemical name properly the first time, and then all subsequent times use the common name, if it is unequivocal.

Some journals either require or request that the experimental design be reported within this subsection, rather than in a separate subsection (O'Connor, 1991). In addition, some journals require the writer to justify use of a particular experimental design if alternate methods are available (O'Connor, 1991).

Steps to Prepare the Organism for Analysis

During your experiments, you may need to prepare the entire organism, or a harvested part of it, for a chemical, biochemical, or enzymatic assay. These steps must be presented in detail. For instance, you may need to freeze-dry your tissue before analysis. Or you may need to determine fresh weight, followed by a determination of dry weight, and then an ashing step to get the tissue ready for a mineral analysis. Or you may need to condition the organism for a low-temperature treatment. These are a few examples of the preparatory steps you may need to present, and almost every assay will have a dedicated protocol for preparing the sample for analysis. Some words that were formerly italicized, because they are not in English, have been anglicized, and these should no longer be italicized (unless your particular journal still prints them in italics.). Examples are in situ, in vivo, in vitro, and in vacuo.

Chemical and Biochemical Analyses and How They Were Conducted

If an established procedure was followed to the letter, simply cite the reference to the procedure. You may need to modify an existing procedure, which requires a description of the changes made. It would be inappropriate for a writer to state simply "the procedure of Smith and Jones (1998) was used with modifications." Instead, supply all details about modifications in procedures, equipment, or materials, and their effects on the analyses. Depending upon the complexity of an analysis, consider a schematic diagram to explain what you did. Some journals allow, or require, that supplementary information be presented via an online source that is described in its entirety so the reader may access it as needed.

Enzymatic Analyses

For some experiments, the researcher will use an enzymatic assay. Sometimes, it may be better to separate the preparation steps from the enzymatic analysis steps, and present them in an earlier section. Preparatory steps are usually presented with the methods for the enzymatic analysis. Assays on enzymes must be presented in enough detail to enable readers to judge and repeat the science conducted in the manuscript. Again, be sure you include descriptions of all modifications, apparatus, and procedures used in the research. In most cases, it is required that you give the Enzyme Commission, or EC, number in your first reference to the enzyme. Later in this chapter you will find a basic description of the EC system. Again, depending upon the complexity of the analysis, you may want to include an explanatory schematic diagram.

Physiological Analyses and Miscellaneous Procedures

Here is where to report any other more complex analyses. These probably involved several steps and/or several sets of apparatus such as systems that permit tests with electrophoresis, photosynthesis, and respiration. As usual, describe in appropriate detail the procedures and protocols, along with chemicals and equipment used. Include the name and location (city, state, and country if international) of any company that made an apparatus or more complex piece of equipment. If you have used a methodology and equipment exactly as stated in a published article, then simply cite that reference and clarify any modifications made. Schematic diagrams can help convey complex methods. If you have developed a new procedure or apparatus, you might insert another paragraph in this subsection to describe it. Consider whether the new procedure or apparatus is important enough to warrant a separate, descriptive publication.

Experimental Design, Data Analysis, and Statistical Presentation

Before you start to write this subsection of the materials and methods, please reread or peruse the section in Chapter 4 on "The special case for statisticians," and Chapter 10 on presentation of statistically analyzed data. These two chapters will help you understand what is needed in this subsection. You may find also that you need to consult with a statistician before proceeding.

Present the reader with an appropriate statistical basis for understanding the experiments reported. Begin with the basics of the experimental design, including the number of treatments, with a full description of how treatments were arranged, and the number of replications and observations. Next, describe how data were collected, if that was not addressed under each type of analysis. This should be followed by the way the database was managed and analyzed. In some journals, presenting ANOVA tables or other statistics in great detail will be expected, or required. Your journal is likely to require citation of the analytical software and methods used for common procedures like mean-separation tests and regression.

A FEW THINGS TO CONSIDER

- Give only a very gentle acknowledgement of any problems, weaknesses, or shortcomings of previously published research. It is inappropriate, and many times, unethical behavior to refuse to include, to misrepresent, or to hurl insulting remarks toward relevant methods and procedures developed by others. This is important because, first, it is unprofessional. Second, you might attack previously published work too strongly, and there is certainly the possibility the researcher you attacked could become a reviewer of your manuscript. In fact, this often happens because editors, in their search for manuscript reviewers, will peruse the literature cited to find researchers working in the same area. It may be happenstance, but your manuscript might be sent for review to the person you attacked in your manuscript. In the same manner, neither overplay nor underplay the value of your methods and procedures, especially if they are truthful and noteworthy additions to our body of knowledge.

- In general, we suggest that you err on the conservative side and cite a reference for virtually all procedures used, and then allow the reviewers or editors to remove the reference, if they feel it is common knowledge. Do not be afraid to reference something, if you feel a citation is necessary.

- If you use human subjects or animals in your research, you must receive permission to commence the research. Report approvals from that committee in your manuscript. If you do not, it may lead to a direct rejection of the manuscript (Peat et al., 2002). See written examples by Macknin et al. (1998) and Mathews and Sukiaini (1997) for research involving humans and other animals, respectively. Informed consent may also be an issue. Peat et al. (2002) and Gustavii (2003) have an excellent treatment of this issue.

- Use only the past tense, because all research methods are now past events.

- Write as much of the text as you can in the active voice. It will take fewer words, and will be more direct and descriptive. Passive voice is common in older literature, and some editors still prefer a mix of active and passive voices (Day and Gastel, 2006).

- Do not use jargon. If it is necessary to use specialized terms, be sure to define them upon first use.

- Use abbreviations sparingly, if at all, and only when necessary and appropriate. Abbreviations can confuse the reader, rather than clarify situations and save space. If you believe you must abbreviate, then consult the style manual or recent issues of your journal. Most journals have a standard list of abbreviations usable without definition.

- Editors or reviewers may require an indication of the accuracy and precision of all measurements made in the conduct of the research (Katz, 1985). Check your journal of choice to learn whether this is required.

- Avoid qualifiers such as "occasionally," "almost always," or "maybe" when you present information in the materials and methods section.

- If the manuscript is a methodology paper, the authors must accurately, clearly, and succinctly describe the method they have developed. In addition, the writers must show data in the results section that were used to validate the procedure. The reader must be able to follow the steps of a new procedure or the use of a new apparatus or equipment without question.

- Accuracy is how close a datum is to the truth of its value. Precision is the degree of reproducibility, whether it is a measurement of length, concentration, and so on. An archer can be very precise in the final location of the arrows shot at the target, but that group of closely spaced arrows may be a long way from the bull's eye of the target, and therefore, not very accurate. Strive to be both accurate and precise in your writing.

- Most readers will ignore the materials and methods section unless they have a deep level of interest in the topic, which explains why the section may be printed in a smaller font size or appear at the end of the article. However, do not be lulled into a false sense of security that reviewers will not pay much attention to your materials and methods. Often, this section is scrutinized intensely because the reviewer can render an opinion to the editor on the quality of the procedures much more easily than they can render a solid opinion on the analysis and interpretation of the data.

THE ENZYME COMMISSION (EC) SYSTEM FOR DESCRIBING ENZYMES

The abbreviation EC stands for the Enzyme Commission, and among other things, this entity assigns a number to each enzyme in biological systems (Webb, 1984). This information is updated on a web site: www.chem.qmul.ac.uk/iubmb/enzyme/.

The following example will explain why you should visit this web site if you are working with any aspects of enzymes. The first three numbers define the major class, the subclass, and the sub-subclass, respectively. The last number is a serial number within the sub-subclass, and it indicates the order in which each enzyme is assigned to the list. For instance, the common plant enzyme triose-phosphate isomerase is designated EC 5.3.1.1. It is an isomerase (Class 5), and it is in the third subclass (3), which contains enzymes that involve oxidation in one part of the substrate molecule during the time another part is being reduced. It is in the first sub-subclass (1), which contains enzymes that interconvert aldoses and ketoses. Finally, it is the first entry in this sub-subclass (of 19, several years ago). When describing an enzyme for the first time in a manuscript, always use the systematic name first, followed by the EC classification number, followed by the trivial name. Subsequent referrals to this enzyme may then be made by using the trivial name (Webb, 1984).

USE OF SIGNIFICANT FIGURES WHEN EXPRESSING DATA

There are several "rules of thumb" for reporting data. Below is a set of eight rules for expressing significant figures in the presentation of data (Holtzclaw and Robinson, 1988), but virtually any resource dealing with basic natural science will address significant figures.

- All measurements will have an uncertainty of at least one unit in the last digit of the quantity measured.
- Results calculated from measurements are as uncertain as the original measurement(s); that is, they are no better than the worst measurement.
- To determine the number of significant figures, follow this sequence. In a given number, find the first nonzero digit on the left and count to the right, including this first nonzero digit and all remaining digits. Unless the last digit is a zero lying immediately to the left of the decimal point, this is the number of significant figures in the number.
- When a number ends in zero(s) that are to the left of a decimal point, the trailing zero(s) may or may not be significant figures.
- When adding or subtracting, report the results with the same number of decimal places as that of the number with the least number of decimal places that are significant.

- When multiplying or dividing, report the product or quotient, respectively, with no more digits than the least number of significant figures in the numbers involved in the computation.

- When you need to round a number at a certain point (i.e., at a certain number of significant figures), drop the digits that follow if the first of them is less than 5. If the first of them to be dropped is greater than 5, or if it is 5 followed by other nonzero digits, increase the preceding digit by 1. If the digit to be dropped is 5 or 5 followed only by zeros, it is common practice to increase the preceding digit by 1 if it is odd and to leave it unchanged if it is even. Thus, dropping a 5 always leaves an even number.

- When you have the correct number of significant figures for your value, then arrange the number so that it is between 1 and 999 and use the prefixes listed later in this chapter to describe your unit of measure. It is no longer acceptable to use scientific notation to express numbers (i. e., 2.34×10^7). Note that prefixes are in multiples of 1000, and that is why you report the number between 1 and 999. For example, a wavelength in the visible range of the spectrum would be 555 nm rather than 5.55×10^{-8} m.

THE SYSTEME INTERNATIONALE (SI)

The breadth and depth of information about the SI system of units for the presentation of data is very large, and a complete treatment of the topic is beyond the scope of this book. However, we will present many of the basic principles you are likely to need. Virtually every journal now requires SI for the presentation of all units of measure. A few journals will permit, or even require, parenthetical insertion of English system units, because of their audience, but in most cases, this should not be done.

The SI consists of the presentation of data in the form of seven base units, two supplementary units, and 17 derived units (directly derived from base or supplementary units) (Table 15.1). Eleven quite common units of measure can no longer be used within SI, but there is an approved set of units that can be used in SI (Table 15.2). There are also seven original non-SI units that have now been converted into units that are fully acceptable within SI (Table 15.3). It is now required to record all measurements as a number from 1 to 999 as a coefficient to a unit of measure, and then use this number with an approved prefix to describe virtually all numbers appearing in scientific literature (Table 15.4).

GUIDELINES FOR MEASURING AND REPORTING ENVIRONMENTAL CONDITIONS

Let us now go on to measuring, monitoring, and reporting the environment for research studies conducted in the life sciences. It is no secret that many classical experiments in the life sciences, and especially the plant sciences, have been found irreproducible. Often, this can be attributed to a lack of controlling and monitoring the controlled environment in which the biological specimen was held during the progress of the experiment. In the 1960s, as this became more and more apparent, groups of scientists formed working groups that studied, discussed, and presented information about how to measure, monitor, and report these environmental parameters.

As a result of these discussions, scientists developed two goals for presenting data in a manuscript. Goal 1 was to develop methods for the measurement of the environmental

TABLE 15.1 Physical Quantities, Units, Symbols, and Derivations of Fully Approved Base, Supplementary, and Derived SI units (Downs, 1988)

Physical quantity	Unit	Symbol	Derivation
Base Units			
Amount of substance	mole	mol	
Electrical current	ampere	A	
Length	meter	m	
Luminous intensity	candela	cd	
Mass	kilogram	kg	
Thermodynamic temperature	kelvin	K	
Time	second	s	
Supplementary Units			
Plane angle	radian	rad	
Solid angle	steradian	sr	
Derived Units			
Absorbed dose	gray	Gy	$J \cdot kg^{-1}$
Capacitance	farad	F	$A \cdot s \cdot V^{-1}$
Conductance	siemens	S	$A \cdot V^{-1}$
Disintegration rate	becquerel	Bq	$1 \cdot s^{-1}$
Electrical charge	coulomb	C	$A \cdot s$
Electrical potential	volt	V	$W \cdot A^{-1}$
Energy	joule	J	$N \cdot m$
Force	newton	N	$kg \cdot m \cdot s^{-2}$
Illumination	lux	lx	$lm \cdot m^{-2}$
Inductance	henry	H	$V \cdot s \cdot A^{-1}$
Luminous flux	lumen	lm	$cd \cdot sr$
Magnetic flux	weber	Wb	$V \cdot s$
Magnetic flux density	teals	T	$Wb \cdot m^{-2}$
Pressure	pascal	Pa	$N \cdot m^{-2}$
Power	watt	W	$J \cdot s^{-1}$
Resistance	ohm	Ω	$V \cdot A^{-1}$
Volume	liter	L	dm^3

conditions such that the reader could consider the experiments repeatable under the same conditions. Goal 2 was to develop a reporting system that could compare the results of one study with another, because the environmental conditions were now a known quantity. Guidelines reported here are used in plant science (Langhans and Tibbitts, 1997) but are applicable to work with any biological organisms. Be sure to determine whether your journal has developed separate guidelines for measuring and reporting the environment in which the experiments were conducted. This will vary among disciplines.

Handling data on environmental conditions of life sciences experiments is similar to handling numbers for significant digits. Rules apply, and the scientist must first learn what are the important parameters to measure in their research system, and then they must measure

TABLE 15.2 Physical Quantities, Corresponding Prohibited Familiar Units, and Corresponding Fully Approved SI Values and Units for Several Familiar Units Formerly Prohibited Within SI (Downs, 1988)

Physical quantity	Prohibited familiar unit	SI value and units
Conductance	mho	1 S
Energy	Btu	1054.35 J
Energy	calorie, gram	4.184 J
Energy	erg	10^{-7} J
Force	dyne	10^{-5} N
Length	ångstrom	0.1 nm
Length	micron	1 μm
Length	millimicron	1 nm
Luminance	stilb	10^4 cd · m^{-2}
Magnetic flux	Maxwell	10^{-8} Wb
Photon flux density	Einstein	1 mol

TABLE 15.3 Quantities, Symbols, and SI-based/Derived Values for Several Non-SI Units that are Acceptable for General Use in SI (Downs, 1988)

Quantity	Symbol	SI-based value
Curie	Ci	37 GBq
Hectare	ha	10^4 m^2
Knot	kn	1.852 km · hr^{-1}
Millibar	mbar	102 Pa
Nautical mile	n · m^{-1}	1852 m
Roentgen	R	2.58×10^{-4} Ci/kg
Ton	t	103 kg

those conditions properly (Langhans and Tibbitts, 1997). A major abuse occurs when the researcher tries to measure or report values for the environmental conditions that are beyond the specifications of the instruments used to measure the environmental parameters. Another form of abuse occurs when the wrong instrument is used to measure the environment. Each instrument used for monitoring the environmental conditions will have an expected accuracy and precision, and researchers should ensure they do not exceed these values (Table 15.5).

For most experiments in the life sciences, the major environmental conditions to be measured and reported may include irradiation as photon flux; irradiation as energy flux; irradiation as spectral flux; air temperature; soil or liquid root zone temperature; atmospheric moisture; air velocity; atmospheric carbon dioxide concentration; watering parameters (how much, how often); substrate; applied nutrition (how much, of what, how often); pH; and electrical conductivity (Langhans and Tibbitts, 1997).

Scientists in the plant sciences have developed a synoptic table that structures all these parameters (Krizek et al., 1997), and it can be used for experiments in all life science disciplines. This table provides information about the environmental parameters to be measured and reported, and it also includes segments on where and when to measure, and what to report. Please note the units (or dimensions) for measuring and reporting. All individuals involved in the publishing process realize you have completed or nearly

TABLE 15.4 Multiplication Factors, Prefixes, and Symbols (All Non-SI Units) Accepted for Use as Adjective Modifiers of SI Units (Downs, 1988)

Multiplication factor	Prefix	Symbol
10^{18}	exa	E
10^{15}	peta	P
10^{12}	tera	T
10^{9}	giga	G
10^{6}	mega	M
10^{3}	kilo	k
10^{2}	hecto	h
10^{1}	deka	da
10^{0}	(value of 1)	—
10^{-1}	deci	d
10^{-2}	centi	c
10^{-3}	milli	m
10^{-6}	micro	μ
10^{-9}	nano	n
10^{-12}	pico	p
10^{-15}	femto	f
10^{-18}	atto	a

completed your research, and you may not have measured and recorded all of these items, simply because you did not know about them. Don't panic! Use this information for reporting your present set of research results wherever they can be used, and use the complete set of information for reporting your future work. It is likely that measuring and reporting all of these environmental conditions will be required for most journal articles in the future.

TABLE 15.5 Expected Instrument Precision and Measurement Accuracy

Parameter	Instrument precision	Measurement accuracy of reading
Radiation		
Flux	$\pm 1\%$	$\pm 10\%$
Spectral flux	$\pm 1\%$	$\pm 5\%$
Temperature		
Air	$\pm 0.1\ °C$	$\pm 0.2\ °C$
Soil/liquid	$\pm 0.1\ °C$	$\pm 0.2\ °C$
Atmospheric moisture		
Relative humidity	$\pm 2\%$	$\pm 5\%$
Dew point temperature	$\pm 0.1\ °C$	$\pm 0.5\ °C$
Water vapor density	$\pm 0.1\ g \cdot m^{-3}$	$\pm 0.1\ g \cdot m^{-3}$
Air velocity	$\pm 2\%$	$\pm 5\%$
Carbon dioxide	$\pm 1\%$	$\pm 3\%$
pH (H^{+} concentration)	$\pm 0.1\ pH$	$\pm 0.1\ pH$
Electrical conductivity		
Salt concentration	$\pm 5\%$	$\pm 5\%$

REFERENCES

BARRASS, R. 2002. *Scientists must write. A guide to better writing for scientists, engineers and students.* Second edition. Routledge, New York, NY.

DAY, R.A. and GASTEL, B. 2006. *How to write and publish a scientific paper.* Sixth edition. Greenwood Press, Westport, CT.

DOWNS, R.J. 1988. Rules for using the International System of units. HortScience 23:811–812.

GUSTAVII, B. 2003. *How to write and illustrate a scientific paper.* Cambridge University Press, New York, NY.

HOLTZCLAW, H.F. JR. and ROBINSON, W.R. 1988. *College chemistry with qualitative analysis.* Eighth edition. D.C. Heath and Co., Lexington, MA.

KATZ, M.J. 1985. *Elements of the scientific paper. A step-by-step guide for students and professionals.* Yale University Press, New Haven, CT.

KATZ, M.J. 2006. *From research to manuscript. A guide to scientific writing.* Springer, Dordrecht, The Netherlands.

KRIZEK, D.T., SAGER, J.C. and TIBBITTS, T.W. 1997. Guidelines for measurement and reporting of environmental conditions. pp. 207–216. In: R.W. LANGHANS and T.W. TIBBITTS (Eds). *Plant growth chamber handbook.* North Central Regional Research Publication No. 340. Agriculture Information Services, Iowa State University, Ames, IA.

LANGHANS, R.W. and TIBBITTS, T.W. (Eds). 1997. *Plant growth chamber handbook.* North Central Regional Research Publication No. 340. Agriculture Information Services, Iowa State University, Ames, IA.

MACKNIN, M.L., PIEDMONTE, M., CALENDINE, C., JANOSKY, J. and WALD, E. 1998. Zinc gluconate lozenges for treating the common cold in children. A randomized controlled trial. J. Am. Med. Assn. 279:1962–1967.

MATHEWS, K.A. and SUKHIANI, H.F. 1997. Randomized controlled trial of cyclosporine for treatment of perianal fistulas in dogs. J. Am. Vet. Med. Assn. 211:1249–1253.

MATTHEWS, J.R., BOWEN, J.M. and MATTHEWS, R.W. 2000. *Successful scientific writing. A step-by-step guide for the biological and medical sciences.* Second edition. Cambridge University Press, New York, NY.

McCABE, L.L. and McCABE, E.R.B. 2000. *How to succeed in academics.* Academic Press, San Diego, CA.

O'CONNOR, M. 1991. *Writing successfully in science.* HarperCollinsAcademic, London, UK.

PEAT, J., ELLIOTT, E., BAUR, L. and KEENA, V. 2002. *Scientific writing. Easy when you know how.* BMJ Books, London, UK.

WEBB, E.C. (Ed.). 1984. *Enzyme nomenclature: Recommendations (1984) of the Nomenclature Committee of the International Union of Biochemistry.* Academic Press, Orlando, FL.

CHAPTER **16**

INTRODUCTION: RATIONALE, REVIEW OF LITERATURE, AND STATEMENT OF OBJECTIVES

A bad beginning makes a bad ending.

—Euripides

DEFINITION AND NEED

The introduction gives readers a sufficient basis to understand why the study was done (Day and Gastel, 2006). Readers should also be in a position to evaluate the scientific merit of the research, under the assumption that the objectives of the research have been met. The introduction is needed because it provides answers to the following questions. Why was this research done? How did our previous knowledge influence this research? Who should be interested in the results of this research? What are the questions being asked? What hypotheses are being tested?

PRINCIPLES

Although it depends upon the format the journal uses, the introduction is often the first section of the manuscript. However, many scientists do not read it unless they read the entire article. The introduction usually consists of three parts that seem somewhat disjunct. However, these parts allow the introduction to flow from point to point as the need to conduct the research is articulated. Usually, the introduction is short, concise, and focused, and it is commonly one to two pages of the typed manuscript (Peat et al., 2002). In addition, it is written almost entirely in the present tense, as the author(s) are developing the need to conduct the research.

There are two basic models for building the introduction, but there are also a few alternate methods, which we will also present. Model number 1 incorporates parts of the materials and methods and results sections into the introduction, whereas model number 2 is a more traditional approach. We prefer model number 2, which we now summarize. The writer should follow the bulleted steps under model number 2 for building the introduction section of the manuscript.

Getting Published in the Life Sciences, First Edition. By Richard J. Gladon, William R. Graves, and J. Michael Kelly
© 2011 Wiley-Blackwell. Published 2011 by John Wiley & Sons, Inc.

The first part is one or two paragraphs long, and provides a rationale for the research. This begins to set the stage for the methodology and data that follow.

The second part comprises one or several paragraphs that contain references that acknowledge the previous work of others. It is mandatory that you have references in the introduction and, therefore, the review of the literature, because you cannot relate the work you will be doing to the previous literature without them. These paragraphs show what is already known about the subject, and they show how this work will relate to, or is needed as a result of, or extension of, this prior work. If you have published previous research in the area covered by this manuscript, and parts of that research are important to the research presented in this manuscript, then be sure to address that work in the review of literature, because it will help orient the reader.

The third part is normally one paragraph long, and is a clear and concise statement of the purpose(s) of the work. This will almost always be in the form of an overall objective and the specific objectives that, taken together, show the reader what hypothesis(es) will be tested and question(s) answered. The introduction should be as short, concise, and to the point as the writer can make it. The first and last paragraphs are especially important, because the writer needs to begin strongly and end strongly. Finally, the authors must ensure they do not promise more than the manuscript can deliver, whether it is methodology or results.

During development of these three parts of the introduction, the writer should get across to the reader why she or he should be interested in and care about this research and its significance. The writer can also explain why previous research results should be reopened or extended, and often this can be accomplished by an appropriately balanced presentation and discussion of controversies in the published literature. Many times, an excellent item to include in the rationale is the identification of a specifically targeted audience and how that audience could apply the results of this research. By incorporation of these items during the development of the introduction, the writer creates links to the discussion section of the manuscript, and this linkage can be a good thing that helps the reader understand the value of the research that was conducted. We encourage writers to take advantage of circumstances that present themselves and allow the writer to link the discussion with the introduction. Sometimes, writers will write and revise the introduction and discussion sections simultaneously, or nearly simultaneously, so they can capitalize on cases where the two sections might be linked.

The introduction is also the appropriate place for the writer to introduce and define any specialized terms pertaining to the research reported in the manuscript and the research of others whose literature you are citing. This also applies to specialized abbreviations. However, do not get into the trap of introducing jargon, as this has no place in a manuscript that is scientifically based.

The introduction should be written in the present tense, because the writer is referring to the development of the research problem and knowledge already established in the area of the research project. Model number 1 for building the introduction contains statements of information from the materials and methods and results sections, and we do not recommend that approach. Authors might refer to some of their methodology or the methodologies used, but this should be done only if they need to state reasons for choosing one methodology over another.

Do not keep major points or major discoveries made by others out of the introduction so that you can "pop" a surprise on the reader in the discussion section. Readers of scientific research papers really do not like to be surprised when they were not led up to the surprise. Keep in mind that this is not a murder mystery you are writing. In the words of Ratnoff (1981), "Reading a scientific article isn't the same as reading a detective story. We want to know from the start that the butler did it."

ITEMS TO AVOID

You should avoid these items when writing the introduction (Peat et al., 2002):

- Do not give the reader an unnecessarily long historical review of the published literature, unless you are writing a review article. In your journey through your research project and your visits to the library, you should become acquainted with all the relevant literature. However, you do not need to reference all of it. As with other sections of the manuscript, authors need to be careful to include only enough information to orient the reader without going overboard by presenting a more exhaustive literature review than is necessary or by making promises to deliver information beyond the scope of the manuscript.
- Authors should also avoid too much analysis of, criticisms of, and comparisons to existing literature. These should be saved for the discussion section of the manuscript.
- Do not analyze the relationship of previous literature to your work more than once.
- The author should not overplay or underplay the value of the work reported in this manuscript or the work reported in the literature.
- Finally, it is of paramount importance that the writer does not choose to refuse presentation of, misrepresent, or insult the relevant work of others who have published their work previously. Few things anger reviewers and editors more than writers who choose to take a solely positive approach to presenting their work while presenting the work of others only in a much less favorable light.

THE TWO BASIC MODELS

Two basic models have been developed for building the introduction (Day and Gastel, 2006). Depending upon the journal, or its editor, one or other of these models will be preferred. Statements of the materials and methods, results, take-home messages, or conclusions are sometimes included, as in model number 1, especially in some journals with younger editors. We prefer the more traditional approach of model number 2, but you should know about both models, as you may, some day, want to publish a paper in a journal that requires use of model number 1. We also present a few alternate methods.

Model Number 1

Model number 1, developed by several writers (Day and Gastel, 2006; Gustavii, 2003), contains the following elements:

- a statement of the nature and scope of the problem to be investigated;
- a review of pertinent literature to orient the reader;
- a statement of the methods used in the research (it also may be appropriate to state the reasons why specific methods were chosen);
- statement(s) of the principal results;
- statement(s) of the principal conclusion(s) suggested by the results.

Notice that there are no statements of objectives included within this model. In addition, note that there are statements of the materials and methods, results, and conclusions. The writer must be careful if they use this model number 1 and include these parts for the presentation of the materials and methods, results, and conclusions in the introduction. Often, the reader does not have enough information to be able to understand how these developed, and they become confused to a point where they cannot clearly understand what has been done by previous researchers and what was accomplished by the author(s) of this manuscript (Peat et al., 2002).

Model Number 2

The origin of this model is unknown, but several people have addressed parts or all of it (Davis, 2005; Lebrun, 2007; Matthews et al., 2000; Yang, 1995). This model may contain up to seven items that are detailed below as items 1 through 7. Item 1 often will be one paragraph that explains the rationale, as the need, to do the research. Items 2 through 6. Constitute the review of the published literature, and this group of items may be anywhere from one to six or seven paragraphs long. Item 7 is the statement(s) of the objective(s) of the research to be conducted and reported in the manuscript.

In practice, most authors concentrate on items one, two, and seven, and they will do a good job of developing them fully. Therefore, a minimal introduction may be composed of only three paragraphs, and these three paragraphs will be one paragraph each for items one, two, and seven, in that order. Many introductions will contain at least three paragraphs, but most will contain four, five, or more paragraphs. Note that this model does not contain any statements that present or discuss information from the materials and methods, results, discussion, take-home messages, or conclusions of the manuscript.

The seven items are as follows:

1. A rationale, or developed set of reasons, for conducting the research. Sometimes, this is called the nature and scope of the problem researched in the article. This portion is extremely important in orienting the reader to the need for the research.

2. A complete, but not exhaustive, review of the literature pertinent to the work in the manuscript. This portion of the introduction provides enough important background information that the reader becomes oriented to the problem and then can understand what the authors wanted to accomplish, and why they wanted to accomplish it, without reading any previous literature. This review of the literature also shows the reader the authors know what has been done previously, and it will not be duplicating work already completed. Establishment of credibility is a major goal of this segment. In writing the review of literature paragraph(s), there are several "traps" that can befall writers, and the writer must be cognizant of them. These "traps" are based on a poor story plot of the manuscript outside the introduction, plagiarism, imprecision in presenting the information of others, and the use of judgmental adjectives in describing the work of others. Lebrun (2007) should be consulted if more information is needed about these "traps."

3. What question(s) is (are) being asked. Sometimes this is incorporated into the statement(s) about the rationale or nature and scope of the article.

4. What hypotheses are being tested? Sometimes this is included in either the statements about the rationale or nature and scope of the article, or they may be part of the objective(s) statement(s).

5. Who should be concerned about, or care about, whether or not this research was conducted and reported.

6. Why the reader should be interested in the results of this study. Why did you choose to do the research and why is it so important?

7. Thorough, clear, and direct statement(s) of the objective(s) of the research. These may also be presented as the hypothesis or hypotheses of the research. Usually, this is located in the last paragraph of the introduction, and it may take several sentences to articulate the overall objective(s) and any specific objectives.

DEVELOPING STATEMENTS OF OBJECTIVES

In many instances, the last paragraph of the introduction is the most important one in the entire manuscript. Many reviewers refer first to this last paragraph to determine the hypothesis or hypotheses for which the research was conducted. The same may be said for the overall and specific objective(s) of the manuscript. If the reviewer cannot determine the intent behind the work reported in the manuscript after reading this last paragraph, they often become lost in the manuscript. On the other hand, if they find a clear, concise presentation of why the research was conducted, the remainder of the manuscript falls into place for them. This should be your goal for the ending paragraph of the introduction section of your manuscript. A clear presentation of what were the aims of the research is paramount to success in publishing your manuscript.

Under normal circumstances, a student working on an advanced degree will define specific and/or overall objectives of their research project as part of the research proposal they developed for their degree program. Likewise, a practicing scientist develops specific and/or overall objectives as a part of the grant proposal that covers their work. However, there are several instances where these statements of objectives need to be developed in the first place or be revised substantively. Sometimes a student chooses an externally funded research project for their advanced degree, and the research project has already been defined by the objectives of the original grant proposal. In this case, the student has no opportunity to experience the development of statements of objectives of the research project. In addition, either a graduate student or a scientist could adjust objectives during the conduct of the research based on new insights or information.

The beginning writer should also have the experience of thinking critically about the information they convey to the reader and how well that information develops a frame of reference for the reader when they read the introduction. In many cases, the take-home messages the writer developed when they started to write the manuscript will relate to one, or more, of the specific objectives of the research. Thus, the take-home messages, rewritten appropriately, might serve as a basis for construction of the specific objectives that appear in the introduction. Theoretically, the research and the preparation of the manuscript should have flowed from the specific objectives of the research proposal to the take-home messages of the manuscript. However, as the specific objectives of the research project changed due to preliminary results or the manner in which the research was conducted, the writer must often reverse this flow so that the take-home messages flow into the specific objectives of the manuscript at the time it is written. In addition, the writer may have a separate take-home message that is related solely to methods development, and this may become a separate specific objective of the manuscript. Writers should not feel insecure about making a specific-objective statement about a new technique that was developed or a published technique that the writer found did not work well and to which substantial improvements were made.

Most manuscripts yield an appropriate published article if the manuscript has two to four specific objectives of the original research, because these specific objectives often become the take-home messages of the manuscript. The writer should keep in mind this typical upper limit of four specific objectives or take-home messages as he or she develops the research proposal and the work that must be completed in the proposal. If the writer finds they have five or more specific objectives, then they should consider breaking the research project into two separate projects, which then might lead to two separate manuscripts.

ALTERNATE METHODS

In some disciplines, neither model 1 nor model 2 will be applicable due to any one of several factors. These factors may be the intended audience, the journal, how an individual writer develops his or her approach to both solving the problem and then reporting the results. Certainly there may also be other factors. Katz (2006) has suggested that a four-part approach will work well in these instances. These four parts are somewhat similar to the standard approaches in models 1 and 2, but, nonetheless, they are different and may find use for certain writers. This approach consists of the following parts, and it takes the introduction from the general to the specific. Many journals and writers in the biomedical sciences use this method.

- The writer should start the introduction with a general statement or statements of what things are known about the item that is the focus of the manuscript. Because the statement or statements are well known facts, there should be use of few, if any, references. A major goal here is to set up the reader for the point later in the introduction where the writer identifies the gap in our knowledge about this situation. Under certain circumstances, this section may be compared with the rationale for doing the work, especially when one also includes statements of the gaps in our knowledge.

- In the same paragraph, or the next paragraph(s), the writer should present a discourse focused on the information already in the literature. This part should review the existing literature appropriately, but it should not be overdone. It is appropriate to use this part of the introduction to point out smaller breaches in our knowledge of the subject, especially when this can be synthesized over several issues.

- In a new paragraph, the writer should now fully develop and present the gaps that are present in our knowledge of the subject. Normally, there should only be one or two gaps presented in a given manuscript, and if there are more than two gaps, then the writer should consider dividing the information into two manuscripts. These statements of the gaps in our knowledge correspond to the rationale for doing the work.

- Next, the writer presents and develops the manner in which he or she will develop a plan of attack for solving the problem(s) associated with this gap in our knowledge. In essence, this plan of attack may be interpreted as the overall objective of the research work, with individual support via the specific objectives that, taken together, produce an answer to the problem.

Another alternate method has been suggested by Peat et al. (2002). Their method is based on the following template, which may work for some journals or some audiences. This method is very short and to the point (three paragraphs only), and this may lead to broad appeal for many writers and reviewers. It does not, however, contain a rationale for conducting the work.

- In the first paragraph, tell the reader what we know about the subject.

- In the second paragraph, tell the reader what is not known about the subject.

- In the final paragraph, tell the reader why this work was done and what it accomplished. One or more sentences in this last paragraph must, out of necessity, develop the content and expectations the reader must be able to understand throughout the remainder of the manuscript. This paragraph should not contain a summary of any of the results that came out of your research, as this breaks the flow of information for the reader.

EXERCISES IN EDITING INTRODUCTIONS AND FORMULATING STATEMENTS OF OBJECTIVES

Three exercises (16.1, 16.2, and 16.3) are found on the following pages. Each contains a paragraph from an introduction to a scientific paper and is devoted to editing an introduction to a manuscript. To complete these exercises, first try and identify the three major parts of the introduction as designated in the information about model number 2. When one or two parts are not present, identify the missing part(s), and then try to make the introduction complete by supplying the missing part(s), as well as you can. Please use model number 2 as your basis for building this introduction. We purposefully double-spaced these paragraphs so it will be easier for you to edit them.

In Exercise 16.4, you should practice formulating statements of objectives. Normally, a student working on a Master of Science or PhD would have defined specific and/or overall objectives of their research project as part of the research proposal they developed at the beginning of their degree program. Likewise, a practicing scientist early in his or her professional career should have developed specific and/or overall objectives as a part of the grant proposal(s) that fund(s) their work. However, there are several instances where these statements of objectives need to be developed in the first place, or revised, at the time a manuscript is being prepared. If this has been the case, it would be helpful for the writer to get some experience writing statements covering the specific objective(s) and/or the overall objective(s) that will be included within the manuscript.

EXERCISE 16.1   Out-of-Class Editing of Introduction I

Taken from Wein, H.C. and Turner, A.D. 1994. Screening fresh-market tomatoes for susceptibility to catfacing with GA_3 foliar sprays. HortScience 29:36–37.

The formation of irregularly shaped fruit with enlarged blossom-end scars, termed catfacing, occurs more frequently in some cultivars than in others (Knavel and Mohr, 1969; Saito and Ito, 1971; Sawhney and Greyson, 1971). Since the disorder can cause significant reductions in marketable yield (Wien and Zhang, 1991), there has been interest in selecting for genotypes less subject to catfacing. Commonly, selection is practiced after the disorder has been induced by exposure to relatively low temperatures in the field (Barten et al., 1992; Elkind et al., 1990) or in a growth chamber. However, the unreliability of weather conditions in the field and space limitations in controlled environments limit the usefulness of these techniques. The discovery that GA_3 foliar sprays at transplanting can induce catfacing in known susceptible cultivars suggests that this technique might be used as an economical and convenient alternative screening tool (Wien and Zhang, 1991). The present work is a test of this hypothesis.

EXERCISE 16.2   Out-of-Class Editing of Introduction II

Taken from Shu, Z.H., Oberly, G.H, Cary, E.E. and Rutzke, M. 1994. Absorption and translocation of boron applied to aerial tissues of fruiting "Reliance" peach trees. HortScience 29:25–27.

Boron deficiency in plants is the most widespread of the known essential micronutrient deficiencies and has been reported for one or more crops in 43 states of the United States (Sparr, 1970). The range of B concentrations between deficiency and toxicity levels can be relatively narrow for some plant species. Boron deficiency caused spring dieback of twigs and branches of peach trees that had grown normally the previous year (McLarty and Woodbridge, 1949). However, due to the sensitivity of different crops to B, increasing the levels of B available to some plants by even small amounts can cause B toxicity and reduce yield (Gupta, 1985). Boron toxicity in peaches induced by soil B application has been reported (Cibes et al., 1955). Since peach has been ranked sensitive to B (Ayers and Westcot, 1976), foliar B application to peaches is currently recommended in many growing areas of the world. Nevertheless, the basic knowledge of B uptake and translocation by peach trees and the factors influencing the uptake of foliar-applied B is limited (Swietlik and Faust, 1984). The objectives of this study were to determine whether B applied to peach tree foliage is absorbed and translocated in adequate amounts to be an effective production practice.

EXERCISE 16.3 Out-of-Class Editing of Introduction III

Taken from Rhea, S.A., Miller, W.W., Blank, R.R. and Palmquist, D.E. 1996. Presence and behavior of colloidal nitrogen and phosphorus in a Sierra Nevada watershed soil. J. Environ. Qual. 25:1449–1451.

Colloid nutrient transport can play a significant role in the migration of organic and inorganic nutrient species (McDowell-Boyer et al., 1986; Degueldre et al., 1989; Ryan and Gschwend, 1990; Chin and Gschwend, 1991; Backhus et al., 1993). Vitousek and Melillo (1979) and Lewis and Grant (1979) have reported mobile organic colloids to be a significant source of water pollution. More recently, Qualls and Haines (1991) determined that the translocation of suspended colloidal organic forms of N and P played a significant role in the nutrient regulation of soil and stream ecosystems. The clarity of Lake Tahoe has decreased an estimated 7 m during the last 20 yr. as a result of increased nutrient loading (Goldman and Byron, 1986). Colloid forms may well represent a previously unrecognized source of mobile nutrients in the Sierra Nevada. This paper reports findings from a small-scale pilot study measuring the soil distribution of colloidal N and P following a prolonged drought (1987–1992), before and after an artificial precipitation event. Our objective was to evaluate the potential for colloidal N and P to mobilize in solution and leach into waters ultimately discharging into Lake Tahoe.

EXERCISE 16.4   Developing Statements of Objectives

The first purpose of this exercise is to give inexperienced writers help in writing statements of specific and overall objectives they can use in the introduction of their manuscript. The second purpose is to have beginning writers learn to think critically about the information they convey to the reader and how well that information develops a frame of reference for the reader upon reading the introduction. In most cases, each take-home message developed as a start to writing the manuscript will relate to one, or more, of the specific objectives of the research project. Thus, your take-home messages might also serve as a basis for construction of the specific objectives in the introduction. Theoretically, the preparation of the original research proposal, the conduct of the research, and the preparation of the manuscript should have flowed from the specific objectives of the research proposal to the take-home messages of the manuscript. However, as often happens in practice, the original specific objectives might have changed as preliminary results were obtained or the manner in which the research was conducted (e.g., a path different from the original path is now followed). Thus, at the time of manuscript production, the writer might often reverse this flow so that the take-home messages now flow into the specific objectives of the manuscript.

For each take-home message you have developed previously, or if you already have an original specific objective from a research proposal, write a *clear and direct statement* that represents a specific objective, either new or revised, that is an appropriate balance of conciseness and detail. Please limit this to four specific objectives. If two or more of these specific objectives can be grouped, then you might develop one or two overall objectives that may become overarching objectives of your entire research project. If this is the case, then complete the portion of the exercise where you can place overall objectives. In *one or more clear and direct statements for each specific objective*, explain how and why this specific objective became a portion of your research project. The writer should practice writing these specific objectives and their supporting statements elsewhere. When you are satisfied they represent what you want to relay to the reader, use this exercise as a model for writing and defending the items that will become the specific objectives, and possibly one or two overall objectives, of your manuscript.

Specific objective #1: _____

_____

_____

How/why became portion of research: _____

_____

_____

Specific objective #2: _____

_____

_____

How/why became portion of research: _____

_____

_____

Specific objective #3: _____

_____

_____

How/why became portion of research: _____

_____

_____

Specific objective #4: _____

_____

_____

How/why became portion of research: _____

_____

_____

Overall objective #1: _____

_____

_____

How/why became portion of research: _____

_____

_____

Overall objective #2: _____

_____

_____

How/why became portion of research: _____

_____

_____

REFERENCES

DAVIS, M. 2005. *Scientific papers and presentations*. Second edition. Academic Press, San Diego, CA.

DAY, R.A. and GASTEL, B. 2006. *How to write and publish a scientific paper*. Sixth edition. Greenwood Press, Westport, CT.

GUSTAVII, B. 2003. *How to write and illustrate a scientific paper*. Cambridge University Press, New York, NY.

KATZ, M.J. 2006. *From research to manuscript. A guide to scientific writing*. Springer, Dordrecht, The Netherlands.

LEBRUN, J.-L. 2007. *Scientific writing. A reader and writer's guide*. World Scientific Publishing Co., Singapore.

MATTHEWS, J.R., BOWEN, J.M. and MATTHEWS, R.W. 2000. *Successful scientific writing. A step-by-step guide for the biological and medical sciences*. Second edition. Cambridge University Press, New York, NY.

PEAT, J., ELLIOTT, E., BAUR, L. and KEENA, V. 2002. *Scientific writing. Easy when you know how*. BMJ Books, London, UK.

RATNOFF, O.D. 1981. How to read a paper. In: K.S. WARREN (Ed.) *Coping with the biomedical literature*, pp. 95–101. Praeger, New York, NY.

YANG, J.T. 1995. *An outline of scientific writing*. World Scientific Publishing Co., Singapore.

CHAPTER 17

DISCUSSION

Do not condemn the opinion of another because it differs from your own. You both may be wrong.

—Dandemis

DEFINITION AND NEED

If the results section represents the heart of your manuscript, then the discussion section may represent its soul. The discussion comprises your analysis and interpretation of the results you obtained from your experiments, and it shows how those results relate to the observations of others, that is, how they are placed into the context of observations made previously by you or someone else. The discussion also provides an opportunity to use facts from your results to build a case that your data have meaning and will impact others.

The discussion section can be the most difficult section of the manuscript to write, especially for inexperienced writers. A major reason for this is that the content and flow of the discussion are difficult to define and integrate properly into the manuscript (Day and Gastel, 2006). When writers do not express clearly the answers to questions that arise among readers, the chances of the manuscript being accepted for publication decrease. A manuscript can be rejected due to a poorly developed discussion. Inclusion of answers to questions likely to be asked by readers of the manuscript often sets a course that leads to rapid acceptance of the manuscript. When a reader finishes the discussion section, she or he should have answers to all, or at least some, of the following questions:

- So what?
- Who should care about this information?
- What new information has been gained?
- How consistent are these results with previous knowledge?
- What are the practical and esoteric implications of these results?
- What is the next step?

KEY ELEMENTS

Generally, each of the following eight elements must be in the discussion section of manuscripts in the life sciences.

Getting Published in the Life Sciences, First Edition. By Richard J. Gladon, William R. Graves, and J. Michael Kelly
© 2011 Wiley-Blackwell. Published 2011 by John Wiley & Sons, Inc.

- The take-home messages and conclusions that have evolved from the research reported in the manuscript must be prominent. Sometimes there is a separate Conclusions section.

- Analysis and interpretations of your data are critical. What do your facts mean, and who will care?

- References should acknowledge previously published research work and show how it relates to your work. Your new findings may either corroborate or refute information contained in articles published previously.

- Provide comparisons and contrasts between the results and interpretations reported in this manuscript and research already published.

- Acknowledge problems, weaknesses, shortcomings, and so on, of the research reported in this manuscript. It is better to address these issues before review of the manuscript than to be placed in a position of complete defense of all the work reported in the manuscript by one or more reviewers.

- Indicate how these results advance the state of knowledge of the field of interest; that is, publish this manuscript because . . .

- Sentences that are in a mix of tenses should be used to establish a timeframe of the research reported. Choose past tense for specific references to results that evolved from this research (specific data and take-home messages). Present tense is appropriate for general, over-arching conclusions that may synthesize concepts from these data and observations made in previous research.

- Resolve all major issues that were the focus of the introduction section of the manuscript. You have been building up to the discussion section, whetting the appetite of your readers for your new information. The discussion is the time and place to give them that satisfaction.

OTHER HELPFUL ELEMENTS

Depending upon the type of manuscript and the intended audience, the following four elements may be essential, beneficial, or unnecessary:

- extended analysis and discussion of controversies;

- identification of specifically targeted audiences and how they could apply these results;

- recommendations that specific types of experiments should be done as a result of this work;

- speculation, provided the writer does not overindulge and it is acceptable to the journal.

ISSUES TO BE AVOIDED

- *Restating results.* A pitfall of many discussion sections is text that does little more than restate the results. Take special care to avoid unnecessary repetition of results presented previously. It is often the tendency to wander in writing the discussion, and most often, reviewers will request that the discussion be reduced (Katz, 1985).

- *Poor style.* Loose, scattered, and/or illogical writing or organizational style has no place.

- *Poor value judgment.* Neither overplay nor underplay the value of the research reported in this manuscript or findings reported in previously published articles you have cited.

- *Unused results.* Results (from the results section of the manuscript) presented but not analyzed, interpreted, and discussed may need to be deleted. If you have written a good draft of the discussion and found you have not broached a category of your data, then you should consider deleting those data from the manuscript. Conversely, you may want to include discussion about those data, and you may need to revise your discussion section considerably.

- *Ignoring others.* Avoid refusing to include, misrepresenting, or insulting the relevant work of other researchers who have published previously. Wrongly disregarding such work is one of the best ways available to ensure those researchers become reviewers of your manuscript.

- *Overemphasis of your work's limits.* Watch for and avoid excessive use of phrases like "additional research is needed to determine . . . ." Reviewers might tell you to do that research work before you attempt to publish, and then reject your manuscript as being incomplete. Occasional use of such phrases can be acceptable; the key is not to over-use them.

RELATIONSHIP WITH THE INTRODUCTION SECTION

As suggested in the eighth entry of the "Key Elements" section, there should be a strong, discernable relationship between the introduction and discussion sections of a manuscript. Some writers believe that every thing in the discussion should have been addressed in the introduction section. The introduction should have built the case that the work needed to be done, during development of the rationale, and that someone would be impacted by it. Also, the introduction should have provided selected references to pertinent literature and direct statements of your research objectives. Building on, resolving, or otherwise dealing with each of these elements of the introduction is now important as you construct your discussion. Readers should finish the discussion section satisfied that you have met all stated objectives, have built on previously introduced literature through a synthesis of your results with that of the previous literature, and have identified audiences who may be impacted by your results. Reviewers should believe that ". . . yes, this is interesting, important, relevant, original work that has been conducted well and reported objectively and honestly, and therefore, should be disseminated to others."

STRUCTURAL METHOD FOR BUILDING THE DISCUSSION

Many authors, especially inexperienced authors, find that preparing the discussion section of a manuscript is particularly challenging. In fact, many first-time writers and inexperienced writers feel this sentiment fits their outlook on preparing the discussion:

> Writing is easy. You just stare at the typewriter until drops of blood appear on your forehead.
>
> —Red Smith

On the brighter side, you have considerable freedom to be creative during your construction of the discussion. For the most part, it is up to you how you frame the discussion, the literature you cite, and the emphases on which you elect to focus. On the flip side of that freedom is the fact that few journals designate specific expectations or formatting requirements for the discussion section. Thus, the burden falls on the writer to develop a discussion that accomplishes its goals in an organized, interesting, and engaging manner. We suggest that inexperienced writers, and especially first-time writers, try the conclude–expand method for building the discussion section. As more experience is gained, and individual writing skills are improved, modifications that will become part of the author's individual style can be incorporated.

THE CONCLUDE–EXPAND METHOD

This method provides an easy way to get started. It also presents you with an instant structure/format to follow. The discussion can certainly be revised to deviate from this structure later, if you wish, but the conclude–expand method should help you start in an organized manner.

- This method dictates that the first paragraph of the discussion is written as a compilation of your take-home messages; that is, simply write your take-home messages, one after another, to make a paragraph of two to four sentences. It can be as simple as that, or better yet, you may try adding some different words at the beginning of some of the take-home messages so that you create smoother transitions from one take-home message to the next (e.g., "... We also conclude that ...", "... Our data also show that ..."). The length of this first paragraph will depend upon the number of take-home messages you address in your manuscript. You may want to consider ending this paragraph with a statement of conclusions to prepare your reader for what will come in the next few paragraphs.

- Each subsequent paragraph is devoted to synthesizing and strengthening each take-home message, at a rate of one per paragraph, in the same order presented in the first paragraph of the Discussion section. Hence, the second paragraph of the discussion is your synthesis and strengthening of the first take-home message you listed in the first paragraph. The third paragraph synthesizes the second take-home message, and so on until you have discussed all take-home messages. The synthesizing and strengthening that occurs in each paragraph for each take-home message is the wordage you use to:

    o show how your results support the take-home message of that paragraph (without directly restating your results);

    o compare and contrast your results with those previously published;

    o state why your results are important; and

    o tell to whom those results are important.

- In some journals, or sometimes as a part of the style of a particular author, a final paragraph will be provided. The purpose of this final paragraph is to link together all of the take-home messages into an over-arching conclusion, or conclusions, or a theory concerning the interpretation of the meaning of the entire project. Often, this brings the significance of the research to the forefront, and in some manner, the reader must be able to relate to the significance of the work, or the manuscript will not be published. Some journals invite this last paragraph, and others show disdain for it. One way to determine whether you should include this last paragraph is to peruse recent issues of your journal of choice and look at the end of the discussion section to see if this last paragraph is there. If you include a final paragraph, it should not be a simple restating of the information in the first paragraph of the discussion but rather should present an over-arching conclusion or theory.

ADDITIONAL SUGGESTIONS TO HELP YOU GET STARTED

If you continue to struggle with your discussion, and you have started and maybe restarted the writing process several times, here are additional suggestions that might help you expedite the process.

- Begin strongly and end strongly, because this is where your most prominent ideas should be (Yang, 1995). Starting with your take-home messages is one way to start your discussion in a strong, forceful, and high impact way. We also suggest that you end with something useful, bold, or particularly significant. This is a case where inclusion of a final paragraph may be very effective, if the journal allows it and your results lend themselves to it. The presence of something particularly compelling at the end of the paper often seals the fate of the manuscript.
- Always move from the specific (your results) to the general, via inferences, as you develop the discussion section. The power of this approach has been discussed by Platt (1964).
- Do not feel compelled to write an exhaustive diatribe. In fact, excessive verbiage detracts from your messages and reflects that you are unfocused or unclear about the meaning of your data/results. As a rough rule of thumb, a discussion section that is well organized should typically not exceed about two and a half times the length of the text of your results.

REFERENCES

DAY, R.A. and GASTEL, B. 2006. *How to write and publish a scientific paper.* Sixth edition. Greenwood Press, Westport, CT.

KATZ, M.J. 1985. *Elements of the scientific paper. A step-by-step guide for students and professionals.* Yale University Press, New Haven, CT.

PLATT, J.R. 1964. Strong inference. Science 146:347–353.

YANG, J.T. 1995. *An outline of scientific writing.* World Scientific Publishing Co., Singapore.

CHAPTER **18**

ABSTRACT

Usually, a good abstract is followed by a good paper; a poor abstract is a harbinger of woes to come.

—Robert A. Day

DEFINITION AND NEED

The abstract presents a clear, concise, and specific synopsis of the manuscript. It is a miniature version, or microcosm, of the manuscript, which aids the reader in making a decision about reading the entire paper (Day and Gastel, 2006). The abstract almost always comes at the start of an article, and it largely replaces the terminal summary found in older literature. Most scientists agree that an abstract is of more value than a terminal summary, because an abstract is a condensation of the entire article, whereas the terminal summary might only recapitulate some of the results and the conclusions. Herein lies the main difference—and it is an important difference—an abstract can substitute for reading the article, but the summary only makes sense when the entire article has been read previously (Matthews et al., 2000).

Basically, there are three types of abstract (Day and Gastel, 2006). The *informative* abstract is a summary of all the work contained in the article. It is the most common form of abstract used today and, in most cases, it is the choice for most scientific journals, as it summarizes all parts of the work that were conducted. Busy scientists like this type of abstract because it can substitute (at least partially) for reading the entire article. The second type, the *indicative, or descriptive*, abstract tells what type of information will be contained in the entire article, but it will not divulge any specific information about what was determined as a result of the work. This type of abstract has less favor with busy scientists, as the scientist cannot glean any information about what was determined in the work without reading the entire article. The third type is the *structured* abstract. The structured abstract is derived from the usage of section headings for each part of the abstract. It can be used on either an informative or an indicative abstract. Few scientific journals use a structured abstract for their articles, but more have been choosing this option recently. The structured abstract is found most often in review articles, conference proceedings, and other, longer documents. These types of documents are cases where there would be too much specific information to relay to the reader, and the writer tries to communicate to the reader what can be found within the article/document.

The abstract is needed as an intermediate link between the title, which is very short, and the larger body of information in the text of the manuscript. In fact, the title and abstract should be intimately intertwined, and it has been suggested that every high-impact word in the title should also appear in the abstract (Lebrun, 2007). As a secondary need, the abstract

---

229

briefly summarizes each section of the article and answers at least some of the following questions. What was the rationale, or justification, for conducting the research? What were the objectives of the research? What hypotheses are being tested? How was the research conducted? What facts have been learned from this research? What conclusion(s) can be drawn from this research? What does the information discovered mean to people other than the author(s)? The abstract provides answers to some or all of these questions, but it must do so in a manner that allows it to "stand alone."

As a microcosm of the manuscript, an abstract should contain, in a logical sequence, short summaries of each of the major sections of the manuscript. Thus, the abstract should comprise summaries of the rationale for conducting the research, including the overall and specific objectives, the materials and methods, the results, and most importantly, a summary of the major results and conclusion(s) evolving from the research. A discussion, per se, and its associated analysis and interpretation, does not necessarily appear in an abstract. The role of the discussion is replaced by a presentation of the conclusion(s), which does not carry with it a lengthy analysis and interpretation of the results, and this yields a shorter piece.

Each summary of a section should comprise one to four sentences, yielding an abstract that contains about 10 to 15 sentences. In almost all cases, an author will struggle to limit the length of the abstract, rather than it becoming an issue of finding information to expand it.

Although we have not confirmed this story, we understand that the shortest abstract ever written was done by one A. Einstein (Day and Gastel, 2006). Albert was having trouble writing an abstract for a very important manuscript. The source of his trouble was that he had too many important words that had to be included, or so he thought. A colleague advised him to revise it several times, and in each revision, he should remove all unimportant words from the revised abstract so the remaining words would have more impact. Well, Albert followed the advice of his colleague, and when he was done revising the abstract, it read as follows: $E = mc^2$.

PRINCIPLES

Next to the title, the abstract is the most read part of an article. Most scientists read the title of the article first, and if the title piques their interest, they read the abstract. Then, if the abstract continues to stir interest, they may read one or more parts of the entire article. In realistic terms, the title is the "bait," and the abstract is the "hook, line, and sinker" that allows authors to "catch" the interest of the reader and "reel them in" (encourage the reader) to search for more information within the article. For this reason, an abstract must "stand alone" and give an appropriate amount of information to the reader. It also must be able to "stand alone," because it will be republished by abstracting services that often use the abstract without condensation or modification. Finally, the abstract is often transmitted separately from the remainder of the manuscript and, thus, again must be able to stand alone. Although we will not judge whether this is good or bad, some readers stop during or after reading the abstract because they have come to a more thorough understanding of the content of the manuscript, and it was not what they thought it was, based on the title.

The title, usually read first, sets an expectation by the reader that lures the reader into reading further. If, upon reading the abstract, the reader cannot connect the title with the abstract, the disconnect will cause the reader to choose not to read the article. However, neither should repeat verbatim what the other is stating, and this coherence should result in a strengthening of both parts. Lebrun (2007) suggests that 30 to 80% of the high impact words in the title should also appear in the abstract, and maybe within the first few sentences, depending upon the format used to construct the abstract. If there is less than 30% coherence, then the reader will probably not read the article, due to lack of coherence and direction. If there is greater than 80% coherence, then there may be too much repetition, and the reader may lose interest due to a lack of conciseness in the abstract. In general, the greater the percentage coherence, the greater the chance the reader will continue and read the entire article, provided the abstract expands upon the title, rather than simply repeating it. Part of Chapter 19 is focused on preparation of the final copy of the title, and there we provide an exercise that helps you determine the percentage coherence between the title and the abstract.

The take-home messages or conclusions that evolve from the work must be developed and presented unequivocally in the abstract. This can be a difficult task, especially for inexperienced authors, because most abstracts must be less than 200 to 250 words in length, and 100 to 200 words is much better. A few journals have a limit of 300 words, whereas others may have a limit of 3 to 5% of the entire manuscript. For some journals, this percentage may be as high as 5 to 10% of the entire article (Alley, 1996). For structured abstracts, which are often used with conference proceedings, the upper limit may be 300, 400, or 500 words (Matthews et al., 2000).

If the depth and breadth of the manuscript is such that a longer abstract is required, then make it as long as needed (O'Connor and Woodford, 1975). However, be judicious in your word selection and what you would like to tell the reader. Keep in mind the final result of Albert Einstein's putative battle to use only the most important words. Authors should not keep the reader guessing, nor should they go into so much detail that it stops the reader from continuing to read the article, unless the reader had an incorrect understanding of the article based on the title.

THIS INFORMATION WILL HELP YOU CONSTRUCT THE ABSTRACT

- Condense each major section of the manuscript into one to four sentences, and weave those sentences into a story that relays the basic idea of that section of the entire manuscript.

- For research reports in most scientific journals, abstracts must be informative, rather than indicative or descriptive, stating particularly major findings, conclusions drawn, and the significance of the work. Some journals now require structured abstracts, so you need to check what is appropriate for your journal of choice.

- The abstract should be written in the past tense only; everything has been done already. However, some writers disagree with this statement and suggest the entire abstract should be written in the present tense so that a higher level of interest can be generated (Lebrun, 2007).

- Use only action verbs, and write in the active voice, as these will yield the same ideas in fewer words.

- Abstracts should be quantitative, self-contained, and self-explanatory, without reference to anything but the title (Day and Gastel, 2006).

- The abstract must conform to the style mandated by the journal, especially in terms of length. For most journals, the maximum allowable unit will be governed by word count (usually 100 to 300 words), number of characters, or a percentage of the entire manuscript.

- The abstract should be written as one paragraph only, and it should consist of complete sentences only. Recently, a few medical journals are permitting appropriately divided abstracts called structured abstracts (Day and Gastel, 2006).

- Abstracts should use only standard terminology that is understood widely.

- The abstract should be the last section of text written. The reason for this is that the abstract needs to be a clear and concise summary of the entire content of the manuscript. It has been suggested that a "working draft" of the abstract should be one of the first items written so that it may serve as a guide in the development of the manuscript (Gustavii, 2003). Often, there will exist a "working draft" of the abstract, as many scientists will have prepared an abstract for presentation at scientific meetings. We feel take-home messages are better at keeping the writer on-target.

- Probably the most common problem with building an abstract is inclusion of too much detail. Often, this causes the abstract to become too long to fit the specifications designated by the journal. Economy of words without loss of accuracy, precision, and clarity is the key to success in writing an abstract. This is true whether the abstract is for a peer-reviewed, scientific, journal article or for an oral or poster presentation at a scientific meeting.

Abstracts should not contain the following items:

- statements that essentially repeat the title;
- illustrations, tables, or figures;
- literature citations, except under the most unique circumstances;
- justification(s) related to why you chose to do this research (this is best approached as part of the rationale portion of the introduction, but many authors prefer to have a very short rationale as part of the abstract);
- abbreviations, jargon, and specialized words (normally, the only permissible abbreviations are those that are extremely common, such as DNA, or extremely lengthy ones, such as the name of a complex chemical; journals vary on this, and they may need to be defined);
- low-impact phrases such as "was studied," "will be discussed," or "are described";
- reporting information not included in the text of the manuscript;
- promises that are more than the manuscript can deliver;
- unnecessary repetition of reporting data and other forms of results;
- loose, scattered, or illogical writing or organizational style.

PURPOSES AND QUALITIES OF AN ABSTRACT

Lebrun (2007) has an excellent treatment of the purposes and qualities of an abstract, and we will summarize those characteristics here. Five of these characteristics are focused on benefits to the reader of the abstract, whereas two are focused on benefits to the writer.

For the Reader

The first purpose of the abstract is to make the title completely clear. This is best accomplished by there being a high percentage of coherent words between the title and the abstract. The second purpose is to provide information about exactly what the author(s) have contributed to our body of scientific knowledge. This is accomplished via a complete description of what the researcher found, without confusing it with information that already exists in the literature. This is the major reason why there should be no references in the abstract. The third purpose is to help the reader decide whether or not they want to invest the time it takes to read the article completely. Unless the information in the title and the abstract are coherent and understandable, the reader will probably opt to not read the entire article because it may waste her or his time. The fourth purpose is that it allows the reader to gain knowledge quickly, and this helps the reader stay informed about new developments in his or her discipline. Finally, it helps the reader assess the level of difficulty of the article. If the reader knows they will need lots of time to read and comprehend the content of the entire article, then they can allocate the time needed. Some time during their career, most scientists have found themselves in the middle of an article, and realized that they need to spend more time on the article than they first believed.

For the Writer

The first purpose to benefit the writer is that a good abstract provides more key words, and this helps broaden the scope of search engines trying to locate new information. This leads to more people "finding" the published article, and some percentage of these will read it. The ultimate hope here is that the reader will learn from the published article and cite it in his or her scientific writings. The second purpose for the writer is that it allows the writer to provide more detail to the reader than just what the reader can glean from looking at the title and key words.

A STRUCTURAL MODEL FOR BUILDING YOUR ABSTRACT

We have found that the easiest way to write an abstract is to divide it into parts, and then allocate the designated number of sentences (approximate, of course) to each part. We present a structural model below that contains six parts. We also suggest that at least five of the parts should be contained in every informative abstract, and under certain circumstances, the sixth part, the analysis and interpretation, might be included. Other authors suggest four parts (Katz, 1985), a different set of four parts (Day and Gastel, 2006), a further different four parts (Lebrun, 2007), and seven parts for biomedical journals (Peat et al., 2002). Again, the most important thing you can do is to look at the instructions to authors document, the style manual of the journal, or recent issues of the journal to determine exactly what should be included in the abstract for your particular journal.

Sentences in most scientific writing average about 15 words. The total number of sentences, summed over all sections of the abstract, will usually be between 10 and 17, so this will give an abstract of 150 to 250 words. This number of words, or the corresponding number of characters, will be acceptable for most journals. Certainly, it behooves the writer to check with the instructions to authors for the journal to determine beforehand what is the limit and how this limit will be measured (e.g., word count, character count, or percentage of the entire manuscript). These sentence allocations may also be addressed as a percentage of the entire abstract, and this should be done on a proportional basis; that is, each sentence constitutes 5 to 10% of the entire abstract.

Our structural model consists of the following parts:

1. Rationale, justification, or need for the research (1 to 3 sentences)
2. Objective(s)/hypothesis (es) (1 to 3 sentences)

3. Materials and methods (2 to 3 sentences)
4. Results (2 to 4 sentences)
5. Analysis/interpretation (1 to 4 sentences—sparingly, only if necessary)
6. Take-home messages/overall conclusions (2 to 4 sentences)

SOME ADDITIONAL INFORMATION

One or more parts of the abstract may be eliminated due to author preference or journal style considerations. In addition, the analysis/interpretation part may be deleted completely or incorporated into either the results or the conclusions part, thus yielding five sections, and maybe correspondingly fewer sentences. In our work teaching manuscript preparation, we have found that the best abstracts contain five or six sections. This approach yields an informative abstract, or one that condenses the entire content of the manuscript. Researchers, and other readers, often find that reading this type of abstract substitutes for reading the entire article. This may be good or bad, depending upon one's perspective.

If you are having difficulty getting started or completing the abstract, try the following.

- Write the final draft of the abstract only after you have written the entire manuscript. (However, some writers prefer to develop a preliminary draft to help guide them in writing the remainder of the manuscript.)
- Focus on your take-home messages in the development of the abstract.
- Write one or two sentences that explain what your research means to others/society.

EXAMPLE TO PUT INTO PRACTICE THIS BUILDING SCHEME FOR THE ABSTRACT

AT THE INTERFACE OF PHYLOGENETICS AND POPULATION GENETICS, THE PHYLOGEOGRAPHY OF *Dirca occidentalis* (THYMELAEACEAE)

William R. Graves and James A. Schrader

(From: Am. J. Bot. 2008. 95:1454–1465; used with modifications)

ABSTRACT

Rationale	*Dirca occidentalis* is a rare shrub indigenous to only six counties near the San Francisco Bay in California. The populations in these six counties seem to show that there are four geographically disjunct populations (East Bay, North Bay, Salmon Creek, and Peninsula) and that these may constitute four divergent species rather than one species.
Objective(s)	Our objective was to determine whether these four geographically disjunct populations were four divergent species, or one species with four forms of ecological variation.
Methods/Materials	We used intersimple sequence repeat markers and automated genotyping to probe 29 colonies of *D. occidentalis* from the four geographically disjunct populations. Subsequently, we used methods of phylogenetics and population genetics to model variation across the four possible species.
Results	Results show the four disjunct populations are genetically isolated and have undergone divergence. Phylogenetic analyses indicate that the East Bay population was the first to diverge, followed by the North Bay, then the Salmon Creek, and then the Peninsula populations.
Analysis/Interpret.	This order of divergence suggests an intriguing natural history for *D. occidentalis* that is explained by the dynamic geological and climatic history of the San Francisco Bay Area. Spatial genetic structure detected for the species suggests an interaction of four factors: limited seed dispersal, clonal regeneration, distances traveled by pollinators, and genetic isolation of the four populations. Genetic diversity within the North Bay and Salmon Creek populations is low, indicating poor ecological fitness and risk of decline.
Conclusions	Intersimple sequence repeat markers resolved phylogeographic structure within *D. occidentalis*, results unattainable with internal transcribed spacer methods, and the integration of tools of phylogenetics and population biology led to an enhanced understanding of the divergence of this endemic species.

EXERCISE 18.1   Find the Missing Abstract Parts (with a Title Exercise)

Instructions

Draw a circle, or a "lasso," around each element of this sample abstract and label the sentences that correspond to each of the five or six elements that constitute a complete abstract.

Environmental problems caused by petroleum-based plastics have led to interest in alternatives made from biodegradable polymers (bioplastics), but little effort has been made to evaluate horticultural pots made from these materials. We hypothesized the stability and longevity of pots made from polymers of the hydrophobic corn protein, zein, is sufficient to make commercial use of zein-based pots feasible. Zein-based, bioplastic pots of two wall thicknesses were filled with either a peat-based, soilless potting mix or with coarse perlite and irrigated every 2 or 4 days. After 10 weeks, weight loss of pots was determined as a measure of their degradation. In a second experiment to simulate the potential practice of installing plants in the landscape without pot removal, bioplastic pots of two sidewall thicknesses were filled with the soilless potting mix and planted in either drained or saturated field soil, and the two substrates were either sterilized (autoclaved) or nonsterilized. Pots previously filled with soilless mix lost nearly twice as much weight as pots that contained perlite, and irrigation every four days led to greater weight loss than irrigation every two days. The pots released nitrogen (N) as they degraded; as much as 208 mg N/L was present in leachate after irrigation with water. After 12 weeks, pots in drained soils had greater weight loss than did pots in saturated soils regardless of whether the substrate had been sterilized. Zein-based, bioplastic pots appear suitable for crops having production cycles of less than three months, and the pots will decompose and release N if installed with plants in the landscape. Further research is needed to increase the longevity of zein-based pots for crops with longer production cycles. In addition, the influence of pots made from zein on plant growth needs to be determined, and potential effects of degrading pots installed in the landscape on the establishment of transplants merit investigation.

Title Exercise

Based on this abstract, try and develop an appropriate manuscript title.

EXERCISE 18.2   Building Your Abstract for Your Manuscript

Instructions

By using the structural model presented earlier in this chapter, and the admonitions of what is and is not appropriate to place in the abstract, construct the abstract for your manuscript. If you are having trouble getting started, refer to the hints we also gave earlier in this chapter. Try and keep your abstract to 150 to 250 words, and be sure to include one to four sentences on each of these items.

Rationale:

Objective(s):

Materials and Methods:

Results:

Analysis/Interpretation:

Take-Home Messages/Conclusions:

Answer objectively: Would you read the complete article, on the basis of this abstract?

EXERCISE 18.3   Would you Read this Article Based on this Abstract?

Instructions

After you have developed your abstract, revised it, and refined it several times, give copies of it to some fellow co-workers in your laboratory, your major professor, or a few other graduate students in your laboratory. These scientists should be chosen for their interest in and knowledge of the research area covered by the abstract. They should have enough interest in the research that they would want to read the article if it were published. Ask them to review it critically, and ask them to mark it with any suggestions they feel appropriate. Finally, and especially if they return it to you in person, ask them to answer this question (honestly!). Would you read this article based on this abstract?

EXERCISE 18.4   Identification of High-Impact Words

Instructions

Sit down with a printed copy of your finished abstract. *Part A*: Draw a strike line through all prepositions, articles, and low-impact words, and a circle, or lasso, around each high-impact word. Next, write each of these high-impact words into the blank lines below. *Part B*: After you have collated all the high-impact words, group them into categories of most important, important, and least important.

Part A:

_____	_____	_____
_____	_____	_____
_____	_____	_____
_____	_____	_____
_____	_____	_____
_____	_____	_____
_____	_____	_____
_____	_____	_____
_____	_____	_____
_____	_____	_____
_____	_____	_____
_____	_____	_____
_____	_____	_____

Part B:

Most important:	*Important*:	*Least important*:
_____	_____	_____
_____	_____	_____
_____	_____	_____
_____	_____	_____
_____	_____	_____
_____	_____	_____
_____	_____	_____
_____	_____	_____
_____	_____	_____
_____	_____	_____
_____	_____	_____
_____	_____	_____
_____	_____	_____
_____	_____	_____
_____	_____	_____

REFERENCES

ALLEY, M. 1996. *The craft of scientific writing.* Third edition. Springer-Verlag New York, Inc., New York, NY.

DAY, R.A. and GASTEL, B. 2006. *How to write and publish a scientific paper.* Sixth edition. Greenwood Press, Westport, CT.

GUSTAVII, B. 2003. *How to write and illustrate a scientific paper.* Cambridge University Press, New York, NY.

KATZ, M.J. 1985. *Elements of the scientific paper. A step-by-step guide for students and professionals.* Yale University Press, New Haven, CT.

LEBRUN, J.-L. 2007. *Scientific writing. A reader and writer's guide.* World Scientific Publishing Co., Singapore.

MATTHEWS, J.R., BOWEN, J.M. and MATTHEWS, R.W. 2000. *Successful scientific writing. A step-by-step guide for the biological and medical sciences.* Second edition. Cambridge University Press, New York, NY.

O'CONNOR, M. and WOODFORD, F.P. 1975. *Writing scientific papers in English.* American Elsevier, New York, NY.

PEAT, J., ELLIOTT, E., BAUR, L. and KEENA, V. 2002. *Scientific writing. Easy when you know how.* BMJ Books, London, UK.

CHAPTER 19

TITLE, BYLINE, KEYWORDS, AND AUTHORSHIP FOOTNOTE ITEMS

First impressions are strong impressions; a title ought therefore to be . . . a definite and concise indication of what is to come.

—T. Clifford Allbutt

DEFINITIONS AND NEED

These are the last four items you will address, or finalize, during the preparation of your manuscript. There are many possible definitions of the title of a manuscript, but the best one we have found may be attributed to Robert A. Day (Day and Gastel, 2006): ". . . the fewest possible words that adequately describe the contents of the paper." The title is needed because it (i) briefly identifies and describes the subject and content of the entire manuscript (Alley, 1996; Davis, 2005; Peat et al., 2002), (ii) separates this document from all others in the field (Alley, 1996), and (iii) helps the reader decide whether to read the entire article (Davis, 2005). In addition, the title must be accurate, specific, complete, meaningful, understandable, unambiguous, and attractive to potential readers on the first reading (Peat et al., 2002). Yang (1995) has presented ten rules for making a title.

The concept of the byline comes from newspaper journalism, and it is the boldface line on the top of the article that says, "This article is by. . . ." The byline is simply a listing of the author(s) of the manuscript and, in some journals, it also contains the location(s) where the research was completed, the affiliation(s) of the author(s), and sometimes, the rank of each coauthor. The byline is needed, at the most basic level, because it identifies who has taken complete responsibility for the entire scope of work that constitutes the manuscript. It also helps to identify the author(s), and it presents information about how to contact the author(s). Use of e-mail as the medium for one scientist to communicate with another scientist has largely replaced written letters, and to some extent, telephone conversations. Therefore, the e-mail addresses are very important parts of the byline or the footnotes on one of the first pages of the manuscript. In developing the byline, answers to two important questions must be determined. Who will be a coauthor? What is the order of coauthors, if there is more than one author?

The keywords (or index words) are words that are not in the title but have significant meaning within the context of the field of research. The keywords may be simple, single words, or they may be an amalgamation of several words (a phrase), and they should not be repeats, or even partial repeats, of the words in the title. "High impact" words that are in the title automatically become keywords in nearly all journals. The keywords are needed because they are used by indexing services to help readers, literature searchers,

Getting Published in the Life Sciences, First Edition. By Richard J. Gladon, William R. Graves, and J. Michael Kelly
© 2011 Wiley-Blackwell. Published 2011 by John Wiley & Sons, Inc.

and other people "find" your particular article among the thousands of articles that are published in a given time period.

The authorship footnote usually appears at the bottom of the title page of the manuscript, and it transmits to the reader several pieces of information that do not have a proper place elsewhere in the manuscript. It will contain the received and accepted dates of the manuscript, an institutional journal paper number (if the sponsoring institution requires it), funding acknowledgments, and an acknowledgment of other forms of aid throughout the conduct of the research and preparation of the manuscript. The authorship footnote also often provides the complete mailing address, e-mail address, and/or telephone number of the corresponding author of the manuscript. These dates of submission and acceptance of the manuscript are necessary, because the first person publishing a discovery often receives credit for the discovery and the discovery may be associated with them, in the form of their name being used as a label for the discovery. The acknowledgments are needed because we must give credit to the people and institutions that made it possible for the work to get to the stage of a manuscript for publication. The format and content of the authorship footnote vary widely among journals. Be certain that you check the instructions to authors, the style manual, or recent issues of your journal to determine the appropriate location, format, and content of the authorship footnote. Journals that use a blind review process by necessity must keep all of the information on the author(s) on a separate title page that can be removed before the manuscript is sent out for review.

PRINCIPLES

The Title

The title you will now finalize should be based on the provisional or working title you derived in Chapter 7. These two titles are almost never exactly the same, because they serve different purposes (Davis, 2005). The provisional title you developed previously was structured to help guide you, *the writer*, through the process of manuscript development. The final title will change for two reasons. First, as a person or research team writes a manuscript, there will be numerous items that are included, deleted, or changed, and this will probably mandate a change in the title, purely on that basis. In addition, and probably even more importantly, the title needs to be changed from a provisional title to a final title so that it can be crafted with *the reader* in mind. The final title must help the reader make a very quick decision about reading at least the abstract, if not the entire article. If it does not help the reader make this decision quickly, more than likely, you will have lost the reader.

The most important thing to remember is that your title will be read by thousands of people either in the original journal or through an indexing or abstracting service. The title should help to facilitate retrieval of the article during a search of the literature associated with a topic that is of interest to the searcher. For each person who reads the entire article, 500 people may have read the title; this means that 499 will not have read anything more of the article (Kerkut, 1983).

The title should contain high impact words. Those words should be chosen with great care, and the coauthor(s) should use them early in the string of words that constitute the title. In most cases, a prospective reader peruses the title for two seconds or less and then makes a decision about reading the paper (Lebrun, 2007). Do not therefore waste space or time on words that do nothing for your article. The sooner the reader gets to a high impact word or words, the better. The words of the title must be able to stand alone, just like the words of the abstract and the keywords, because the reader will not have already read your article.

They must also be able to stand alone because indexing and abstracting services will separate them from the entire title and "pigeon-hole" them into the database that will be the basis for a search by someone looking for information (Katz, 2006).

The title may be *indicative*, which means that the title indicates or hints to the reader that the manuscript will contain certain information. Alternately, it may be *informative, or declarative*, and in this case, it informs the reader about the content of the manuscript. The best titles are informative or declarative ones that are specific, because they are meaningful to the reader at first sight. They have also completed their mission of answering the main question of the reader, "What is the content of this article?" Often, the best title, or at least a good first draft of a title, is developed by arranging, rearranging, and stringing together the keywords that the author(s) accumulated while writing the manuscript.

The following are probably the most common faults of a bad title: (i) it is too long and the reader becomes disconnected by the time he or she gets to the end of it; (ii) there is a poor choice of a word, which does not mean what the writer intended; (iii) it extends beyond what the article actually contains (i.e., it makes promises it cannot keep); and (iv) there is faulty syntax, or word order, that leads the reader to an incorrect meaning of the content of the article (Davis, 2005; Day and Gastel, 2006). In almost every case, the title should not be a sentence, but instead is a carefully chosen set of words that are both understandable and retrievable by some form of indexing service. The first several words of the title are the most important ones (Davis, 2005).

There are four general categories or types of titles: the headline; the sentence (sometimes with a colon); the question; and the name and verb. In the end, a good title, regardless of its type, indicates the purpose or an overall take-home message of the manuscript. The words that constitute the title should be accurate, informative, clear, concise, lend themselves to indexing, and command attention, even to the casual observer (Davis, 2005). In Chapter 7, you completed an exercise in which you developed a provisional title so that you would have a better focus for developing the take-home messages of your manuscript. You should return to that provisional title, re-read it, maybe several times, and then use it to begin formulation of the final draft of the title you will use for your completed manuscript.

Keep in mind that the provisional title was for your own use, and it should have been used as an aid in helping you write the manuscript. Lebrun (2007) has presented four purposes of the provisional title for the writer; the main one is that it helps the writer to write the manuscript concisely. However, the purpose of the final title is to aid the reader. Its purposes are to inform the reader about the content of the article, to help the reader determine whether she or he should read the entire article, and to allow the reader to come to an understanding of what type of article it is, its specificity, its theoretical level, and its nature (Lebrun, 2007). These different purposes almost always lead to entirely different titles. An exercise on developing the final draft of the title is at the end of this chapter.

Lebrun (2007) lists and discusses several qualities of a title. By nature, these qualities are best described as adjectives that modify a good title, and you may want to compare what quality level your title achieves. These qualities are as follows:

- The title should be unique, so that it separates your title from others, both past and present.
- It should be lasting, rather than new, as items are new only for a short period of time.
- It should be concise (are the keywords/high impact words overly detailed, or just right?).
- It should be clear. The words should have only one meaning and cannot be misunderstood.

- It should be easy to find, by scientists you want to "find" your article and read it.

- It should be honest and representative of the contributions made by you and the coauthors; that is, does the title set expectations to be made by the reader, and are they met?

- It should be as catchy as you can make it, because in two seconds the reader reads the title and makes a decision about whether or not she or he will read the abstract or maybe the entire article. Lebrun (2007) lists seven ways to make a title catchy.

Some Items to Consider in Building a Title

- Avoid using low impact expressions like "Effect of," "Impact of," "Study of," "The nature of," "On the," "Studies on," or "Contributions to." In addition, do not use unnecessary prepositions such as "of," "in," and "on," or unnecessary articles such as "a," "an," or "the." These words are often at the start of the title, and they get the title off to a bad start (O'Connor, 1991).

- Whether you should or should not use a verb in the title is debatable. Some feel verbs should not be in the title (Davis, 2005), whereas others feel it is satisfactory or maybe good to use a verb in the title (Lebrun, 2007). Lebrun (2007) also suggests that numbers and adjectives should be used to accentuate a strong point of contribution to our field of knowledge.

- Use the past tense, if you use a verb in the title. The work you are reporting was done previously. It has been suggested that when a verb is used in the title, the title should then be converted into a sentence (Gustavii, 2003). Such a sentence often yields a stronger title, but it may become too strong and indicate more than the article can deliver (Gustavii, 2003). Recently, this has become an issue of debate, as some people believe that use of the present tense makes the title more "alive" (Katz, 2006). In addition, some object to sentences as titles (Day and Gastel, 2006).

- Little or no punctuation should be used (Peat et al., 2002).

- Most good titles contain 5 to 12 words, with 10 to 12 words being most common (Davis, 2005; Matthews et al., 2000; O'Connor, 1991). Many journals have a word limit, usually 12 to 15 words, or a line limit in the printed article (e.g., two printed lines (Katz, 2006), or they will have a character limit, usually 80 to 100 total characters, including spaces (Matthews et al., 2000). However, some journals have no limit on words, lines, or characters. Titles shorter than five or six words are usually too general, and therefore problematic (Day and Gastel, 2006), and titles longer than 12 to 15 words are usually too detailed, which can cause confusion in the interpretation of the title by the reader. Most readers can grasp only about four items from a title, so choose your words carefully (Alley, 1996).

- In some journals, the title, or a condensation of the title, also serves as a running title, which occurs on each page or alternate pages throughout the printed article. If the running title is a running head of the article, it will be a "headline" on the top of each page or alternate printed pages of the printed article. Alternately, it may serve as a running foot, which will be located at the bottom of each page or alternate pages in the printed article. In most cases, this will be a condensation of the title (most commonly 40 to 60%), rather than the complete title (Gustavii, 2003).

- The title should not contain abbreviations, chemical names, or jargon unless, for instance, the chemical is associated with the central purpose of the manuscript (Day and Gastel, 2006; Yang, 1995).

- If a chemical is associated with the central purpose of the manuscript, then you should use the common name of the chemical, or any other proprietary item, for that matter, so that you do not use a large percentage of the characters available for the title for the name of the chemical. Chemical names are often extremely long, and they can contain many unusual symbols.

- Unless there is an unequivocal reason to use the binomial of the species used in the research in the title, you should use the common name. However, some journals may require its use.

- Avoid the use of a title that presents itself as a question, as most times these do not get published (Matthews et al., 2000). Titles that end in question marks often confuse the reader, or worse, the content of the article may not answer the question satisfactorily (O'Connor, 1991). Often, reviewers believe the author(s) overstepped their bounds and are very presumptuous in the claims made in the article. The same may be said for an overly assertive title, as this will often be too egotistical (Peat et al., 2002).

- Series titles are generally discouraged by editors, because they often imply that the author(s) have "staked a claim" to the research in that area (Day and Gastel, 2006; Matthews et al., 2000). Most journals that allow serial titles will require that at least the first two articles in the series be submitted together so that they go through the review process together. (There have been instances of the second article in the series being accepted for publication, but not the first, and this creates an awkward situation for both the author(s) and the journal.)

- Hanging titles (titles broken by a colon) may not be desirable, because the colon segments the title, and the segments, taken together, may not have the meaning the author(s) originally desired (Matthews et al., 2000).

The Byline

Let us start this section by repeating the quotation at the beginning of Chapter 4.

> If you have coauthors, problems about authorship can range from the trivial to the catastrophic.
>
> —M. O'Connor

We hope you have encountered few difficulties determining who should be listed as a coauthor and what should be the order of the coauthors in the byline. There are no sets of rules or any conventions that can be used to guide author(s) of the manuscript to a logical, acceptable statement of who should be an author and what will be the order of coauthors on the manuscript. Some professional societies are working toward the development of a code for authorship, some may have guidelines in place, and the Vancouver Guidelines for biomedical research have come closest to broader acceptance. At this point, however, there are no consequences that result from not following any of these guidelines, as they are guidelines, not rules.

At times, decisions about authorship can be an issue of ethics, and Chapters 3 and 4 are devoted to helping you make appropriate choices of your coauthors and the ethical considerations associated with authorship issues. Now that you are approaching completion of the manuscript, you may want to return to Chapters 3 and 4, re-read them, and change some things in what will become the byline of your manuscript. We hope you have already completed the exercise on authorship in Chapter 4, and that you have learned important information from that exercise. In addition, we hope you will complete the one at the end of this chapter, as it is an expanded and strengthened look at authorship of your manuscript.

This is critical, because this is a very real issue that must be faced by all writers of manuscripts, because it is the most sensitive part of manuscript preparation.

In an attempt to avoid conflicts and other unpleasant scenarios, some journals are moving toward incorporation of a credit line or a contributors line, which presents and discusses each coauthor's contribution to the article (Gustavii, 2003; Peat et al., 2002). In our opinion, most journals will move in this direction, because this type of presentation focuses the attribution of authorship credit on objective, tangible issues, rather than those of grace and favor (Peat et al., 2002).

In most journals, another part of the byline is the complete mailing address of all authors, so readers with questions can reach any one of the coauthors. If this is the case, then the addresses should be listed in as complete a form as possible for all individuals who participated in the research, including postal codes for directing mail. However, and perhaps for security reasons, some journals now list only the e-mail address of the corresponding author, and the address associated with the byline is the institution where the work was completed. As with the names in the byline, there is no set of rules or conventions about whether the addresses should define the place(s) of work of the author(s) or the place where the research work was actually performed. The important issue here is that the reader should have access in one form or another to at least one author so that questions may be answered appropriately. An exercise that revisits your first attempt at defining the authors and the order of the authors is included in this chapter. Please complete this exercise before submission of your manuscript.

Some Items to Consider in Building the Byline

- The format for the byline will vary from journal to journal, and there is no set format used by many or all journals. You should duplicate the format used in the byline of an article recently published in your chosen journal. Alternately, you should consult the instructions for authors document or the style manual of the journal to help you determine the appropriate format for the byline.

- Some journals do not allow addresses to be contained in the byline (only the author(s) names), and they mandate that the address(es) be in a footnote or as footnotes on the first page of the printed article. Again, you should duplicate the format used in the byline of an article recently published in your chosen journal.

- Most journals no longer print degrees or the rank of the author(s), although some medical journals still do (Day and Gastel, 2006). Again, see the instructions to authors document, the style manual, or recent issues of the journal to ascertain what you should do.

- In every case we know, the institution where the work was conducted must be in the byline. This may also be true if several institutions are collaborating on the project. If more than one unit at one institution or if multiple institutions participate, or if multiple units at multiple institutions contribute to the project, then the author writing the manuscript must be sure that each coauthor is properly and unequivocally linked to the correct unit at the correct institution.

- In all cases we know, if a coauthor moves from the institution where the work was conducted, that should be added as a footnote in the proper place for your journal. This will vary from journal to journal, and you should see a recent issue to determine how to handle this. A prime example of this is when a graduate student completes a degree and moves to another institution for continued studies or for a full-time professional position.

- Exercises that you attempted or completed earlier should have started you thinking about any possible problems that might arise in selecting who should or should not be a coauthor, as well as the order of the coauthors. Please return to the beginning of this section of this chapter and re-read the quote for the byline, because you should not enter into this part of manuscript preparation without some basic understanding of what can happen.

The Keywords

Keywords have significant meaning within the context of the research field associated with the manuscript. These words may also be known as the "index words," "additional index words," or other monikers, depending upon the journal. Indexing services that help journal-article readers "find" a paper use the high impact words in the title and the keywords. These words are entered into a database and, then, when a match is made, the searcher can get information about where the information has been published. In most cases, all of the high impact words in the title automatically become keywords for the purposes of indexing the published article, and, therefore, in most instances, the keywords should not be a repeat, or even a partial repeat, of the words in the title. The keywords may also be multiword units such as "gas chromatography," "liquid scintillation counter," and so on. In all cases, a keyword should be quite specific, and a general term should never be used. Often, the keywords you used to complete your review of literature before you began to work on the research project will function well as keywords for your manuscript (Gustavii, 2003; Katz, 1985). Another rich source of keywords for your article is the set of keywords contained in journal articles you read before you commenced work on your research project (Lebrun, 2007).

Depending upon your journal, several issues may need to be addressed before your manuscript can be submitted. In all of the instances presented later in this paragraph, you should check with your journal's instructions for authors document, the style manual of the journal, or a recent set of issues of the journal to determine how you should handle each of these items. First, you will want to determine the term used for the keywords, because this varies in different journals. Second, you should determine where in the manuscript, or printed article, the keywords are placed. Again, depending upon the journal, this can be in one of several places, but they are usually near the beginning of the manuscript. You may have to choose your keywords from a list published by the journal. To help keep information more broadly accessible, some journals only allow words they want used as keywords, or they may allow you to take them only from a list such as *Index Medicus*. Although this depends on the journal, many journals will require you to submit from three to ten keywords, and in almost all cases, these are keywords that are not in the title of the article. If two keywords have the same meaning, then use one for the title and one for the listing of keywords (Lebrun, 2007). Lebrun (2007) has a particularly good treatment of choosing keywords, and he divides them into three levels of their capability to differentiate and specify your research work.

Some Items to Consider in Building the Keywords

- One of the best ways to obtain keywords for the manuscript is to search through your take-home messages for high impact words.
- Another source may be perusal of the materials and methods section, especially if the main purpose of the work was to develop methods.

The Authorship Footnote

The authorship footnote, which is sometimes called the lead footnote, usually appears at the bottom of the first page, on the title page, or on the cover page of the manuscript, depending upon what your journal calls the first page of the manuscript. Depending upon the journal, it may be located elsewhere in the manuscript, but in most cases, it will be near the beginning of the manuscript because it contains several pieces of information that are important, but not part of the scientific manuscript.

The authorship footnote, under normal conditions, will contain these several components, usually in this order. However, as always, the author should consult the instructions to authors document, the style manual, or a recent copy of the journal to which the manuscript will be submitted to determine the content and order of the authorship footnote for that particular journal. Sometimes, this footnote may be designated with a superscripted "1" at the end of the title of the manuscript on the cover page of the manuscript. There may then be a superscripted "1" at the start of the footnote or, in many cases, the footnote may not be superscripted because the reader is expected to understand that the superscripted "1" at the end of the title designates the authorship footnote.

The authorship footnote may first contain the phrases "Received for publication _____" and "Accepted for publication _____." (These exact phrases may not be encountered in every journal, but most journals will have some designation equivalent to this pair of phrases.) These blanks are completed when each of these days has arrived, and the editorial office will complete the blanks in almost all cases. Occasionally, the corresponding author completes the dates from the hand-written date entered by the editorial office. These entries are important for identification of the scientist(s) who first discovered the scientific information that is part of the manuscript. There have been instances when two completely independent laboratories have discovered the same principle or other piece of scientific information, and these phrases designate who was first and second to record the phenomenon. Probably the most stellar example of this is the work on the helical structure of DNA by Rosalind Franklin, James Watson, Francis Crick, and Linus Pauling. In most cases, however, when two or more groups have identified and published information about the same concept within about one year of one another, scientists generally acknowledge both groups as co-discoverers.

The next item is the journal paper number of the institution where the work was completed. This number is issued mainly to keep track of each manuscript so that page charges and reprint/offprint charges can be associated with the correct published article. For many years, scientists believed that receiving this number from their institution constituted "permission to publish" the work. This was not at all the case, as virtually no institution of higher learning or nonindustrial research center can control what a scientist chooses to publish. However, this may be the case for an industrial organization, because the corporation may want to control what information leaves the organization and also when, so that the work of the organization can be patented or at least kept from competitors. Some institutions no longer assign this number, and it will not therefore appear in the authorship footnote. However, authors should make sure they understand and follow the protocol of their institution and the journal where the manuscript will be published.

Identification of the institution that has taken responsibility for the body of work that constitutes the publication will follow the journal paper number, if that is used, or will follow the accepted for publication date. This is critical when the authors are from more than one institution, because this becomes the institution that has responsibility for all of the costs associated with the publication of the manuscript. In most cases that involve agricultural research done at land-grant schools in the United States, this will be the experiment station that is associated with the university where the researcher works. When this is a publication not completed through an experiment station, the university is acknowledged. In either case, the location of the institution is designated, because some states and some universities have multiple campuses or centers.

The next items will be various forms of acknowledgment of contributions to the body of work reported. This has been summarized nicely by Peat et al. (2002), but we have expanded certain parts of their discourse. Care and judicious exercise of choice is important here, because before long we can be thanking everyone, without real regard to his or her contribution. Alastair Spence states this poignantly: "By all means recognize secretaries, wives or husbands, lovers and parents—but not in the manuscript." Be sure that you check the proper location of the acknowledgments, because some journals have them here, but many have them in another part of the manuscript.

Normally, the first form of acknowledgment is for the main financial support for the conduct of the work. Traditionally, this is accorded to the institution(s) of the author(s) because those institutions are the workplace for the scientists involved with the research work that led to the manuscript. The institution has provided many forms of financial support in the form of the salaries and benefits of the researchers along with access to the laboratories, other facilities, and the space and utilities used during the conduct of the work. A large grant may have covered most or all of the outward expenses associated with the conduct of the research, but the amount of money contributed by the external grant will still only be a fraction of the entire amount of money needed to complete the work.

Immediately after that acknowledgment, the source of external funding organization(s) will be designated. These may be foundations such as the National Science Foundation, National Institutes of Health, or the Rockefeller Foundation. These external sources of funds may be provided as grants, contracts, fellowships, or other forms of financial support. In many cases, the authors will designate, or will be required to designate, the grant number of the external granting source. This primary acknowledgment of the funding source should then be followed by acknowledgment of funding sources that gave smaller amounts of money or other forms of in-kind grants such as chemicals and biological or other materials needed to conduct the research.

The author(s) should next acknowledge all forms of technical help that contributed to the conduct of the research in the manuscript, but not to the point where those contributions warranted that the person become a coauthor on the manuscript. Often, this is the place where a researcher will acknowledge the aid of a technician or other laborer who made major contributions to the work, usually physical. Sometimes, this person may not be in the laboratory of the person publishing the manuscript, but they may have made one or more substantial contributions to the conduct of the research. A good example of this may be a technician who operates a piece of equipment critical to the research (e.g., HPLC or DNA sequencer) in another laboratory, and the data received from that technician, and his or her equipment, may have been the basis for the entire manuscript. In some cases, the acknowledgment is solely for the use of another scientist's equipment (by one of the coauthors), and this is perfectly in keeping with an acknowledgment. In most cases, this should be a simple courtesy extended to the person supplying the help.

Another good example of this type of acknowledgment is for help with the design of the experiments and the statistical analyses and interpretations that moved the data from raw data into a form that was suitable for presentation in a manuscript. This has been covered in Chapter 4, and you should revisit the information about statisticians in that chapter if you or your coauthors are considering authorship for a statistician.

This is also the place where the author(s) can acknowledge ideas that came out of discussions with another scientist or group of scientists. It is also used to acknowledge those individuals who reviewed and made suggestions that improved the overall quality of the manuscript.

The most important item that should guide you in the construction of this portion of the manuscript is the courtesies that you should extend to the people and institutions that helped you. There are few things done in science today that are done by individuals acting alone. We all receive help at some time, for some reason, and many of us seek help when we know that something can be done more simply and quickly if we just enlist the help of another person. As you extend these courtesies to others, be sure you have acknowledged them properly, by accurately describing the contribution made by that individual. This may be done best by having them read the acknowledgment of their contribution(s) before the manuscript is submitted (Day and Gastel, 2006; Yang, 1995). It is better to know that you have acknowledged them improperly before the manuscript has been submitted or accepted for publication and you cannot make any changes in it. In this way, you can establish whether those you have acknowledged are comfortable with what you have said about them (Day and Gastel, 2006). A gram of prevention is worth a kilogram of cure.

Some Items to Consider in Building the Authorship Footnote

- Be sure to follow the standard style and format of the journal to which the manuscript will be submitted. The necessary information for this should be contained in the instructions for authors document or the style manual of the journal.
- Be aware that some journals have authors prepare a separate acknowledgment paragraph that is located in a completely different section of the manuscript, rather than in the authorship paragraph.
- Be generous and inclusive in acknowledging all forms of assistance you have received from others—when appropriate.

EXERCISE 19.1   Transition from Provisional Title to Final Title (Step 1)

Now that your entire manuscript is nearly completed, it is time for you to revisit the title. In Chapter 7, you developed your take-home messages and a provisional title for your manuscript, and they should have helped guide you through the development of your manuscript. Hopefully, your take-home messages kept you on-target as you wrote, and they are evident throughout your manuscript. At this time, however, you must at least revisit your provisional title and make sure it is truly an appropriate title now that you have completed the manuscript. If, upon perusal and thinking about it, it is not, then you need to revisit the title and develop a newer, more appropriate title for your manuscript. Remember, the title will be read by perhaps thousands of people, and it should therefore be meaningful at first sight and represent the entire contents of the manuscript. You should make all attempts to develop an *informative* or *declarative* title, rather than an *indicative* title. Do not use any abbreviations, chemical names, or jargon in the title, and be aware of faulty syntax that produces a title that leads to an incorrect understanding of the content of the manuscript. Most journals have either a word limit or a character limit for the title. The maximum in either of these cases will correspond to a title that has a maximum of about 12 to 15 words. Finally, remember that this edition of the title should be developed for the *reader* of the article, *not the writer*.

Exercise

Now that you have nearly completed the first draft of your manuscript, let the manuscript get "cold" for a few days, if you have the luxury of giving it that much time. Come back to it after some time and try to "tighten" the title to a maximum of 12 to 15 words for a penultimate draft of the title. After you have let this penultimate draft of the title get "cold" for a period of time, again "tighten" it to a final manuscript title that will contain between 5 and 12 words. This may be easier said than done, but you must try and develop as succinct a title as you can.

Penultimate draft of the title—maximum of 12 to 15 words:

_____

_____

_____

_____

Final draft of the title—between 5 and 12 words:

_____

_____

_____

_____

EXERCISE 19.2 Transition from Provisional Title to Final Title (Step 2—Coherence Between the Title and the Abstract)

In Chapter 18 you developed the abstract for your manuscript, and in this chapter, you have developed the final version of your title. Now it is time to check the coherence between these two very important sections of your manuscript. On a separate piece of paper, write or print the final version of your title. Strike a line through these words, in the order presented: all prepositions (of, in, on, etc.); all articles (a, an, the); and then other low impact words that are not either prepositions or articles. Now, you should be left with only the high impact words in your title. List those words on these blank lines. Remember, your title sets a level of expectation for the reader, and nothing frustrates a reader more than to have the title set an expectation not fulfilled by the abstract.

_____   _____   _____

_____   _____   _____

_____   _____   _____

Now, return to your final copy of your abstract, and strike a line through all prepositions, articles, and low impact words. List the remaining words on these blank lines. (You may want to call this the "Albert Einstein Exercise.")

_____   _____   _____

_____   _____   _____

_____   _____   _____

_____   _____   _____

_____   _____   _____

_____   _____   _____

_____   _____   _____

_____   _____   _____

With one color of marker, circle all high impact words that appear in both the title and the first sentence of your abstract. Lebrun (2007) suggests that at least 33% of the words in the title should also appear in this first sentence of the abstract. If the coherence here is less than 33%, then you should investigate and maybe restructure your title to incorporate more key words. With another color of marker, circle all words that appear in both the title and the entire abstract. If 90% or more of the words in the title appear in both places, then there is good coherence between your title and abstract.

EXERCISE 19.3   Development of Keywords

List the high impact words found in your title.

_____   _____   _____

_____   _____   _____

_____   _____   _____

Now, list the high impact words in your abstract that are not repeats from your title.

_____   _____   _____

_____   _____   _____

_____   _____   _____

_____   _____   _____

_____   _____   _____

_____   _____   _____

Peruse your rough draft of your manuscript, and list the high impact words that are not in your title or your abstract. In addition, try and find synonyms for phrases or chemical names that might be good keywords, and list them below.

_____   _____   _____

_____   _____   _____

_____   _____   _____

_____   _____   _____

_____   _____   _____

_____   _____   _____

Consult your journal of choice, and determine whether your keywords will or will not include high impact words from the title. If you are allowed to have some number of keywords in addition to those in the title, then pick another ten or so keywords for inclusion into your manuscript. If the number of keywords your journal allows encompasses the title, then select the appropriate number of keywords from both the title and these lists you just constructed. Be judicious in your use of keywords, and do not go over the limit stated by your journal.

EXERCISE 19.4   Strengthening Your Manuscript Byline

Instructions

Now that you are in the final stages of preparation of your manuscript, it is appropriate for you to revisit, and complete again, an authorship issues grid so that you can be sure you have the appropriate people as coauthors, and that they are in the appropriate order. This grid has been modified from Peat et al. (2002). Please complete the grid by ranking each possible coauthor for each of the activities needed to complete the research project. Use a pencil, as you will probably have to erase some items, maybe even several times. Use the scales below to assign a point score for each of these activities to yourself, as the writer, as well as to each potential coauthor. Remember, when you started writing this manuscript, the contributions by a particular author, or authors, may have been regarded as minimal or of great magnitude. However, upon completion of the manuscript, a potential coauthor may have made many significant contributions to these activities, and may now need to be considered as a bona fide coauthor when earlier they did not qualify for authorship. The opposite may also be true. Start by assigning the initials of possible coauthors to each column, including the one for you. *The maximum points allocation possible for each activity is listed below, and it also represents the sum of all the contributions by all coauthors for that activity.* If there are more than five potential coauthors, then duplicate the grid and add the additional columns to make a grid that encompasses all possible coauthors at these final stages of manuscript preparation. It is likely you will need at least two columns, as few units of research are completed with only one person "wearing all the hats" that are contained in the grid.

In general, a coauthor should have contributed at least 15 to 20 points worth of effort to the cause of the project. If an individual accumulates less than 15 to 20 points, you and any other coauthors that have greater scores must determine whether this potential coauthor deserves to be accorded the privilege of coauthorship. This is not a trivial issue, because decisions on authorship are intimately related to promotion, tenure, and overall ethics issues. Be sure you have not breached an ethical issue by erring to either the conservative side or the liberal side when making the authorship issue decision. Are you comfortable or uncomfortable with the results? Why? Let us assume you now have completed this authorship grid, and you are essentially confident of, if not also pleased by, the results of this exercise. You now have a list of the individuals you should include as coauthors, and you now can list the order of the authors in the byline by ordering them into position as a result of the total score for each individual.

Max. points Possible	Section of work
10	Research idea conception
5	Literature searches and background information assembly
10	Experimental design
10	Conducting the experiments
5	Data collection and database management
10	Data analysis
10	Interpretation of analyzed data
15	Writing of manuscript drafts
15	Critically revising and reviewing manuscript
10	Additional input to intellectual content

	Total possible points		
Contribution	Five	Ten	Fifteen
Minimal	0,1	0,1,2	0,1,2,3
Minor	2	3,4	4,5,6
Moderate	3	5,6	7,8,9
Major	4	7,8	10,11,12
Entire	5	9,10	13,14,15

Activity	Yourself	Potential coauthor 1	Potential coauthor 2	Potential coauthor 3	Potential coauthor 4	Activity total
(Initials)	( )	( )	( )	( )	( )	
Conception	_____	_____	_____	_____	_____	<u>10</u>
Lit. search	_____	_____	_____	_____	_____	<u>5</u>
Expt. des.	_____	_____	_____	_____	_____	<u>10</u>
Conducting	_____	_____	_____	_____	_____	<u>10</u>
Data coll.	_____	_____	_____	_____	_____	<u>5</u>
Data anal.	_____	_____	_____	_____	_____	<u>10</u>
Data interp.	_____	_____	_____	_____	_____	<u>10</u>
Writing	_____	_____	_____	_____	_____	<u>15</u>
Revise/rev.	_____	_____	_____	_____	_____	<u>15</u>
Intell. cont.	_____	_____	_____	_____	_____	<u>10</u>
TOTAL	_____	_____	_____	_____	_____	<u>100</u>

Are you comfortable/uncomfortable with the results? Why?

_____

_____

_____

_____

List of coauthors as they should appear in the byline:

_____

_____

_____

_____

REFERENCES

ALLEY, M. 1996. *The craft of scientific writing*. Third edition. Springer-Verlag, Inc., New York, NY.

DAVIS, M. 2005. *Scientific papers and presentations*. Second edition. Academic Press, San Diego, CA.

DAY, R.A. and GASTEL, B. 2006. *How to write and publish a scientific paper*. Sixth edition. Greenwood Press, Westport, CT.

GUSTAVII, B. 2003. *How to write and illustrate a scientific paper*. Cambridge University Press, New York, NY.

KATZ, M.J. 1985. *Elements of the scientific paper. A step-by-step guide for students and professionals*. Yale University Press, New Haven, CT.

KATZ, M.J. 2006. *From research to manuscript. A guide to scientific writing*. Springer, Dordrecht, The Netherlands.

KERKUT, G.A. 1983. Choosing a title for a paper. Comp. Biochem. Physiol. 74A: 1.

LEBRUN, J.-L. 2007. *Scientific writing. A reader and writer's guide*. World Scientific Publishing Co., Singapore.

MATTHEWS, J.R., BOWEN, J.M. and MATTHEWS, R.W. 2000. *Successful scientific writing. A step-by-step guide for the biological and medical sciences*. Second edition. Cambridge University Press, New York, NY.

O'CONNOR, M. 1991. *Writing successfully in science*. HarperCollinsAcademic, London, UK.

PEAT, J., ELLIOTT, E., BAUR, L. and KEENA, V. 2002. *Scientific writing. Easy when you know how*. BMJ Books, London, UK.

YANG, J.T. 1995. *An outline of scientific writing*. World Scientific Publishing Co., Singapore.

CHAPTER 20

POLISHING YOUR MANUSCRIPT WITH GOOD WORD USAGE

The greatest possible merit of style is, of course, to make the words absolutely disappear into the thought.

—Nathaniel Hawthorne

CHOICE OF WORDS

Each word used in a scientific manuscript should be chosen deliberately. Its selection should be based on the accuracy, precision, clarity, and economy it fosters. And it should be necessary to the point that its removal causes a loss of meaning.

All words should be chosen explicitly for the accuracy of their meaning because inaccurate words cause misunderstandings. Vague and ambiguous words create problems of interpretation for all readers. Each word must represent, not just imply, a single meaning. Avoid words that can have multiple meanings depending upon the context of the sentence. Clarity can be promoted by using simple, common words instead of unusual ones; scientific writing is not the place to show off one's vocabulary. A writer should select the most direct, simple word to convey the intended meaning, as captured by this quotation from Sir Winston Churchill:

Old words are best, and old words when short are best of all.

An additional, similar quotation by one of the coauthors of this book, one evening after a few beers, is also appropriate, although the authors of these quotations are certainly not in the same class:

Understanding and depth of thought are manifest through simplicity of communication.

—Richard J. Gladon

Many of the suggestions contained in this chapter came to us as comments by reviewers of our manuscripts or other sources (Luellen, 2001, 2007).

SPECIFIC WORD CHOICES FOR ACCURACY AND PRECISION

- Nouns and verbs are the key building blocks of sentences. Use qualifiers such as adjectives and adverbs only when their inclusion makes a sentence clearer, more accurate, or more informative. Avoid excessive qualifiers that clutter and confuse the reader. Often the most straightforward and readily understood sentence consists of a subject, verb, and an object.

- The writer must be cognizant of the correct tense to use. Generally, the abstract and the new results presented in the manuscript are written in the present tense, whereas the introduction, materials and methods, and previously established results are in the past tense. Some authors and editors, however, believe the entire manuscript should be written in the past tense, as all of the work was done previously. You should determine this when you choose your journal.

- Keep a dictionary nearby, and use it often. Sometimes we learn a new word that works in one context, but we use it incorrectly in another context. If you have any doubt about the correct meaning or usage of a word, consult a dictionary. Likewise, use a dictionary to compare synonyms for subtle differences in meaning and choose the word that captures what you mean most accurately and precisely.

- Avoid using a word or phrase repeatedly in a short space.

- Some scientists want to make statements about an issue when they should not, because the statement is not fully correct in all circumstances. A good analogy to use here is that of a fence. When a person walks on a fence, they may get off the fence on either side, depending upon what works better in a given circumstance. The writer may walk the fence by using what we call "weasel words." Use words that occupy the top of the fence sparingly, if at all, and only when absolutely necessary. Some examples of "weasel words" are may, seem(s), might, appear(s), could, possibly, suggest(s), probably, indicate(s), apparently, maybe, and perhaps. Never use multiple "weasel words" within the same sentence: "The data suggest that our hypothesis probably could be valid." Reviewers, editors, and readers will soon become alarmed by the choice of words, and they will look for reasons why a definitive statement has not been used. At this point, the manuscript may be in trouble.

- Do not coin new terms unless absolutely necessary.

- Eliminate vague qualifiers like fairly, quite, certain, rather, many, very, much, several, adequate, complete, substantial, small, little, considerable, relatively, much more, and much less. If a word expresses an absolutely unequivocal quality or condition, then do not use a comparative or qualifier in front of it (e.g., very complete, more or less pregnant). However, if the condition is not absolute, then it may be appropriate to qualify it (nearly complete, very strong). When a qualifier must be used, be sure the reader understands that the meaning of the qualifier is for comparative purposes. The degree of accuracy should be reflected in the writing. If accuracy is required, then give numbers, such as "the growth was 7 cm," and do not say "... rather large, 6 to 8 cm."

- Learn to recognize low impact words and eliminate them from the manuscript, including the following low impact verbs: "attained," "carried out," "indicated," "facilitated," "implemented," and "experienced."

- Also be wary of dangling words like infinitives and words that end in "ing" or "ed."

- In scientific writing, some words are perennially troublesome or misused completely. Consult a dictionary or another resource (Luellen, 2007) to determine which spelling is correct for your use of that word. In some cases, the words will be pronounced similarly, but the meanings of the words are very different (see below). In some cases, one form is a verb and the other form is an adjective or adverb.

allude vs. elude   can vs. may   allusion vs. illusion
anybody vs. any body   effect vs. affect   farther vs. further
like vs. as   imply vs. infer   lay vs. lie
accent vs. accentuate   principal vs. principle   while vs. although
maximum vs. optimum   alternative vs. alternate   enable vs. permit
where vs. in which   absorbed vs. adsorbed   due vs. do
elicit vs. illicit   whether vs. weather   greater vs. higher
partly vs. partially   provided vs. providing   comprise vs. compose
predominant vs. predominate
complimentary vs. complementary
loss vs. lose vs. loose
sight vs. site vs. cite
varying vs. various vs. variable vs. varied vs. different
assure vs. ensure vs. insure
minimal vs. negligible vs. slight amount
concentration vs. content vs. level

- Eliminate use of slang terms, as they usually evolve from jargon. This list could go on forever, but a few examples of common scientific slang and jargon include "supernate," "centrifugate" (or its counterpart, "fuged"), "electrophoresed," "scope" (instead of "microscope"), "reps" (instead of "replications"), "boosted" (instead of "increased").

- Improper use of singular versus plural can create problems for many writers, and then the reader. Here are some examples that can be problematic:

datum vs. data   genus vs. genera   alga vs. algae
fungus vs. fungi   medium vs. media   sheep vs. sheep
spectrum vs. spectra   deer vs. deer   fish vs. fish
radius vs. radii   or
phenomenon vs. phenomena   fish vs. fishes

- Eliminate unneeded prefixes. For example, there is usually no need to say something was "precooled"; it was simply "cooled."

- Eliminate use of double prepositions. For example, "outside" and "of" are prepositions, but we do not need to say something is "outside of the box"; it is "outside the box."

- Eliminate use of two or more words when fewer will get the job done more clearly. Some examples:

before, rather than, prior to
short, instead of, not long
too many, rather than, a greater number of
in a couple weeks, rather than, in a couple of weeks

SPECIFIC WORD CHOICES FOR CONCISENESS

- Write in direct sentences that are short, simple, and factual.
- Use only meaningful words.

SPECIFIC WORD CHOICES FOR CLARITY

Active Voice

The active voice is direct, brief, and interesting, and it brings the writing to life. Use verbs actively, and do not use verbs hidden later in the sentence. However, some authors suggest that the writer can make the words more interesting if she or he moves back and forth between the active and the passive voices. Consider these examples:

- "The growth of plants is rapid." vs. "Plants grow rapidly."
- "Isolation of the toxin was achieved." vs. "We isolated the toxin."
- "The results of the present study are explanatory of the reason . . ." vs. "Our results explain why . . ."

First Person

We suggest you use the first person in all your scientific writing. When most middle-aged and older scientists today were in primary and secondary school, we were taught not to write anything in the first person. Exclusive use of first person may irritate some reviewers, so incorporate it gradually into your writing style. Why use the first person? It takes less space (words) than using third person ("I" vs. "the investigator"). It is honest. You did the work and writing, so why write as if someone else did? The first person is clear to the reader, particularly in the discussion, where use of the first person distinguishes what your data are and what those data mean versus your statements about the data of others.

Avoid Stacked Nouns and Modifiers

These are sometimes called railroad nouns or boxcars. When the stack is ridiculously long, we call them "boxxxcarrrs." They are cumbersome, awkward to understand, and all too common. Note that stacking often requires the use of hyphenation, which can be difficult to express properly and can be confusing to the reader. Rewriting to eliminate stacking may result in more words, but that is OK if clarity is increased. As a general rule, do not stack more than two nouns (i.e., limit your writing to only one hyphen). If you must use modifiers, then make sure they modify correctly; in order to do this, you may need to sacrifice brevity for clarity. Modifiers may include adjectives, adverbs, misused nouns, and misused participles, and a good scientific writer is always on guard against their use. Adjectives used in series and separated by commas may be stacked. For instance, "His writing style was ridiculous, cumbersome, imprecise, wordy, and dull" is an appropriate use of a series of modifiers. Here are several examples of modifiers that should be written differently.

- Instead of "ATP-activated actinomysin contraction" say "Actomysin contraction activated by ATP."
- Instead of "Silica gel-coated glass-fiber paper chromatography" say "Chromatography on paper made from glass fibers and coated with silica gel."
- Instead of "light glucose cells" say "cells grown in light in the presence of glucose."

More Words than Necessary—Excessive Words

In many situations, writers use more words than they need to drive home a point. Usually, the group of words they use can be replaced by only one or two words, as most of these examples show.

all of, both of—just use "all" or "both"

at the present time, at this point in time—use "at present," or better yet, "now"

by means of—use "by"

during the course of, in the course of—just use "during" or "in"

in order to—use "to"

prior to, previous to—use "before," "preceding," or "ahead of"

small in size, rectangular in shape, red in color, tenuous in nature—use "small," "rectangular," "red," "tenuous," respectively

to be—almost never needed, and therefore, just eliminate

reason why—omit "why" when reason is used as a noun. "The reason is . . ."

Miscellaneous Words that are Very Important

Many words used commonly in scientific writing have a very specific meaning that relates to only one set of circumstances. Writers must often choose between two seemingly paired words in which both could have the intended meaning. It is important to know which word is the appropriate word for each circumstance. This listing is certainly not exhaustive, but it will help the writer understand why confusion arises. Many of these examples have been presented in a handout from the publications office in the College of Agriculture at Iowa State University, and it has been used with permission of the author, who is not actually listed on the handout. Many of these examples may also be found in books devoted to better scientific writing (Luellen, 2001; Luellen, 2007; Malmfors et al., 2000).

Word(s)	Explanations of uses
Above, below	Writer is referring to something preceding or following, rather than physically over or under; loose reference; convenient for writer but confusing to reader.
Affect, effect	Affect is a verb, and it means to influence. Effect, used as a verb, means to bring about. Used as a noun, it means result.
And, but	OK to use these words at the beginning of a complete sentence. Useful transition words between related or contrasting statements, respectively. Use can vary with journal.
Apparent, apparently	Also means obvious(ly), clear(ly), plain(ly) evident(ly), seeming(ly), ostensib(ly), or observab(ly). Reader may not understand your meaning, and ambiguities arise. Use obviously, clearly, seemingly, and so on, to remove all doubt.
Appear, seem	Also appears/seems and appeared/seemed. Appear means to be physically present and observable, whereas seems is a state of mind or presence of mind.
Case	Can be ambiguous, misleading, or ludicrous. Could be confused with a case of, say, fine wine. Use in this instance.

Word(s)	Explanations of uses
Compare with, compare to	Compare with means to examine differences and similarities between two items, but compare to means to represent as similar. This example is excellent: The music of Brahms compares to that of Beethoven, but to do that, one must first compare the music of Brahms with the music of Beethoven.
Comprise	Means to contain, include, or encompass, but does not mean to constitute or compose. Has been misused so much that constitute and compose are now accepted by some as viable meanings, albeit incorrectly. Use and meaning so confused, it is best to not use it.
Correlated with, corr. to	Items may be related to one another, but these items are correlated with one another.
Different from, diff. than	Always different from. One thing differs from another thing, but you may differ with your colleagues.
Due to	Does not mean because of. Due is an adjective modifier and must be related directly to noun, not a concept or series of ideas from the rest of a statement. Often used to "weasel out" of a definitive statement.
Either ... or, neither ... nor	Apply to no more than two items or categories.
Etc.	Use at least two items, categories, or illustrations before using and so forth or etc. Do not use an "and" before etc., as the et of etc. means "and."
Experience(d)	If something is experienced, it must be sensed, or be sensory. Inanimate, unsensing objects cannot experience something; only animate objects can experience it.
Former, latter	Refer only to the first and second, respectively, of only two items or categories.
Following	After is more accurate if after is the intended meaning. After the parade, the festival ended.
High(er), low(er)	Used too often and too inaccurately and imprecisely. Reader may interpret an item is above or below something, when really it is greater than something. Much better to use the more accurate words greater, lesser, larger, smaller, fewer, and so on.
However	Place it more often inside a sentence or major element, rather than at the beginning or end. "But" is a better choice at the beginning of the sentence or element, but can vary with publisher.
Irregardless	Not an appropriate word. Use either regardless or irrespective.
It was found, decided, determined, felt, etc.	Word usage is too evasive. Be more direct.
Less(er), few(er)	Less(er) refers to quantity or amount only; few(er) refers to number.

Majority	See if most will do. There are several types of majority—simple, 2/3, 3/4, and so on.
Percent, percentage	Not synonyms. Use percent or the percent symbol only with a number. In all other cases, use percentage.
Varying, various	Varying something means you are changing certain things individually. Various, different, or differing means you may select from a number of items that are not necessarily changing and are not the same.
Where	Should not be used when the meaning is in which or for which. Use when you mean a physical location.
Which is, that were, who are, etc.	Almost never needed. Instead of data that were analyzed by analysis of variance, use data analyzed by analysis of variance. Often, rephrasing helps. "A survey, which was conducted in 2000, . . . , can become "A survey conducted in 2000, or better yet, "A 2000 survey."

Hyphens

The presence or absence of hyphens can have very dramatic effects on clarity.

- For many years, when a suffix was added to a root word, a hyphen was not used between the root word and the suffix. This situation continues today. Examples using the suffixes -ette, -ism, and -ist, are pipette, capitalism, and naturalist, respectively.

- Conversely, when a prefix was attached to its root word, a hyphen was inserted between the prefix and the root word. We do not know how this practice originated, but it is no longer used for most prefixes. Most of these words that combine a prefix and a root word will be formed by simply attaching the prefix to the beginning of the root word. Exceptions include when the prefix is used with a proper noun or numerals (e.g., pro-Iowa or mid-80s) or when the lack of a hyphen creates ambiguity or awkwardness. Another example is when different prefixes precede the same root word, and the writer wishes not to repeat the root word (e.g., predoctoral and post-doctoral become pre- and post-doctoral). In other instances, the words can simply be predoctoral and postdoctoral.

- Many dictionaries will list some of the common words that no longer require a hyphen, and the writer should check if there is any doubt. Here are several examples of prefixes that no longer require a hyphen when attached to a root word (Mish, 2004): anti, bi, bio, con, counter, de, di, dis, ed, ele, elec, epi, ex, full, hard, head, heat, home, homo, hyper, im, in, inter, ir, mega, micro, mis, mono, multi, non, ortho, over, out, para, ped, per, peri, photo, poly, pseudo, pyro, quad, qual, quant, quin, quit, rain, re, semi, sub, super, sym, syn, tele, tetra, tri, ultra, un, under, uni, up, van, ven, veni, veno, ver. Surely there are other examples not included here, but if there is any doubt, the writer should check in a dictionary.

- Several prefixes are used without hyphenation and are attached directly to the root word (the most frequent way), or they are hyphenated at certain times (Mish, 2004). Here are several examples of this case: co, cross, down, neo.

- The prefix well is generally not hyphenated and is not attached directly to the root word (Mish, 2004).

- Several prefixes are not hyphenated, except when they precede a proper noun (Mish, 2004), and examples are: post, pre, pro.

- Several prefixes are hyphenated and directly attached to the root word in each instance we observed (Mish, 2004), and here are several examples: one-, quasi-, self-, two-.

- Several prefixes are randomly found either as a detached word immediately before the root word, unhyphenated and attached directly to the root word, or hyphenated and attached directly to the root word (Mish, 2004). Here are several examples: half, hand, hands, high, life, long, off, open, out, quarter, queen, quick, race, radio, red, right, rough, run, sand, school, set, short, test, time, trans, turn, vice, vis, war, wet, wild, and wind.

- Sometimes, a hyphen is needed so that ambiguity can be eliminated. A professional athlete might re-sign a contract with his or her original team, but the athlete would not want to resign from that team. Likewise, if a FAX did not come through the first time, the receiver might resend it if you had not re-sent it to him or her.

- Compound modifiers must often include hyphens to clarify what is modifying what. Under normal circumstances, adjectives modify nouns. The problems arise when adjectives and nouns are used to modify other adjectives or nouns. For instance, a man-eating fish could be very dangerous, but a man eating fish, whether they were shark, swordfish, perch, or trout, probably would not be dangerous. Another example is that a fast acting dean may not be nearly as effective as a fast-acting dean, especially when grant money is involved. Adverbs, which usually end in -ly, are modifiers of adjectives, and their use rarely causes problems.

- A hyphen or en dash should not be used to indicate a range of values. When we want to say that the lengths of a group of axons was 7 to 14 mm, we often would say "axon length was 7–14 mm." In some cases, reporting data in this form may be unequivocal, but in many instances, confusion arises. Let us use the following example. Let us say a researcher was studying exposure to cold temperatures and how those exposures affected a physiological process. In addition, let us assume this researcher used exposures that started at -5 °C, the highest temperature tested was 10 °C, and the testing was done in increments of 3 °C. Would it be unequivocal for this researcher to report the range of exposure temperatures was -5–10 °C? Were these exposure temperatures from -5 °C to -10 °C, or were they from -5 °C to $+10$ °C? How about a range of exposure temperatures starting at -20 °C and going to -5 °C? Would it be clear if it were reported that the range of exposure temperatures was -20–5 °C or -20 $(-)$ -5 °C? Presentation of this information can be clarified completely by using the word "to" rather than the hyphen. In our first example, the researcher would have used temperatures of "-5 to 10 °C," and in the second example, she or he would report the range as "-20 to -5 °C." In addition, many journals require that the dimension, in this case °C, be appended to each reported value. Thus, for some journals, these examples would be "-5 °C to 10 °C," and the second example would be "-20 °C to -5 °C."

- There are instances in which a researcher, for one reason or another, has a missing set of data – the beaker of test solution spilled, exposure to oxygen degraded a sample, and so on. Most journals represent missing data in tables by three hyphens, without regard to the number of significant figures reported. Some journals will use other symbols to indicate missing data (Katz, 2006). It is best to consult the style manual of the journal or a recent issue of the journal because this information will probably not be in the instructions to authors document. An example follows:

17.4	16.8	17.1
16.6	- - -	16.9
17.2	16.9	17.0

- When a qualifying adverb is used as an adjective with a noun, the qualifying adverb is not attached to the noun with a hyphen. The ly of many adverbs used to modify an adjective or a noun replaces a hyphen. Thus, we previously said that we took a real-nice trip, and we should now say it was a really nice trip (Luellen, 2001).

- This example is a special case for the use of a hyphen. When the writer must write out, for instance, a time or age that is exactly between two consecutive whole numbers, it is written in a special way. If you took a long examination, it should be written as "three and one-half hours long," rather than "three and a half hours long." This latter usage is not correct, although instances abound where it is used.

Proper Word Order (Syntax) Denotes a Correct Meaning

Observe these two examples. In both cases, the exact same words are used, and only the order of appearance of the words changes. If you read these carefully, as a reviewer would, you will notice that each example exhibits several meanings depending upon the syntax.

Example 1:   Only his assistant can sign that check.
His only assistant can sign that check.
His assistant only can sign that check.
His assistant can sign only that check.
His assistant can sign that check only.

Example 2:   Not everything in print is founded on fact.
Everything not in print is founded on fact.
Everything in print is not founded on fact.

Strive for Positive Sentences

Avoid using the word "not." It is more informative to say what is than what is not. Instead of saying "the plants did not continue to grow," report "the plants stopped growing." In situations where a positive and a negative statement are in the same sentence, express the positive first: "The goal is to produce rather than to pollute." This same idea also applies to the use of conjunctions. Try and always use "and" rather than "but" or other conjunctions. Certainly, there are instances where use of the conjunctions other than "and" are appropriate, and in those cases, you should feel free to use them.

That Versus Which

The word "that" is overused. Every time you use "that," or which for that matter, imagine the sentence without it, and delete it when you can without loss of clarity or inappropriate change of meaning. Be cognizant of the difference between that and which, and the meaning each implies. Try and use that unless the meaning becomes ambiguous or the structure becomes awkward. Example 1 implies all plants in their study were defoliated, whereas Example 2 implies only some plants were defoliated, and only the defoliated plants survived. As demonstrated in Example 1, proper use of "which" often requires a comma immediately before it, and in many cases, another comma at the end of the dependent clause, if one is used after the word which. The word "that" often specifies a particular item or set of circumstances, whereas the word "which" often has a sole function of adding more information. Issues related to use of punctuation could be explored if necessary (Kirkman, 2006).

Example 1:   The plants, which were defoliated during their study, survived.

Example 2:   The plants that were defoliated during their study survived.

Avoid Beginning Sentences with "There is" and "It is"

It is clear chlorophyll degraded. vs. Chlorophyll degraded.

Sentences Containing Parenthetical Expressions Should be Complete if the Expression is Omitted

The range of plant dry mass (17 to 44 g) was greater than in other studies.

Punctuation

Yes, punctuation matters (Kirkman, 2006). Keep it simple, and when in doubt, consult an appropriate reference book for clarification (CSE, 2006; Day, 1995; Fowler, 1986; Luellen, 2001; Lunsford, 2003; Malmfors et al., 2000; Strunk and White, 2000). Mish (2004) gives a list and presents the symbol of about 30 punctuation marks that are used commonly in English writing. We suggest that you consult your dictionary for descriptions of any marks you do not understand. Use as few marks as possible, because they tend to confuse the reader rather than provide clarity. The warnings above carry about the same weight as we discussed under the heading word choice. If you are writing sentences that require colons, semicolons, exclamation marks, or other punctuation marks, try to simplify your sentences so those forms of punctuation may be eliminated. If you are in doubt regarding the use of a certain punctuation mark, consult a standard style guide. The most elegant of scientific articles often contain only periods and commas as the punctuation marks. Of course, they also contain words.

Commas are especially troublesome because they are needed often and are misused or erroneously not used. We have observed that many writers do not use commas to the extent required, and their manuscript often suffers clarity problems when commas should have been used, but were not.

One basic rule to remember involves serial lists. Use a comma after each term, except for the last term, and be sure to include the comma immediately before the word "and." The comma just before the word "and" is called the terminal comma, and almost every journal we know requires its use. Here are a few examples to help you understand. Place a terminal comma immediately before the word "and" when designating a serial list of some items.

The plants were tall, pyramidal, and uniform.

The flag of the United States is red, white, and blue.

Commas, and periods, are enclosed in quotation marks and parentheses, if your manuscript requires the use of either of these punctuation marks. An exception to this rule occurs in plant science when the writer is discussing a cultivar of a species. A cultivar is a contraction of the words cultivated variety, and many commercially used species have several to many cultivars within one species. In this case, the cultivar name is enclosed in single quotation marks, and the single quotation marks are part of the name, rather than a punctuation mark. Thus, a comma or period would be placed outside the single quotation mark.

Use "weighed properly," rather than "weighed properly",

Use (He is mischievous.), rather than (He is mischievous).

Use chrysanthemum 'Orange Blush', rather than chrysanthemum 'Orange Blush,'

Place a comma before a conjunction that introduces an independent clause. The two independent clauses should be related to one another, and if they are not, the writer might consider making them two separate sentences. In this example, each clause should be written as a separate sentence. The independence of these clauses dictates the need for a comma between them.

The field was sprayed regularly, and the plants appeared healthy.

Abbreviations

Although we encourage economy of words, brevity should not take precedence over clarity. Abbreviations that are not standard should be avoided, at almost all costs. It is better to write "root dry mass" repeatedly than to write "RDM." One such abbreviation might not lead to much confusion. However, authors who tend to use one abbreviation use many, and repeated use of numerous abbreviations detracts from clarity. Keep the reader focused on your data and their meaning. Do not force readers to decipher a host of abbreviations. If, despite this advice, you elect to use one or more abbreviations that are not standard, you must define them in the manuscript. Usually this is done upon their first use. The word(s) are written out and the complete word(s) are followed by the new abbreviation in parentheses, for example, root dry mass (RDM). Some journals have a special heading under which all abbreviations in the manuscript are defined, and then there is no need for the abbreviation to be defined, even the first time. Be sure you check the instructions to authors document, the style manual, or recent issues of your journal to determine how your journal addresses use of abbreviations.

Standard abbreviations are defined by each journal, for use in that journal only. The journal lists the standard abbreviations in a style manual or an instructions to authors document, or the journal will refer the writer to a standard reference source of abbreviations. You can also survey recent articles in the journal for examples. If a journal adopts a certain abbreviation as part of its standard format, you should not re-define it, nor should you create a new, different abbreviation. Some common abbreviations that are standard in many journals include "et al.," "diam," "temp." "°C," "RH," "max.," and "min." The style guide for the journal you select will define such abbreviations.

Avoid Nonsensical Statements and Statements with Multiple Meanings

Here are several examples of sentences that need to be reworked. Often, the clearer sentence contains fewer words.

The most elegant scientific articles might only contain periods and commas. (Wow, the author(s) forgot to add words?)

We need a list of biologists broken down by specialization.

We need a list of biologists classified by specialties.

A porcupine was brought to the laboratory by a biologist in a moribund condition.
A biologist brought a moribund porcupine to the laboratory.

Being in a dilapidated condition, I was able to buy the house for a low price.
I bought the house for a low cost because it was dilapidated.

Ensure that Repeated Verbs are Needed

In this example, there is no need to use "were" twice.

The plots were sprayed twice and were dusted once.
The plots were sprayed twice and dusted once.

But in this example, not using "were" twice implies seedlings can dust. (We all need these type of plants around our home and office.)

The plots were sprayed twice and the seedlings dusted once.
The plots were sprayed twice, and the seedlings were dusted once.

Nouns/Subjects and Verbs Should Agree in Number

Words like "each" and "none" are singular. "None" is singular because it may represent a contraction, without an apostrophe, of the words "not one."

Each of the following experiments was done five times.
None of the plants was deficient in nitrogen.

Do Not Split Infinitives or Parts of a Compound Verb

Infinitive forms of verbs are associated with the word "to." Some examples include "to see," "to weigh," and "to assign." It is incorrect word usage to split the infinitive with an adverb. Thus, instead of writing "It is essential to accurately weigh the samples," write "It is essential to weigh the samples accurately."

We encourage you to write sentences that are simple and declarative, but we realize that many situations do not lend themselves to a straightforward sentence structure. When the verb of a sentence is composed of more than one verb form (participle or infinitive), do not split the verb forms with adverbs. The adverb may precede the verb forms, or it may succeed the verb forms depending upon which place allows the sentence to read more clearly or more easily. Thus, instead of writing "The broccoli was only blanched once," write "The broccoli was blanched only once." The only generally accepted exception to this rule is insertion of the word "not." "Not" may be inserted between any of the verb forms.

Beware of Danglers

A verb form (participle or infinitive) "dangles" when the subject of the verb in question is not the subject of the main clause of the sentence. This type of awkward construction often leads to antecedent problems that render the sentence ambiguous, meaningless, misleading, or downright funny. Establish the subject of every verb, and ensure it is present in the correct context. You know what you are writing, but the reader may not interpret it correctly. Distrust words ending in "...ing" and always check their context for correctness.

"After closing the incision, the animal was placed in a restraining cage." (Who closed the incision? According to this sentence, the animal did it. "Closing" is a dangler.)

"Considering the climate of the region, the plants produced flower buds early." (First the plants considered the climate, and then they decided to flower early?)

Here are a couple favorites of ours:

"Using multiple-regression techniques, the animals in Experiment I were. . ."
"Settling in the collected effluent, we observed what we thought was. . ."

Following vs. After

Avoid using the word "following" when you really mean "after." The word "following" may be mistaken for a form of the verb "to follow" in space rather than in time (". . .the dog following his master.")

Some Introductory Phrases to Omit

Inexperienced writers sometimes write in the same manner they speak because they recite the information to themselves as they write. When many people speak, they add "stuffing" to their words because they need to keep their mouth going while they are thinking. The "stuffing" often appears as introductory phrases or interjection of words like "and stuff" in the middle of a sentence. This is not good spoken grammar, nor is it good written grammar, and all of these "stuffers" should be eliminated. Here are some examples of introductory phrases found usually at the beginning of sentences. In almost all cases they should be eliminated from the manuscript.

As already stated,

It goes without saying that,

Needless to say,

Concerning this matter, it may be borne in mind that,

In this connection, the statement may be made that,

It is interesting to note that,

It has long been known that,

It may be said that,

Typical results are shown

It is worth mentioning that,

Obviously,

It should be mentioned (or noted, or pointed out, or emphasized).

Avoid Statements of Intent

Often, authors are tempted to state what they plan to do in the future. The most common place we see this is in the discussion. Statements of intent can reflect weakness, and they should never be used. A reviewer may reject your paper on the basis that your planned work should have been included before publication of what you have so far. The reality of this entire scenario is that you may never accomplish what you state. "The road to Hell is paved with good intentions." In that way, you may forever be remembered as the person who intended to accomplish something, but never met your goal. It is better to keep the focus on your data and their meaning, rather than drawing attention to what you have not accomplished yet.

Subtle Implications (Nuances) of Word Choice

Sometimes, there is just not quite enough information. Either your data or the data of someone else are incomplete, and you would like to "hedge" on a statement. If you choose your words carefully, words can be used that allow you to make subtle implications about your stand on something without using weasel words that will stick out like a sore thumb. If you say that Smith described something, you place the responsibility for Smith's work squarely on Smith's shoulders. However, if you say that Smith showed something or demonstrated a phenomenon, then you go on record as agreeing with Smith, and you assume responsibility accordingly. If you plan to challenge a finding or interpretation of someone else, then do not first make a statement from which the reader can infer that you agree with it.

To imply that you agree with Smith, you could say:

Smith found . . .

Smith demonstrated . . . or

Smith showed . . .

To imply that you know Smith said something (i.e., you have acknowledged it), but the jury, and maybe you, have not concluded yet whether it is true, you could say:

According to Smith . . .

Smith claimed . . .

Smith reported . . . or

Smith described . . .

These last four statements, particularly the choice of words in the first of them, also could imply that you disagree with Smith and intend to challenge her/his data with new findings.

Explain Your Conclusions

Allow the reader to follow your train of thought, and they should come to the same conclusion you did. This is the only way the reader can judge your conclusions properly.

Do Not Use Possessives

Scientific writing should not contain possessives because, often, "the owner" of the particular item is not known or is ambiguous. Although we encourage brevity and economy of words, this is a time where we should not use a possessive to reduce the number of words. In creative writing, it would be OK to say "She walked by the water's edge." However, in scientific writing, it would be difficult to identify that the water owns an edge, and this would be stated better as "She walked by the edge of the water."

The only really accepted possessive to be used in scientific writing is the word its, which is used without an apostrophe, to avoid confusion with the contraction it's, which means "it is."

Numbers

The numbers one through ten are normally written out, as shown in this sentence. Numbers greater than or equal to 11 will normally be written as the number. An important exception to this when the number is a datum followed by a dimension, such as cm or kg. In this case, all numbers are written as the Arabic numeral. For example, two laboratory rats weighed 2 kg each, and they were 30 cm from nose to tip of the tail. Katz (2006) suggests that all numbers should be written out, except when they modify a dimension.

Anglicized Words

Many Latin words have been anglicized and, therefore, no longer require the use of italics associated with a foreign word. Here are several quite common examples in the life sciences, and a dictionary should be consulted to clarify questions you may have. Several years ago the Latin words *et cetera* were anglicized and abbreviated to etc. Consult a dictionary if there is any doubt. Also, consult your journal of choice, as this should not be considered universal over all journals.

> *et alia* has become et al.
>
> *in vivo* has become in vivo
>
> *in vitro* has become in vitro
>
> *in situ* has become in situ

Be Selective

Include only what the reader needs to understand your main ideas, which we have called the take-home messages. Do not go overboard either by omitting essential information or supplying too much information.

The Word "significant"

It is understood by most life scientists that statistical analyses have been applied to the data presented in the manuscript. Therefore, all references to data that are different from one another should not carry the word significant, because the reader should understand that the data were analyzed and only data that were shown to be different statistically were presented and interpreted. If a set of data exhibits no differences between any of the individual pieces of data, then this information should not be presented, interpreted, and discussed, except in unusual instances.

Dates of the Year

In the United States and several other countries, the day of the year is presented numerically as XX/YY/ZZ, where XX is the month, YY is the date of that month, and ZZ is the year. If a person were born on October 7, 1947, it could be presented numerically as 10/7/47 or 10/07/47. In many other countries, the order of the month and day are reversed. Thus, in these countries, numerical presentation of the date of the year would be YY/XX/ZZ, where YY represents the day of the following month, XX represents the month of the year, and ZZ represents the year. The above example of an October 7, 1947, birthday would be 7/10/47 or 07/10/47. This example of 7/10/47 or 07/10/47 could be read correctly as the seventh of October in 1947, or it could be read incorrectly as the tenth of July in 1947 by other people. To clarify this mess, most journals have opted to use the following format. The day of the month comes first, followed by the month of the year written out, followed by the year. Our example now would be the unequivocal 7 October 1947, and the day, month, and year are not separated by a solidus. Some journals require that the month of the year is abbreviated, and this should be clarified by consulting the journal. In this case, our example now might be 7 Oct. 1947.

The Word Since

The word "since" should not be used to relay the meaning of "because," and it only should be used as a time connotation; for example, "Since 1973, we studied. . . ." Use "because," or maybe "in as much as," when no time connotation is used and the writer means the effect is due to the influence of something else.

While Versus Whereas, Although, But, And

For many years, writers used the word "while" to imply that one thing happened at the same time another event happened, and this is a correct usage of the word. However, some writers have begun using the word "while" to indicate a contrary, comparative situation, that is, a "not if" situation, and this is incorrect usage of the word. The word of choice in this latter instance is "whereas" because it implies two events are occurring simultaneously but they are contrary events. For example, correct usage of while is "I watched for the indicator to change color while I titrated the acid." Incorrect usage of while might be "I walked into the spectrophotometer room while she walked out of it." Correct usage would be "I walked into of the spectrophotometer room, whereas she walked out of it." These same principles can be applied to the words although, but, and and.

One-Sentence Paragraphs

Newspapers and other forms of creative writing often use a one-sentence paragraph to call attention to a specific item the writer wishes to emphasize. Most editors and copy editors of scientific journals will suggest that the one-sentence paragraph simply be incorporated into the previous paragraph unless its incorporation causes ambiguity or a situation where the single sentence is a "dangler" at the end of the paragraph.

Anticipate Questions and Concerns

Look for weaknesses in your data and in the logic used in your arguments. Try to deal with weaknesses up front by acknowledging them in the manuscript and providing a credible rationale for why the issues should not preclude acceptance of the manuscript. Dealing with any types of weaknesses and potential concerns is an important aspect of the discussion section of many manuscripts.

Stimulate the Interest of the Reader

Write the manuscript so that attentive readers (e.g., the reviewers) will formulate questions of interest. Later in the manuscript you can deal with the answers to those questions so that the reader has a sense of satisfaction.

What's New?

Point out directly what is new in your journal article. Acceptance often hinges on originality and on your ability to convey that you have discovered new, valuable information. Do not be afraid to state the novelty of your findings directly.

Text vs. Tables or Figures

Free the text of data that has been presented in a similar manner in tables or figures. However, be aware that it is completely acceptable to present data, or other forms of information, in the text when those data or information are simple and straightforward.

Write Informative, Not Indicative, Sentences

Do not hint at something in a vague way, which is called indicative because it indicates something to the reader, rather than informing him or her. State facts directly (i.e., be informative) so the reader comes away with new and useful information. Often this involves quantification, and it is easy to do, because you simply refer to your data.

> *Indicative*: Plants gained dry weight during the middle of the treatment period.
>
> versus
>
> *Informative*: Plant dry weight increased by 32% between days 24 and 45.

Make the Writing Flow

Avoid cumbersome sentences. Short and simple declarative sentences using common words will serve you well. Research by the American Press Institute, as reported by Bolsky (1988), has shown that a person loses the ability to comprehend a sentence clearly as the number of words in the sentence increases. We present here a sample from the figure presented in Bolsky (1988). If a sentence has eight words or fewer, then readers have 100% comprehension. When sentences have 15 words, comprehension drops to 90%. When sentences have 19 words, comprehension drops to 79%. When there are 28 to 29 words in a sentence, comprehension drops to 50%. Comprehension is about 20% when there are 39 words, and it drops to less than 10% when there are 44 words. Although more intricate, a similar treatment of word length has been presented and shows similar trends of difficulty to understand written works (Flesch, 1960, as cited in Mitchell, 1968). Also, strive for smooth transitions from one idea to the next in a paragraph or from paragraph to paragraph. Short sentences are potent, but you should avoid a succession of short choppy sentences by varying the length of the sentences and occasionally adding a compound sentence. Try to achieve euphony in your writing, so that the text sounds pleasing when read aloud.

REFERENCES

BOLSKY, M.I. 1988. *Better scientific and technical writing*. Prentice Hall, Englewood Cliffs, NJ.

[CSE] COUNCIL OF SCIENCE EDITORS, STYLE MANUAL COMMITTEE. 2006. *Scientific style and format: the CSE manual for authors, editors, and publishers*. Seventh edition. The Council, Reston, VA.

DAY, R.A. 1995. *Scientific English: A guide for scientists and other professionals*. Second edition. The Oryx Press, Phoenix, AZ.

FLESCH, R. 1960. *How to write, speak, and think more effectively*. Harper and Bros., New York, NY.

FOWLER, H.R. 1986. *The Little, Brown handbook*. Third edition. Little, Brown and Co., Boston, MA.

KATZ, M.J. 2006. *From research to manuscript. A guide to scientific writing*. Springer, Dordrecht, The Netherlands.

KIRKMAN, J. 2006. *Punctuation matters*. Fourth edition. Routledge, New York, NY.

LUELLEN, W.R. 2001. *Fine-tuning your writing*. Wise Owl Publishing Co., Madison, WI.

LUELLEN, W.R. 2007. *Dictionary of similar-but-separate words*. Wise Owl Publishing Co., Madison, WI.

LUNSFORD, A.A. 2003. *The St. Martin's handbook*. Fifth edition. Bedford/St. Martin's, Boston, MA.

MALMFORS, B., GARNSWORTHY, P. and GROSSMAN, M. 2000. *Writing and presenting scientific papers*. Nottingham University Press, Nottingham, UK.

MISH, F.C. 2004. *Merriam-Webster's collegiate dictionary*. Eleventh edition. Merriam-Webster, Inc., Springfield, MA.

MITCHELL, J.H. 1968. *Writing for technical and professional journals*. John Wiley & Sons, Inc., New York, NY.

STRUNK, W. JR, and WHITE, E.B. 2000. *The elements of style*. Fourth edition. Allyn and Bacon, Boston, MA.

PART III

REVIEW, SUBMISSION, AND POSTSUBMISSION ISSUES

CHAPTER 21

IN-HOUSE MANUSCRIPT REVIEW PROCESS

The test of a man's or woman's breeding is how they behave in a quarrel. Anybody can behave well when things are going smoothly.

—George Bernard Shaw

PRINCIPLES AND NEED FOR PEER REVIEWS

Research is not complete until the researcher disseminates the results of the accomplished work (Day and Gastel, 2006; Matthews et al., 2000). Thus, most life scientists will become authors sooner or later, as they write to describe what they have accomplished. In the words of Euripides, "If you wish to be a writer, write." As scientists, we also have a responsibility to review manuscripts and offer an opinion on the quality of the writing and science contained in a manuscript that has been or will be submitted for publication by other scientists. Authors have a tremendous opportunity to learn how to write from critical analyses of what reviewers have said about the manuscript they are working to complete. Authors should take advantage of the opportunities afforded by the review process every time they can (Day and Gastel, 2006; Matthews et al., 2000). Thus, Euripides was correct in stating that we need to write to learn to become a writer. However, we also stand to make great gains in our communication skills when we review a manuscript of other scientists and when other scientists have critically reviewed our manuscript. We can learn not only what not to do when writing a manuscript, but also what "tricks of the trade" we can use in our writing. This starts at the point where authors have their first opportunity to have other people and other scientists review and critically analyze the manuscript under preparation. Normally, this occurs when authors seek or are required to seek internal reviews of their manuscript before it is submitted for publication. This is an excellent time to obtain a review so that problems are corrected before the manuscript has been submitted. Truly, peer review is one of the most important foundations of science, particularly for the publication of research results, and the peer review process has been with us for years (Day and Gastel, 2006; Matthews et al., 2000). All parts of science benefit from peer review, because better written and clearer journal articles allow science to move forward more quickly than if scientists had to work their way through a mountain of ambiguous writing that made no sense (Davis, 2005; Day and Gastel, 2006; Matthews et al., 2000).

Getting Published in the Life Sciences, First Edition. By Richard J. Gladon, William R. Graves, and J. Michael Kelly
© 2011 Wiley-Blackwell. Published 2011 by John Wiley & Sons, Inc.

STEPS OCCURRING IMMEDIATELY AFTER COMPLETION OF DRAFTS OF THE MANUSCRIPT

The lead author should thoroughly revise the most recent copy of the manuscript he or she has been developing and should submit copies of it to the other coauthors for their final comments and opinions. This includes one's major professor if still a student. This is an extremely important step that must be followed, because all authors share all of the responsibility for the entire content and accuracy of the manuscript (Rennie et al., 1997; Weltzin et al., 2006). Therefore, they all must approve of its content before it is submitted (Rennie et al., 1997; Weltzin et al., 2006). Incorporate all final suggestions made by every author, as is appropriate, and be sure all authors are completely satisfied with the content of the manuscript. When each coauthor has accepted the manuscript content and form, then the lead author is ready to obtain presubmission reviews from appropriate scientists either in-house or at another institution.

Normally, authors will use, or will be required to use, internal peer reviewers at their home institution. Suggestions from the internal peer reviewers should be taken into consideration by the authors, and appropriate revisions should be made before submission to the journal. After submission, the journal will find several, usually one to three, peer reviewers. These peer reviewers make a judgment on the quality of the science and communication in the manuscript. Normally, reviewers report to an Associate Editor, Consulting Editor, or similar authority. This Associate Editor or Consulting Editor will take into account the judgments of the reviewers, and make a recommendation to the Editor-in-Chief, the Science Editor, or someone with a similar moniker, who makes a final decision. This entire process has been described generically in Chapters 2 and 24.

The main responsibilities of reviewers, both internal and external, are to determine the verifiability of the results and to evaluate the merit or quality of the science. These reviewers do not have the additional responsibility of correcting English usage and grammar indiscretions made by the author(s). All problems associated with poor English usage and grammar should have been resolved by the time reviewers receive the manuscript for either an internal or external review. In addition, reviewers should attempt to detect the rare instances of fraud, in the form of data fabrication, data falsification, or plagiarism by the author(s). It is better to uncover these acts of scientific misconduct at this point, rather than after the manuscript has been published. At that point, retractions must be made, which lead to immense trouble for all involved. It is important that reviews be completed promptly. Reviewers who have agreed to an assignment should arrange their workload to make time available for a thorough review.

HANDLING THE FIRST REVIEW OF YOUR MANUSCRIPT NOT MADE BY COAUTHORS

Waser et al. (1992) have written an excellent piece on writing effective manuscript reviews. We have adjusted their suggestions so that an author who may not have been through a review of a manuscript, and has little or no understanding of the review process, can use the information. We suggest you read this entire section as it contains information useful to both experienced and novice writers and reviewers. Waser et al. (1992) offer advice in two general areas that are critical to high quality reviews—forming and keeping a nonconfrontational attitude and specific suggestions about how to approach a review so that the time spent on the review process is productive, for both the reviewer and author(s).

Forming and Keeping a Nonconfrontational Attitude

- Assume you, as the author, and the reviewer, share a common enterprise—you both like your scientific discipline and want to make it as good as it can be.

- As an author, you should approach the review process in a state of mind where you are convinced the reviewers are trying to help you make the manuscript better and a good contribution to your common scientific discipline.

- Most research work and the resultant manuscripts do not contain "fatal flaws," so be cognizant of the possibility that a reviewer may be inventing one to make him- or herself feel like Sherlock Holmes. If there is any doubt, explore the suggestion and come to a point of understanding regarding the proposed "fatal flaw" before you submit the manuscript.

- As author(s), be sure you know the procedures, materials, systems, and statistical procedures needed for accurate completion of the research. If something does not seem right, then write about it using appropriate language. Later you might find out you had the wrong understanding of the system, and the reviewer was correct in their understanding of the work.

- Be aware that some reviewers will not cut authors any slack, and they will hold the authors to a standard no-one can achieve. No-one can get everything, including methodology, analysis, and experimental design, perfect every time. In addition, what is perfect in one system may not work well in another system.

- You should receive comments from the reviewer(s) about what is both good and bad about the manuscript. Although we scientists are trained to be critical, and we should critically analyze all aspects of science, the criticism(s) leveled by a reviewer should be constructive. The reviewer should point out both strengths and weaknesses in the manuscript.

Some Specific Suggestions on How the Reviewer Should Have Approached the Review

- The reviewer should have focused solely on the manuscript parts he or she could judge competently. If your reviewer cannot judge the quality of your manuscript, then she or he should say so and return it promptly.

- The contents of your manuscript are privileged, confidential information, and if your reviewer consults with a colleague, be sure he or she also understands that further discussions about the contents of the manuscript must be held in confidence.

- Make sure the reviewer did not get herself or himself so focused on nit-picking details that they did not see the forest for the trees when they looked at the manuscript.

- Ask that the reviewer(s) do not do a slipshod job on the in-text citations and the listing of resources in the literature cited section. Most editors state that more factual errors occur in the citation of resources than in any other part of the manuscript. An important issue here is that the author(s) have an appropriate number of citations that are proportional to the magnitude of the study.

- You should hope your reviewers would write the review with economy of words and unparalleled clarity.

- Your manuscript should be reviewed and returned to the writer as quickly as the reviewer can complete it. If your reviewer cannot do the review in a reasonable amount of time, then, without any apologies, ask that she or he return the manuscript so that another reviewer can be located.

- Do not feel threatened by competition, because competition makes all of us function better as professionals. You should welcome new and sometimes controversial ideas and suggestions from the reviewer that may run counter to your own.

Diplomacy in review criticisms is encouraged strongly. As an author, you will be more receptive to comments made constructively than nasty, derogatory comments that come across as a personal attack by the reviewer. The expenditure of time and effort by the reviewer is best spent and appreciated by the author(s) if the author(s) feel the reviewers are on their side, trying to make the manuscript better. However, do keep in mind that the responsibility of the author(s) is to present to the reviewers a well written manuscript. Good reviews should be clear, straightforward, and objective in their criticisms, without putting the author in a defensive posture.

An excessively critical review can easily devastate a young author or an author completing her or his first few manuscripts, and this is counterproductive to the advancement of science. Sarcasm and acrimony are unprofessional, and they have no place in a review. The reviewer should consider herself or himself a colleague of the author, and all criticisms should be helpful, constructive, and explicit, and in the interest of making the manuscript better.

If the manuscript organization is poor or illogical, or if the manuscript is too long, for example, then specific suggestions for improvement should be given by the reviewer. Comments should be relevant to the manuscript under review only, and they should not be an attack on an individual author or a group of authors. Reviewers should have criticized omissions only if the author's conclusions were untenable without them. If there are obvious areas for future study that you have omitted from the discussion, then the reviewer more than likely will note these areas on the "reviewer's comment sheet." If data contradict your text, then the reviewers should have documented them so the author(s) may address them. If specific, published, relevant works were not cited, you can expect that the reviewer will call these works to the author's attention. However, remember that no manuscript is exhaustive.

If your reviewer suspects plagiarism, and makes direct or indirect statements to that effect, then follow the suggestions so that you are not accused of committing plagiarism. Simultaneity occurs often in science, so be alert that a similar paper by another author may have been submitted or published elsewhere.

As an author, you should expect thoroughness from your reviewers as they complete the review of the manuscript or its parts. They should not have skimmed the piece and listed criticisms based on a superficial reading.

Who do you pick for internal reviews—friends or foes? Although it may speed up the time to get to submission, there is little benefit in having a friendly internal reviewer edit your manuscript. There certainly is not much hope for you to get an expert internal reviewer for your manuscript, as most organizations have only one expert in a given discipline or subdiscipline. Therefore, you will probably not get an internal reviewer who will be able to give the strong, expert review from internal sources that we all would like to have. However, many scientists within your organization will have published enough to give you a strong critique of your English usage and grammar, and they can offer suggestions about the science conducted in the manuscript. They will probably have a broad enough scope of knowledge that they can judge how well the science has been done. This is exactly what you need from an internal reviewer. Do not waste your time with a friendly review that tells you all is well, but, rather, seek a strong review for at least the English usage and grammar portion of the manuscript, and hope they can also suggest improvements in the presentation of the science.

You will also have some influence over the choice of external reviewers, as most journals either require or ask that authors submit a list of possible reviewers when the manuscript is submitted. It is not ethically appropriate for you to choose only those people who you know will give you an easy, and most likely positive, review. Your recommendations of possible reviewers for your submitted manuscript should be based on the technical expertise of the reviewer(s) you suggest, not their ability to make you happy. Remember, the manuscript will have your name on it and will be your responsibility for the remainder of your career, so you want a thorough, excellent review before it is published.

GENERAL EXPECTATIONS FOR PARTS OR THE ENTIRE MANUSCRIPT BY THE REVIEWER

As you complete preparation of different sections of your manuscript, there will be opportunities for you to have these portions of your manuscript reviewed by your coauthors, colleagues, classmates, or others. Take advantage of every opportunity to have your manuscript reviewed by anyone who will look at it. If you have been asked to provide a review for someone, or if you are asking someone to review your manuscript, ask the reviewer to focus their energy on providing a critique by addressing the items below, as they are the most basic characteristics of good scientific writing (see Chapter 6). The italic words should be the focal points of this review.

- *Accuracy and precision* must be in everything written.
- *Organization* leads to a properly structured piece.
- A *logical flow* of words and sentences moves the writing from point to point with understanding and smoothness.
- *Clarity* is the basis for understanding, so your reviewer should question everything that could be misunderstood. Write so that you cannot be misunderstood.
- *Economy of words* eliminates words that are neither functional nor concise.
- *Informative words* are more powerful and tell more than indicative words.
- *Proper choice of words* leads to understanding, because many words have multiple meanings and can be interpreted in several ways.
- *Write objectively* so that readers do not suspect bias in any of your statements.
- *Consistency* in your writing allows your work to flow from word to word, sentence to sentence, paragraph to paragraph, and section to section. If the writing is not consistent throughout the manuscript, then the words and structure become a distraction to the reader.
- *Interesting words* keep the reader attentive and focused on the messages you wish to communicate through your manuscript. However, these words should be simple and straightforward, not "flowery" or jargon few can understand.

At the end of this chapter, we have provided what we consider a generic manuscript evaluation form, a generic, nonelectronic reviewer's comment sheet, and a generic electronic reviewer's comment sheet. Each journal will probably have its own form and format for obtaining the opinions of the reviewers, but these will serve as a model (Anon., 1997).

CHECKLIST OF ITEMS REVIEWERS MAY EXPECT TO SEE

When you have completed a rough draft of your entire manuscript, the time for a critical review is upon you, and your coauthors, colleagues, classmates, and so on should participate in these reviews. We provide an extensive checklist of items that some/many manuscript reviewers might expect to see in your manuscript. You should make certain you have incorporated as many as possible of these items.

Some scientists review a manuscript by first reading it from start to finish, maybe even twice, without making any comments on the manuscript or reviewer's comment sheet. After the reviewer allows the manuscript to get "cold" for a while, they will return to it. During this initial reading, reviewers often begin to record items the author(s) must address in order to make the manuscript suitable for publication. Below, we provide, for authors, a checklist of items that should have been addressed during preparation of the manuscript. If they have not been addressed properly, then now is the time to make adjustments. The items within this checklist have been organized by topic so that similar items will be addressed at the same time. (The American Society for Horticultural Science originally designed this checklist for use during the review process of manuscripts submitted to journals the Society publishes, and we authors appreciate permission from the Society to use it with this acknowledgment.)

Rationale for Conducting the Research

- Is this manuscript relevant to the life sciences?
- Are the rationale and objectives of the manuscript stated clearly and concisely?
- Are the objectives important to the discipline within the life sciences?
- Do the facts (data) given relate to the rationale and/or objectives?
- Is the manuscript widely applicable, or is it solely of local interest?

Suitability of Format and Journal

- Is the manuscript suited for the journal to which it will be submitted?
- Is the manuscript better suited for a different journal?
- Is the manuscript a research article, or would it be more appropriate if it were reformatted into another form of communication such as a feature article, an editorial viewpoint, and so on?

Originality

- Has this information (data) been published previously by the author(s) in the same or a similar form?
- Does the manuscript duplicate substantial portions of text or data from a previous work of the same or other authors?

Knowledge of the Field

- Does the author recognize by documentation relevant contributions of predecessors in the discipline?
- If the data and/or text of the manuscript contradict other work, does the author deal with the contradiction(s) appropriately and honestly?

Readability

- Is the writing style clear, accurate, and precise? Ambiguous or vague statements will be challenged appropriately.
- Are the style rules for abbreviations, italics, punctuation, and so on followed properly?
- Are special terms used and defined properly?
- Is there minimal or no use of jargon in the manuscript?

Organization

- Is the manuscript too short or too long in relation to the importance of the subject and amount of data presented?
- Do all parts of the manuscript warrant publication? Does the author introduce irrelevant material?
- Are topics presented and discussed in a logical sequence, following the format set forth in the instructions to authors or style manual of the journal?
- Does the underlying thesis of the manuscript carry throughout the manuscript?

Technical and Factual Accuracy

- Are there errors of fact, interpretation, or calculation?
- Are measurements reported in SI units in the appropriate style of the journal?
- Have all biological, chemical, and other materials been identified properly by their scientific or common names as required by the journal?
- Is the presentation biased in any way, for example, data presentation, analysis, or, interpretation?
- Have fact(s) been distinguished from supposition(s) during presentation?

Title, Byline, Additional Index Words, and Abstract

- Is the title appropriate for the material presented and descriptive of the work?
- Has the byline been constructed correctly for the journal?
- Are additional index words or keywords given? Do they represent the content of the manuscript adequately, without repeating words in the title?
- Is the abstract concise and representative of the manuscript? Does the abstract exceed 5% of the length of the manuscript, 200 words, or whatever is the limit imposed by the journal?

Materials and Methods

- Are the technical and experimental methods appropriate for the study, adequately described, and readily replicable?
- Are the controls adequate and described well?
- Are all forms of the statistical treatment appropriate for the study and journal?

Results and Discussion

- Are the data presented appropriately and honestly?
- Is the discussion relevant to and within the bounds of the study?

- Have ideas, concepts, or interpretations been over- or under-emphasized?
- Are data reported in a format consistent with the experimental design?
- Are the reasoning and deductions of the author(s) logical and well organized, and can they be followed without loss of understanding?
- Are practical applications presented and discussed, especially if they are within the scope of the journal?
- Did the author compare, contrast, and explain differences in results within the experiment or contrary to previous studies?

Conclusions

- Do conclusions follow logically from the experiments conducted and data generated?
- Can the reader verify the conclusion(s) from the data given and resources cited?

Literature Citations (References)

- Has every resource listed in the literature cited section been cited in the text at least once?
- Are all in-text citations included in the literature-cited section?
- Are there too few or too many citations? Are some citations superfluous?
- Are citations of important resources missing?
- Are resources listed in the literature-cited section formatted properly for the journal? Are they complete?
- Are all resources traceable?

Tables and Figures

- Are all tables and figures cited at least once in the text and introduced in the proper sequence required by the format of the journal?
- Do tables and figures duplicate information in the text, or vice versa?
- Are there too many or too few tables and figures?
- Can any tables or figures be condensed or eliminated?
- Are the form, arrangement, and format of tables and figures satisfactory for the journal? Are they too crowded? Should they be combined with others?
- Can the reader understand table titles and figure captions without referring to the text?
- Do tables and figures present data accurately, clearly, and concisely?
- Are the bodies and footnotes of tables formatted properly for the journal?

Figures Only

- Is the technical quality of the artwork, photographs, and so on adequate?
- Are all figures identified properly?
- Do all illustrations show what they purport to show?
- Are legends, keys, scales, and magnification given, where necessary?
- Does the scaling of axes on graphs and histograms distort the information presented and its interpretation?

MANUSCRIPT EVALUATION FORM FOR PEER REVIEW

The information reported in this manuscript is mainly (check one):

Basic Research _____ Applied Research _____ Technology Transfer_____

	Yes	No
1) Rationale and objectives are clearly stated and supported	___	___
2) Methods and procedures are clearly understandable, complete, and create confidence in the results	___	___
3) Results are presented logically and coherently	___	___
4) Discussion of results is clear and adequate	___	___
5) Statistical procedures are appropriate for the type of data presented . . .	___	___
6) Lettering on illustrations is easily legible	___	___
7) Captions to figures are clear and sufficient	___	___

8) Literature citations are (check one):

Inadequate _____   Appropriate _____   Excessive _____

The manuscript could be reduced _____ slightly, by _____% , or reduced _____

substantially, by _____% , without loss of meaning or information.

Summary of review:	Excellent	Good	Fair	Poor
Originality	___	___	___	___
Technical quality	___	___	___	___
Clarity of presentation	___	___	___	___
Potential significance	___	___	___	___

I recommend the manuscript be (check one):

_____ Accepted as submitted

_____ Accepted, contingent upon minor revision

_____ Accepted, contingent upon major revision

_____ Declined, but recommended for revision and resubmission

Reviewer's name _____   Date _____

(Do Not Sign This Form If You Wish to Remain Anonymous.)

NONELECTRONIC REVIEWER'S COMMENT SHEET

Often, a writer can receive helpful, direct comments from a reviewer, whether it is during the various stages of preparation, the internal review stage, or the external review stage. These comments often allow author(s) to avoid mistakes and improve the overall quality of the manuscript. Please enter your comments on this nonelectronic version of the reviewer's comment sheet, and return it to the author(s) when you have completed your review of the manuscript. Comments on this sheet should not be simple editorial marks, which should be placed directly on the manuscript, but rather, these comments should be broader and more substantive suggestions for the author(s). In most cases, some form of information the author(s) can use to correct the problem should substantiate these comments. For example, use of Method ABC is incorrect in the context of this research, and you should use Method XYZ. And when it is possible, you should designate the reference(s) that contains Method XYZ. Please be fair, direct, and honest in your comments to the author(s), but at the same time, please help them by giving them useful, constructive suggestions for improving the manuscript. *Malicious, destructive criticism benefits no one.* Please attach this sheet (with a paper clip) to your reviewed copy of the manuscript, and the author(s) will get the completed form. If you wish to remain anonymous, please do not complete the reviewer's name line on the form.

Author(s)' name(s):

Title of manuscript:

Reviewer's name:

One to two sentences by the reviewer that summarize the essence of the manuscript:

Additional comments and suggestions for improvement:

_____

_____

_____

_____

_____

_____

_____

_____

_____

_____

_____

_____

_____

_____

_____

_____

_____

_____

_____

_____

_____

_____

_____

_____

_____

_____

_____

_____

_____

_____

_____

ELECTRONIC REVIEWER'S COMMENT SHEET

Often, a writer can receive helpful, direct comments from a reviewer, whether it is during the various stages of preparation, the internal review stage, or the external review stage. These comments often allow author(s) to avoid mistakes and improve the overall quality of the manuscript. Please enter your comments on this electronic version of the reviewer's comment sheet, and return it to the author(s) when you have completed your review of the manuscript. You may want to print a copy of this completed comment sheet and then return it to the author(s), or you may want to send an electronic version of it through the editorial office to the author(s). Comments on this sheet should not be simple editorial marks, which should be placed directly on the manuscript, but, rather, these comments should be broader and more substantive suggestions for the author(s). In most cases, some form of information the author(s) can use to correct the problem should substantiate these comments. For example, use of Method ABC is incorrect in the context of this research, and you should use Method XYZ. And when it is possible, you should designate the reference(s) that contains Method XYZ. Please be fair, direct, and honest in your comments to the author(s), but at the same time, please help them by giving them useful, constructive suggestions for improving the manuscript. *Malicious, destructive criticism benefits no one.* If you wish to remain anonymous, please do not complete the reviewer's name line on the form.

Author(s)' name(s):

Title of manuscript:

Reviewer's name:

One to two sentences by the reviewer that summarizes the essence of the paper:

Additional comments and suggestions for improvement:

REFERENCES

ANON. 1997. *American Society for Horticultural Science publications style manual.* American Society for Horticultural Science Press, Alexandria, VA.

DAVIS, M. 2005. *Scientific papers and presentations.* Second edition. Academic Press, San Diego, CA.

DAY, R.A. and GASTEL, B. 2006. *How to write and publish a scientific paper.* Sixth edition. Greenwood Press, Westport, CT.

MATTHEWS, J.R., BOWEN, J.M. and MATTHEWS, R.W. 2000. *Successful scientific writing. A step-by-step guide for the biological and medical sciences.* Cambridge University Press, New York, NY.

RENNIE, D., YANK, V. and EMMANUEL, L. 1997. *When authorship fails: A proposal to make contributors accountable.* J. Am. Med. Assoc. 278:579–585.

WASER, N.M., PRICE, M.V. and GROSBERG, R.K. 1992. *Writing an effective manuscript review.* BioScience 42: 621–623.

WELTZIN, J.F., BELOTE, R.T., WILLIAMS, L.T., KELLER, J.K. and ENGEL, E.C. 2006. *Authorship in ecology: Attribution, accountability, and responsibility.* Front. Ecol. Environ 4:435–441.

CHAPTER 22

COVER LETTER AND SUBMISSION FORM PREPARATION

> Never be afraid to try something new. Remember that a lone amateur built the Ark. A large group of professionals built the Titanic.
>
> —Dave Barry

DEFINITION AND NEED

The cover letter is a document of transmission that conveys vital information about the submission of a manuscript to the person in authority at the editorial office of the journal. A submission form is an alternate document of transmission that requests similar information from the corresponding author. Normally, the submission form also functions as a mechanism to obtain more specific information about the content and status of the manuscript. The purposes of the cover letter and the submission form are to present your manuscript formally to the editor or other person of authority and to provide him or her with information that clarifies the status and content of the manuscript submitted. These documents are needed so that it is clear what the editor and editorial office are to do with the manuscript. In general, no manuscript should be submitted for publication, in either hardcopy format or electronically, without an accompanying cover letter and/or the completed submission form (Day and Gastel, 2006). Be sure you check the type of document of transmission required so you do not expend a lot of time and energy composing a cover letter, only to discover that a completed submission form takes the place of the cover letter. An example of a cover letter is furnished in Appendix A.22.1 as an answer to Exercise 22.1, and a completed submission form is provided at the end of this chapter.

When one is required, it is essential that the cover letter be editorially flawless (Matthews et al., 2000). By virtue of its purpose, the cover letter "covers" the submitted manuscript, and it is therefore the first item of your work the editor sees. You do not have a second chance to make a good first impression. If the quality of your cover letter is substandard, the editor may infer bad things about your manuscript—such as the notion that the work reported in your manuscript has been done poorly and without concern for its image.

PRINCIPLES

Construct the cover letter and/or submission form when a manuscript has been revised several times, reviewed in-house by colleagues, edited into final form, approved for submission by all authors, approved for submission by the appropriate institution(s) and their appropriate

Getting Published in the Life Sciences, First Edition. By Richard J. Gladon, William R. Graves, and J. Michael Kelly
© 2011 Wiley-Blackwell. Published 2011 by John Wiley & Sons, Inc.

offices, when necessary, and prepared in final form for the intended journal. Normally, the cover letter will be constructed and signed by the co-author designated as the corresponding author for the manuscript. The corresponding author has been defined in Chapter 4, and that chapter also details the specific responsibilities of the corresponding author during the entire publication process.

THE COVER LETTER

Keep the cover letter as brief as possible. It should not be longer than one typed page. Be businesslike, friendly, and polite. Indicate that you appreciate consideration of your manuscript for publication in their journal. Do not make any requests for special treatment, such as a fast review. A crisis situation on your part (e.g., a promotion/tenure package that needs to be submitted immediately) does not necessarily constitute a crisis situation for the journal. This is the time to cultivate good will.

The cover letter should address, but not be limited to, the following areas.

- It should have all parts of a formal business letter, such as the date, addressee information, salutation, closing, and so on;
- The editor's name, or the name of another person in charge at the journal, should be contained in the address and salutation of the letter. Be sure it is the current person in charge. Do not use "To Whom It May Concern" or "Dear Editor." It looks very tacky if you do not include the person's name. That person may be left with the impression you are not even interested enough in the journal to determine to whom you should address the cover letter (Gustavii, 2003; Matthews et al., 2000).
- The complete title should be included.
- It should contain the complete identification of all author(s) of the manuscript.
- The place(s) where the author(s) are employed should be identified.
- It should state the number of copies of the manuscript and figures you are including, if you are submitting a hard copy (e.g., an original plus two copies for reviewers).
- The complete name and address of the corresponding author should be provided, including voice and fax numbers, along with the e-mail and place of business addresses of the corresponding author.
- The name of the journal to which you are submitting the manuscript should be stated. Some editors are associated with more than one journal, and some publishers issue several journals. It is easy for an editor to think the author(s) want to submit to one journal, when in reality, they wanted to submit the manuscript to another journal published by the same organization.
- The type or category of manuscript should be identified (e.g., brief note, research report, journal article, letter to the editor, technical comment, review article, feature article, etc.).
- The discipline or category used to classify the content of the manuscript should be identified (e.g., plant physiology, veterinary anatomy, animal genetics, etc.).
- Special features or requirements of the manuscript, such as layout, special reproduction of figures, unusual length, etc., should be mentioned. There may be no unusual or special requests, and if so, that is fine.
- State that this manuscript is new and is not being considered for publication anywhere else. This includes any and all parts of the manuscript (Matthews et al., 2000). The

only permissible exception to this rule is for an abstract and the associated information to be presented (either orally or as a poster presentation) at a scientific meeting.

- Consider commenting on the desired reproduction of the figure(s), especially if you are providing line drawings or photographs. Indicate whether you have prepared them at the same size they are to appear in the journal or whether they are to be enlarged or reduced, and by how much. It is disconcerting to prepare figures for full-page reproduction only to find the journal has shrunk them to quarter-page size and clarity has been lost completely (O'Connor, 1991).

- Indicate your commitment (or your institution's) to pay for page charges and reprint/offprint charges, if any. You may have to clear this with departmental or university offices before, rather than after, manuscript submission. If institution officials only, and not the corresponding author, may make a written commitment to pay, then indicate this to the editor and request the appropriate officials to provide the commitment. Some journal publishing houses will not even process a manuscript until they have written assurance that at least page charges will be met. Some journals waive page charges in cases of hardship, such as for an author from a developing country. Some journals have no page charges, some have no reprint/offprint charges, and some do not charge for either pages or reprints/offprints. Be sure you know the policy of the journal before manuscript submission, because this portion of the publication process is very costly. Many times, this information is available on the journal website, and many of your questions will be answered if you simply check the website of the journal before proceeding.

- Indicate your (or your institution's) commitment to pay for any additional charges such as excess illustrations, colored illustrations, unusual layouts, and so on, if these are part of your manuscript.

- Include any pressing questions you have about the processing of manuscripts, journal policies, or other items that need clarification. However, if you have any questions, check the website of the journal before you begin to ask straightforward questions that already have been answered (e.g., the instructions to authors, the instructions to contributors, the journal website, the style manual of the journal, and so on). Limit such questions to very important matters only. Often, no such ancillary issues arise.

Your cover letter also might include the following, depending upon your journal.

- You might state that all authors have contributed significantly or appropriately to the conduct of the research, have reviewed and understand the manuscript, and have approved its content for submission (Katz, 2006). With ethical situations as they are these days, many journals require this information either in the cover letter or a form specifically designed for this function, and it may be that all authors must sign the form (Matthews et al., 2000).

- Depending upon the journal, you may be able to suggest that certain scientists should not, or you prefer that they will not, review your manuscript. As we all know, there are a variety of reasons for allowance of this request, one of which may be a conflict of interest on the part of the reviewer (Day and Gastel, 2006).

- Some journals allow you, or may even encourage you, to suggest an Associate Editor or other form of adjudicator who will oversee the review and decision-making process for your submission (Day and Gastel, 2006).

- Some journals require copies of the forms cleared through the institution when work with animals or humans has been conducted (O'Connor, 1991; O'Connor and Woodford, 1975).

- A one-paragraph summary that relays the essence of the entire article to the editor, and subsequently to the reviewers, might be appropriate (Gustavii, 2003).
- Some journals consider the act of submission of the manuscript as an assignment of copyright to the journal, if the manuscript subsequently becomes a published article. Be sure you understand whether this is the case, and be sure you, and your co-authors, understand and approve of this system (O'Connor, 1991).
- Alert the editor to the fact that you have cited in your manuscript any of the following: unpublished work by you or other researchers, a personal communication, or an article that has been accepted for publication or is in press (Gustavii, 2003; Matthews et al., 2000). This includes citation of persons in the acknowledgements. Some journals may require that all individuals cited in this way must sign a permission form that allows the author(s), journal, and publisher to use their name (Matthews et al., 2000).
- You may need to declare if you have been, are currently, or may well become involved in a financial conflict of interest if the manuscript is published (Gustavii, 2003).
- It may be good to state that none of the authors has any relationship to the manufacturers or distributors of materials tested or used in the conduct of the research (Katz, 2006; Matthews et al., 2000).
- You may need to announce that this manuscript is a continuation of a series of work that already has been published elsewhere (Gustavii, 2003; O'Connor, 1991).
- If this manuscript is one of two or several that you are submitting for publication simultaneously, you should explain to the editor all pertinent details about the other manuscript(s) so the editor can act accordingly with this manuscript and the other manuscript(s) (Gustavii, 2003; O'Connor, 1991).
- You might include some sentences about why you have chosen this particular journal and how the journal's scope matches the content of the manuscript (Matthews et al., 2000).
- If a submission form that contains your suggestions for reviewers of the manuscript is not part of the submission package, then you might add a list of several of these to the content of the cover letter (Matthews et al., 2000; O'Connor, 1991).

THE SUBMISSION FORM

If you have determined your journal of choice requires a submission form, then locate it and complete all required parts of it. Most likely you will find it on the website of the journal, and it will probably be in a format that allows you to complete and submit the form electronically. The form also may be found in the style manual for that journal, its instructions to authors, its instructions to contributors, or in a recent issue of the journal. The journal has decided they need this information, so do not leave areas blank. Sometimes it contains information similar to what you have included with your cover letter; other times, it will request a completely different set of information. A common extra set of information in submission forms is a request for the names of colleagues who have reviewed the manuscript before submission (in-house reviewers) (O'Connor, 1991). The request may include both in-house and external reviewers. Another common set of required information is the names, telephone numbers, and e-mail addresses of three to five possible reviewers the editor may enlist for review of this manuscript.

EXERCISE 22.1   Find the Missing Parts of this Cover Letter

(An appropriate cover letter that exhibits the missing parts in this letter may be found in Appendix A.22.1. There are ten items that need to be corrected.)

Mr. Michael Neff
Publications Director
Publications Department
American Society for Horticultural Science
600 Cameron Street
Alexandria, VA 22314-2562

Dear Mr. Neff:

Enclosed are copies of our manuscript titled "Seed Germination and Seedling Growth of *Alnus maritima* from Its Three Disjunct Populations." We have retained the original of the manuscript and the originals of the figures. This manuscript is being submitted for possible publication in the Seed Physiology section of the *Journal of the American Society for Horticultural Science*. William R. Graves will function as the corresponding author for this manuscript, and he can be reached at the addresses below. We have provided the figures at a size that we believe can be used without any reduction during the printing process. We also will pay for any special fees for sizing and printing the two photographic figures, if that is necessary due to something we have overlooked. This manuscript is new. We will arrange for payment of the reprint charges at the appropriate time. Mr. Schrader has begun his search for a fulltime, permanent position, and we would really appreciate a fast review and decision on the fate of this manuscript. Thank you for your consideration of this manuscript.

William R. Graves   James A. Schrader
Professor   Graduate Research Assistant

Iowa State University
Department of Horticulture
129 Horticulture Hall
Ames, IA 50011-1100
FAX: 515-294-0730

APPENDIX A.22.1   Example of a well written cover letter that is the answer to Exercise I:

(Portions of this letter in boldface type are those missing in the letter of Exercise I.)

March 21, 2003

Mr. Michael Neff
Publications Director
Publications Department
American Society for Horticultural Science
600 Cameron Street
Alexandria, VA 22314-2562

Dear Mr. Neff:

Enclosed are **three complete** copies of a manuscript by **James A. Schrader and** William R. Graves titled "Seed Germination and Seedling Growth of *Alnus maritima* from Its Three Disjunct Populations." We have retained the original of the manuscript and the originals of the figures. This manuscript is being submitted for possible publication **as a research report** in the Seed Physiology section of the *Journal of the American Society for Horticultural Science*. William R. Graves will function as the corresponding author for this manuscript, and he can be reached at the addresses below. We have provided the figures at a size that we believe can be used without any reduction during the printing process. We also will pay for any special fees for sizing and printing the two photographic figures, if that is necessary due to something we have overlooked. This manuscript is new **and is not being considered for publication elsewhere**. We will arrange for payment of the **publishing fee ($100 per printed page) and** reprint charges at the appropriate time. **Mr. Schrader has begun his search for a fulltime, permanent position, and we would really appreciate a fast review and decision on the fate of this manuscript.** Thank you for your consideration of this manuscript.

Sincerely,

William R. Graves   James A. Schrader
Professor   Graduate Research Assistant

Iowa State University
Department of Horticulture
129 Horticulture Hall
Ames, IA 50011-1100
Voice: 515-294-0034
FAX: 515-294-0730
E-mail: graves@iastate.edu

EXERCISE 22.2   Writing Your Own Cover Letter

Now that you have finished your manuscript, and you are ready to submit it to an editorial office, you should write your cover letter. If you are not familiar with the standard formats in which business letters appear, you should find a reference book that shows the appropriate format(s) of a business letter (Fowler, 1986; Lunsford, 2003). Use a format to develop the letter, and include all appropriate items listed earlier in this chapter. If you are having trouble synthesizing your cover letter, look at the corrected cover letter example we have provided in Appendix A.22.1, or look elsewhere for example letters you can use as a model (Gustavii, 2003; Katz, 2006; O'Connor, 1991). Do not make this document part of your manuscript, because you may not be successful with the journal you choose initially.

If you are forced to submit your manuscript to another journal because your first choice has released or rejected it, be sure to revise your cover letter to meet the requirements of the new journal. Take care to remove anything from your cover letter and manuscript that might suggest the manuscript had been submitted previously to another journal, and was not accepted for publication. A couple of exceptions to this rule might be if your manuscript was released or rejected by the journal in its original form and you have now revised the manuscript extensively in accordance with the reviewer's suggestions, added new data, or reanalyzed the data based on new insights. A second situation where it would be appropriate to mention a prior submission would be the unlikely case of not being able to return the revised manuscript within the appropriate timeframe once the manuscript has been reviewed and returned to the author(s) for revisions. Failure to respond in the time required by the journal usually means the manuscript is handled as a new submission rather than a revision. With knowledge of a prior submission and revision, the editor might choose to assign your manuscript to the same associate editor and reviewers, and this may possibly accelerate the review and acceptance of the manuscript.

JOURNAL OF PUBLISHING IN LIFE SCIENCES
JOURNALS
Manuscript Submission Form

To:
Editors-in-Chief
Journal of Publishing in Life Science Journals
Department of _____
The State University, City, State 12345-6789

From:
Corresponding author: _____

Address: _____

Voice: ____-____-_____ FAX: ____-____-_____ E-mail: _____

Manuscript title: _____

Subject Category:

- ❏ Biotechnology, Molecular Biology
- ❏ Breeding, Cultivars, Rootstocks
- ❏ Crop Production/Cropping Efficiency
- ❏ Developmental Physiology
- ❏ Environmental Stress Physiology
- ❏ Genetics and Breeding
- ❏ Marketing and Economics
- ❏ Pest Management
- ❏ Photosynthesis/Source-Sink Physiology
- ❏ Plant Growth Regulators
- ❏ Plant Pathology
- ❏ Postharvest Biology and Technology
- ❏ Food Processing
- ❏ Propagation and Tissue Culture
- ❏ Seed Physiology and Technology
- ❏ Soil-Plant-Water Relationships
- ❏ Soil management, Fertilization, Irrigation
- ❏ Other

Names of Two or More Colleagues Who Have Reviewed the Manuscript (Required):

_____

Name, Voice Number, and E-mail Address of Three to Five People Outside Author(s)' Institution(s) Qualified to Act as a Reviewer of This Manuscript:

_____

_____

Statement: **I certify that the information presented in this manuscript has not been published elsewhere and is not under consideration for publication in other journals at this time. I will arrange for payment of the publishing fee of $100 per printed page, upon acceptance of the manuscript for publication.**

_____   _____
Date   **Signature of corresponding author**

JOURNAL OF PUBLISHING IN LIFE SCIENCES
JOURNALS
Manuscript Submission Form

To:
Editors-in-Chief
Journal of Publishing in Life Science Journals
Departments of Horticulture and Forestry
Iowa State University, Ames, IA 50011-1100

From:
Corresponding author: James A. Schrader

Address: 106 Horticulture Hall, Iowa State University, Department of Horticulture, Ames, IA 50011-1100

Voice: 515-294-9940   FAX: 515-294-0730   E-mail: jschrade@iastate.edu

Manuscript title: Seed Germination and Seedling Growth of *Alnus maritima* from Its Three Disjunct Populations

Subject Category:

- ❑ Biotechnology, Molecular Biology
- ❑ Breeding, Cultivars, Rootstocks
- ❑ Crop Production/Cropping Efficiency
- ❑ Developmental Physiology
- ❑ Environmental Stress Physiology
- ❑ Genetics and Breeding
- ❑ Marketing and Economics
- ❑ Pest Management
- ❑ Photosynthesis/Source-Sink Physiology
- ❑ Plant Growth Regulators
- ❑ Plant Pathology
- ❑ Postharvest Biology and Technology
- ❑ Food Processing
- ❑ Propagation and Tissue Culture
- X Seed Physiology and Technology
- ❑ Soil-Plant-Water Relationships
- ❑ Soil management, Fertilization, Irrigation
- ❑ Other

Names of Two or More Colleagues Who Have Reviewed the Manuscript (Required):
Jeffery K. Iles, Mark P. Widrlechner

Name, Voice Number, and E-mail Address of Three to Five People Outside Author(s)' Institution(s) Qualified to Act as a Reviewer of This Manuscript:
Michael Dirr, 706-542-2471, hor370@arches.uga.edu; Harold Pellett, 612-443-2460, no e-mail address available; Laura Jull, 608-262-1450, lgjull@facstaff.wisc.edu

Statement: I certify that the information presented in this manuscript has not been published elsewhere and is not under consideration for publication in other journals at this time. I will arrange for payment of the publishing fee of $100 per printed page, upon acceptance of the manuscript for publication.

August 11, 2004 _____
Date   **Signature of corresponding author**

REFERENCES

DAY, R.A. and GASTEL, B. 2006. *How to write and publish a scientific paper.* Sixth edition. Greenwood Press, Westport, CT.

FOWLER, H.R. 1986. *The Little, Brown handbook.* Third edition. Little, Brown and Co., Boston, MA.

GUSTAVII, B. 2003. *How to write and illustrate a scientific paper.* Cambridge University Press, New York, NY.

KATZ, M.J. 2006. *From research to manuscript. A guide to scientific writing.* Springer, Dordrecht, The Netherlands.

LUNSFORD, A.A. 2003. *The St. Martin's handbook.* Fifth edition. Bedford/St. Martin's, Boston, MA.

MATTHEWS, J.R., BOWEN, J.M. and MATTHEWS, R.W. 2000. *Successful scientific writing. A step-by-step guide for the biological and medical sciences.* Cambridge University Press, New York, NY.

O'CONNOR, M. 1991. *Writing successfully in science.* HarperCollinsAcademic, London, UK.

O'CONNOR, M. and WOODFORD, F.P. 1975. *Writing scientific papers in English.* American Elsevier, New York, NY.

CHAPTER 23

FINAL CHECKLIST AND INITIAL MANUSCRIPT SUBMISSION

Know that your most worthy efforts will be scorned by your peers, for it is they who suffer most when you excel. If your actions and ambitions do not threaten them, you are simply striving toward the insignificant.

—Wess Roberts

NEED

You are now at the brink of making the initial submission of your manuscript. The following checklist will allow you to make sure you have included all of the important items in your submission packet (CSE, 2006). Remember, you do not have a second chance to make a good first impression, and first impressions are very powerful, so make them work for you. You should proceed through this checklist, and address each item. Because there are hundreds of journals in the life sciences, we can only give you a more or less generic listing. The best place for you to look for specific items that need to be addressed in the submission packet is the instructions to authors or the instructions to contributors documents for your journal.

FINAL CHECKLIST OF ITEMS TO BE INCLUDED IN INITIAL MANUSCRIPT SUBMISSION

- Be sure you have reread the instructions to authors or instructions to contributors document for your journal of choice to make sure you have met all special requirements spelled out therein.
- Carefully proofread your entire manuscript to ensure there are no major errors in it and that all pages and partial pages are present and you have accounted for them in the proper order.
- Be sure you have made the required number of photocopies of your manuscript required by your journal of choice, if you are submitting a hard copy.
- Be sure you have numbered the pages consecutively by whatever system is used by your journal (numbers only, senior author name—number, corresponding author name – number, etc.). Be sure you have started to number the pages on what your journal considers the first page of the manuscript, and place consecutive page numbers on all pages throughout the manuscript.

Getting Published in the Life Sciences, First Edition. By Richard J. Gladon, William R. Graves, and J. Michael Kelly
© 2011 Wiley-Blackwell. Published 2011 by John Wiley & Sons, Inc.

- For nearly all journals, compile the entire written text of the manuscript (texts of the title, byline, authorship footnote, abstract, introduction, materials and methods, results, and discussion), and order these pages as the first ones in the manuscript.

- Immediately after the written text, you should follow with the references section, followed by the tables, one per page, as governed by table size, in consecutive order of appearance in the written text.

- The last table should be followed by the listing of the figure captions consecutively on succeeding pages, followed by the figures in consecutive order, one per page, as defined by the page(s) that contain the figure captions.

- As if you were proofreading, read through each table title and each figure caption, while simultaneously observing each respective form of data presentation. Be sure each form of data presentation can stand alone, without need to seek clarification by reading the written text. Do the same for all table headings and identifiers in each figure.

- Make absolutely sure that every literature citation in the written text, table(s), and figure(s) is matched to a citation (or resource) in the references section of the manuscript, and vice versa.

- Make sure all identifying data for the source of each literature citation is contained in the references section of the manuscript and is accurate. This includes correct spelling of the name(s) of all authors on each resource, the correct year, the correct citation title, the correct locating information such as the journal name, its volume number, issue number (if needed), and inclusive pages. This is accomplished by comparing each word and number with its source on a copy of the original publication. Exercise great care with this item as any editor will tell you the greatest number of errors in any manuscript occurs in this area.

- Make sure every table and figure contained in the manuscript is used at least once (via a citation of that table or figure in the written text).

- Make sure every citation of a table or figure in the text is referring to the appropriate table or figure in the manuscript.

- Make sure your title, byline, abstract, and keywords are appropriate for the content of your manuscript and your target audience. Add or delete items as necessary.

- Be sure you include copies of releases via letter or form for all items for which the publisher requires a release.

- Use 12-point Times, Times Roman, Tahoma, or Arial font. Palatino may be an acceptable alternate for some journals.

- Use margins of at least 25 mm on all four sides of the paper.

- Double-space *everything* unless instructed otherwise.

- Use spell-checking software for all parts of the manuscript. However, proofreading is still necessary, because the spelling of a word may be correct, but its context can be very wrong, even embarrassingly wrong (Day and Gastel, 2006).

- All copies of the manuscript must be generated in their entirety by using a laser printer, or one of similar high quality.

- In North American, use only white paper that is 21.6 cm by 27.9 cm. For some countries, use of A4 paper will be satisfactory, if 21.6 cm by 27.9 cm paper is not required or used.

- Use paper clips or binder clips, not staples, to bind the pages of each copy of your manuscript.

- Remember that tables, figures, and photographs are not placed in the body of the paper and that figure captions and table titles are treated differently. Most journals still require placement of the tables and figures at the back of the manuscript. However, do be aware that some journals now require authors to place the tables and figures within the text at the appropriate place.

- Use plain-paper pages, line-numbered pages or boxed-in, line-numbered pages, as specified by your journal.

- Do not use right justification, but you can use left justification and centering, as necessary.

- Use italic font in place of underlining. Normally, there are no instances of underlining in any part of a manuscript.

- Print your manuscript on one side of your paper; do not duplex.

- Two spaces should separate sentences throughout your manuscript.

- All of the written text of your manuscript should be printed in portrait format. However, you may use landscape format for tables and figures only if this allows you to convey the information more clearly.

- Prepare and enclose a cover letter and/or submission form, as required by your journal.

INITIAL SUBMISSION AND POSTSUBMISSION EVENTS

After all corrections have been made and all authors are pleased with the content of the manuscript, the corresponding author is just about ready to submit the article to the proper place for consideration by the publisher. At this point, probably the best thing the corresponding author can do is to reread the instructions to authors just before the manuscript is submitted. This gives the corresponding author one more chance to make sure everything is correct, and there is not a massive blunder in the manuscript that was not caught by the people who revised and reviewed it (Davis, 2005). The corresponding author must include a cover letter or completed submission form, or both, depending upon the journal. The corresponding author will submit the manuscript and its associated items to the Publications Director, Editor-in-chief, or other person in authority at the journal office. *Submit the manuscript to only one journal at a time, as multiple submissions are forbidden by virtually every journal. We know of no journal where multiple submissions are an acceptable practice.*

Some journals continue to accept only hard copies of the manuscript and its associated items. This method is called conventional submission (Day and Gastel, 2006). Although it may be slow when compared with electronic submission, there is nothing inherently wrong with this mode, as it has worked well for about 350 years. More recently, however, many journals have started to accept the manuscript in electronic format only, or in instances where the manuscript cannot be transmitted electronically, a hard copy may be submitted. In addition, many journals will now accept submissions only in electronic form. Electronic submission usually occurs through the website of the journal (Day and Gastel, 2006). For many authors, the first several electronic submissions they make are fraught with errors, sometimes humorous, and sometimes embarrassing. Both authors and journals are getting better at this process, and fewer problems are occurring. Benefits of electronic submission are that it speeds up the publication process and decreases the number of copies of the manuscript and its associated artwork (Day and Gastel, 2006). In nearly all cases, the journal will need to receive the final, edited and corrected version electronically (more than likely, this will be required). In this case, the journal will not have to typeset the entire manuscript, and this will speed the publication process and minimize errors.

If you choose to submit conventionally, or are required to submit it that way, then be sure to enclose the manuscript and its associated artwork in packaging that will ensure the manuscript(s) arrive in good condition. Often, padded or bubble-wrap envelopes are used to help absorb the shocks the parcel will receive during transit. Never staple a manuscript, as the staples will tear the paper during transit and subsequent removal of the staples also decreases the appearance of the manuscript. Use a paper clip or some other form of a binder clip to keep each manuscript together. Always retain at least one complete, original copy of the manuscript and its associated artwork. Postal services throughout the world have no guarantees that a given package will not get lost forever. Do not use oversized paper or oversized photographic materials, as both of these have a greater chance of becoming damaged in transit. Be sure you apply the proper amount of postage to get the submission envelope where it is going, whether in the same country or overseas. Use at least first-class postage, or better yet, use priority mail, express mail, airmail, or a courier service that delivers either overnight or in a few days. The peace of mind will be worth it. All over-seas mail should be sent airmail, at the least, or it may take one to two months to arrive by surface mail. Courier services to foreign countries are still quite expensive, and they should be used only when there are no other options.

Authors in countries with limited access to electronic mail and the Internet should contact the editorial office of the journal to determine their course of action for submitting a manuscript when these capabilities are not readily accessible. Often, the editor will be will-ing to work with authors who have this problem. If the submission is still electronic, but not via electronic mail, then wrap the disk, diskette, or compact disc in an appropriate mailing envelope or special disk mailer, and you might consider a double wrapping system to help ensure the carrier does not get damaged during transit.

Immediate Postsubmission Events

Shortly after the journal publication office has received the manuscript, a person in charge will acknowledge receipt of the manuscript. You probably will receive a postcard, letter, or e-mail message acknowledging receipt of your manuscript. Along with that acknowledge-ment, a manuscript number, or some other form of moniker, will be assigned to the manu-script. This number, or moniker, is a journal office item only, and it never appears on the published article. In addition, many journals also divulge the name and contact information of the Associate Editor or other adjudicator in charge of the manuscript. The corresponding author will also be told to address all subsequent communications to this person, who will be in charge of the manuscript. This acknowledgment may also indicate the timeframe normally required to complete the review and evaluation of your manuscript.

The manuscript is then sent to the appropriate Associate Editor or other adjudicator. Either a hard copy or an electronic copy may be sent. Although these next steps or actions are major parts of publishing a manuscript, there are really only two more things the author(s) must do before the manuscript is published. These are (i) to respond to the com-ments of those who have reviewed the manuscript, and then, if the manuscript has been accepted, (ii) proofread and accept the galley proof of the published article (Matthews et al., 2000).

Now your patience will be tested because you will have to wait more or less patiently until a decision is rendered. Depending upon the journal, it may take several weeks to com-plete the review and evaluation process for your manuscript. The next communication you receive will bear the good/bad news, unless some major flaw in formatting the manuscript has been uncovered. In this instance, the editor will return your manuscript quickly for correction before review. If you do not hear from someone at the journal within five to six months after receiving notification of receipt of the manuscript, it is appropriate for you to inquire about the status of your manuscript. The timeframe for the review process has been criticized greatly over the years, and most journals continue to explore ways to make the processing time shorter, so that authors can get on with the revisions needed for final acceptance or submission to another journal.

REFERENCES

[CSE] COUNCIL OF SCIENCE EDITORS, STYLE MANUAL COMMITTEE. 2006. *Scientific style and format: the CSE manual for authors, editors, and publishers.* Seventh edition. The Council, Reston, VA.

DAVIS, M. 2005. *Scientific papers and presentations.* Second edition. Academic Press, San Diego, CA.

DAY, R.A. and GASTEL, B. 2006. *How to write and publish a scientific paper.* Sixth edition. Greenwood Press, Westport, CT.

MATTHEWS, J.R., BOWEN, J.M. and MATTHEWS, R.W. 2000. *Successful scientific writing. A step-by-step guide for the biological and medical sciences.* Cambridge University Press, New York, NY.

CHAPTER 24

THE PEER-REVIEW PROCESS OF YOUR INITIAL SUBMISSION

Genius is one percent inspiration and ninety-nine percent perspiration.
— Thomas A. Edison

Your manuscript is both good and original, but the part that is good is not original, and the part that is original is not good.
— Samuel Johnson

DEFINITION AND NEED

The peer-review process for adjudicating the quality of reported scientific work is an especially important part of the system by which our scientific knowledge base grows and matures (Shatz, 2004). This peer review system constitutes a check and balance system that has worked admirably for many years. The peer-review system consists of authors submitting manuscripts to journals, as publishers, who then acquire the services of one to several experts in the specific field of the reported research. These experts serve as peer reviewers of the manuscript, and their charge is to thoroughly read and understand the manuscript. After suitable deliberation, the reviewer makes a judgment in the form of a recommendation to an adjudicator (henceforth called Associate Editor) on the quality of the science in the manuscript. Normally, the peer reviewers forward their recommendations on the fate of the manuscript and their associated comments to the Associate Editor assigned to that manuscript. This Associate Editor comes to a recommendation or a decision on the acceptability of the manuscript for publication, after taking into account the opinions and comments of all reviewers.

Neither the peer reviewers nor the Associate Editor should have to spend time fixing the English, grammar, sentence structure, and so on, of the manuscript. Such problems should have been avoided in the first place or corrected during revisions by the coauthors. This is especially important because many journals now handle submitted manuscripts electronically, and it is difficult for reviewers or the Associate Editor to edit for proper English usage. *The job of reviewers and the Associate Editor is to use their expertise to judge the quality of the science completed in the manuscript. Their time should not be spent correcting grammar and other communication problems that should have been cleaned from the manuscript long before submission.*

Let us now look at a generalized path a manuscript may take after it has been written, revised, reviewed internally, and submitted. There are many journals publishing scientific work in all disciplines. Every one has developed its own system for peer reviewing and adjudicating the quality of a manuscript submitted for publication. Here we present a basal understanding of the events that occur during the peer-review process associated with publishing a manuscript in a peer-reviewed journal (for overviews, see Davis, 2005; Day and Gastel, 2006; Matthews et al., 2000; O'Connor and Woodford, 1975). Depending upon your specific journal, the overall process may be more or less complex than the following example.

PEER REVIEW

Upon receipt of the manuscript, the Associate Editor sends a paper or electronic copy of your manuscript to two to four independent reviewers with expertise in the subject of the work. Many journals now require the corresponding author to provide the names of several potential reviewers, and the Associate Editor may use one or more of those individuals along with his or her own choice of reviewers. Often, the Associate Editor will contact potential reviewers by e-mail to determine whether they can work a review into their busy routine. If the Associate Editor receives a positive response from the potential reviewer, then the process is accelerated greatly. If a negative response is received, then the Associate Editor must find another reviewer. Subsequently, the manuscript, either in hard copy or electronic form, is sent to the necessary number of reviewers, and the review process begins. This portion of the process usually takes several weeks, as each reviewer needs to work uninterrupted time for his or her review into her or his schedule. Some time later, the Associate Editor will receive all the reviews from the peer reviewers. At this point, the Associate Editor takes into account the opinions and comments of these independent reviewers, in addition to his or her own evaluation, in coming to a decision about the fate of the manuscript. Normally, the Associate Editor makes a recommendation to the Science Editor, Technical Editor, Editor-in-chief, or other person with the authority to make the final decision about the fate of the manuscript. The decision by the Associate Editor returned to the corresponding author usually falls into one of these five categories:

1. Accept as is; very minor or no revision necessary.
2. Accept with minor revisions; author must revise, but only a small amount of revision.
3. Accept with major revisions; author must revise, but it is a major job.
4. Release/decline/reject with encouragement to revise and resubmit.
5. Release/decline/reject with no suggestion to revise and resubmit.

Recently, some journals have allowed only options 1, 4, and 5. The manuscript is returned to the corresponding author, and the corresponding author and the other coauthor(s) act upon the decision of the Associate Editor.

REVISION

If the manuscript has been accepted as is or with such minor revisions that they can be done by the Associate Editor without returning the manuscript to the corresponding author, then the manuscript is sent to the Science Editor, Technical Editor, Editor-in-chief, or other person with final authority on acceptance of the manuscript. This is extremely rare. It is also important for the writer to understand that the Associate Editor only makes a recommendation to the person who has the final say on the fate of the manuscript. Normally, the Science Editor, Technical Editor, Editor-in-chief, and so on, makes the final decision on the fate of the manuscript. If the manuscript is not accepted as is, then it is returned to the corresponding author, and he or she will work with the coauthor(s) to revise the manuscript as directed by the comments and opinions of the independent reviewers, the Associate Editor, and maybe the Science Editor, Technical Editor, Editor-in-Chief, and so on. Many Associate Editors/adjudicators suggest this is where there is the greatest hold-up in moving the manuscript through the publication process. Often, the corresponding author or the coauthor(s) "sit" on the manuscript and do not revise it in a timely manner. This has caused most journals to place a time limit on the revision and return, usually six or eight weeks. If the corresponding author does not complete the revisions by then, the manuscript must be resubmitted as a new manuscript, rather than a revision.

RESUBMISSION

When the revision has been completed, the corresponding author returns the revised manuscript to the Associate Editor for a recommendation for acceptance by the Associate Editor or possible further review by the original reviewers (this is the most common scenario) or a new set of reviewers. In most instances, not all suggestions made by the Associate Editor and the reviewers are accepted and incorporated into the revised manuscript. In those cases, the corresponding author must include a response letter in which he or she delineates what changes they have accepted and incorporated and which changes they decline to make and their justification(s) for not making those changes. It is understood that all other comments and opinions by the independent, expert reviewers and the Associate Editor will be incorporated into the revised manuscript, unless described and justified in this response letter. Details about the content and construction of the response letter are to be found in Chapter 25.

In some cases, the author(s) may decide the reviewers and Associate Editor were so hard on the manuscript that they would be better off submitting it elsewhere or simply trashing the manuscript without a further attempt to publish it. Under the assumption the manuscript is neither accepted as is nor completely released/declined/rejected, but rather returned for revision and resubmission, the author(s) then will complete a revision of the manuscript and resubmit to the Associate Editor.

RECOMMENDATION FOR ACCEPTANCE OR RELEASE BY THE ASSOCIATE EDITOR

The Associate Editor now studies the revised manuscript and determines whether the manuscript should be accepted as it has been revised, be returned to the corresponding author for further revising, or sent out for a completely new review because of the great amount of revision done to the original manuscript. This completely new review may occur with the same reviewers as used the first time the manuscript was submitted, or another set of reviewers may be used. If the Associate Editor recommends acceptance of the manuscript, it is forwarded to the Science Editor, Technical Editor, Editor-in-chief, or other person with final authority to accept the manuscript. If, at this time, the manuscript is returned to the corresponding author and coauthor(s) for further revision, then the coauthor(s) must do as the Associate Editor mandates. Normally, there are no options to bargain one's way to an acceptance of the manuscript at this point. If the Associate Editor deems the manuscript must be reviewed again, then the process of independent review is reinitiated, and the coauthor(s) must wait for the review process to occur again. Usually, the Associate Editor will get two or three new, independent reviews of the paper and again make a decision on their recommendation of the fate of the manuscript.

If the Associate Editor, taking into account the opinions and comments from the independent reviewers, decides the manuscript is not worthy of publication, then the author(s) can choose between submitting the revised manuscript to another journal or simply suspending the goal of publishing the manuscript. Another option is to request that a committee (e.g., a society's Publications Committee) look at the manuscript to determine whether the author(s) have been treated justly in the review process. Some journals and societies have such a committee that hears complaints from authors who believe they have been mistreated during the review of a manuscript.

If, however, the Associate Editor feels the manuscript now merits publication, the Associate Editor sends a letter to the corresponding author stating that she or he will recommend to the Science Editor, Technical Editor, Editor-in-chief, or other person in authority that the manuscript should be accepted for publication. At the same time, the Associate Editor forwards the manuscript to the Science Editor, Technical Editor, Editor-in-chief, or other person in authority, for their review of the manuscript and a final recommendation on the publication of the manuscript.

Davis (2005) and O'Connor and Woodford (1975) have given several generalized reasons why a manuscript may be deemed unacceptable for publication in a given journal. It is good for writers, especially inexperienced writers, to understand these reasons. Authors can use an understanding of these reasons in advance to prevent a negative decision on their manuscript. Also, it is insightful to see that it may not have been the style or abilities of the writer, or the quality of the science conducted, that caused release or rejection of the manuscript. Here are several of these reasons as detailed in Davis (2005) and O'Connor and Woodford (1975):

- The research reported was not within the scope of the journal.
- The research was neither conceived nor executed appropriately.
- The manuscript was written poorly.
- The manuscript did not follow the protocol established by the journal.
- The research results were inconclusive, including an insufficient set of data.
- There are incorrect interpretations of the analyzed data.
- Analysis and interpretation of the results are missing or incorrect.
- Discussion length (speculation, interpretation) is not warranted by the data.
- The research was trivial.
- The research was not new or may have repeated work published previously.
- Too much material was presented, and it was unfocused.
- The manuscript was too long (very common, even for skilled writers).
- The manuscript contained unimportant data.

REFERENCES

DAVIS, M. 2005. *Scientific papers and presentations*. Second edition. Academic Press, San Diego, CA.

DAY, R.A. and GASTEL, B. 2006. *How to write and publish a scientific paper*. Sixth edition. Greenwood Press, Westport, CT.

MATTHEWS, J.R., BOWEN, J.M. and MATTHEWS, R.W. 2000. *Successful scientific writing. A step-by-step guide for the biological and medical sciences*. Cambridge University Press, New York.

O'CONNOR, M. and WOODFORD, F.P. 1975. *Writing scientific papers in English*. American Elsevier, New York, NY.

SHATZ, D. 2004. *Peer review. A critical inquiry*. Rowman & Littlefield Publishers, Inc., New York, NY.

CHAPTER 25

RESPONSE LETTER PREPARATION AND MANUSCRIPT RESUBMISSION

Success is the ability to go from one failure to another with no loss of enthusiasm.

—Sir Winston Churchill

DEFINITION AND NEED

Congratulations! If you are ready to prepare a response letter, chances are good that you have nearly reached your goal of acceptance of your manuscript for publication. Here we will summarize the information from a section of the previous chapter. In this case, you have been advised that (i) your manuscript has been accepted without revision (an extremely rare occurrence) or (ii) if you complete all modifications as designated by the Associate Editor, your manuscript will be accepted for publication. Alternately, you may have been asked to revise and resubmit your manuscript, *but without the assurance that it will be accepted upon your revision and resubmission*. In this case, one of several things can occur. First, the Associate Editor accepts your revisions and forwards the manuscript (to the person who makes the final decision) with a recommendation to accept for publication. Second, the Associate Editor accepts your revision, but feels it was too extensive to approve for publication without the reviewers seeing it again. Third, the Associate Editor does not want to make a decision about it until the initial reviewers have reviewed the manuscript again. Fourth, the Associate Editor does not want to make a decision on acceptance without a review by a different set of reviewers from those who did the original review. If your manuscript was reviewed too critically, there would be no response letter because the manuscript would not have been accepted pending revision. You would have been told your manuscript no longer is under consideration for publication, and the editor handling your manuscript may offer several options for your manuscript.

As you move into this critical stage in the publication process, you should not overlook an important fact—*your manuscript has not been accepted yet*. Careless attention to detail or an inappropriate attitude at this stage of the game can lead to release/decline/rejection of your manuscript. Proceed with care!

Let us assume your communication from the Associate Editor carries a message that you need to revise and resubmit your manuscript, without regard to the assurances you have been granted about its acceptance. In this instance, you will be required to revise the manuscript and then resubmit it. The main difference here from the original submission you made previously is that this resubmission will be to the Associate Editor and not to the editorial office. In addition, you will need to prepare another letter that will cover the resubmission packet, but this time it will be called a response letter, rather than the cover letter of the original submission.

Getting Published in the Life Sciences, First Edition. By Richard J. Gladon, William R. Graves, and J. Michael Kelly
© 2011 Wiley-Blackwell. Published 2011 by John Wiley & Sons, Inc.

THE RESPONSE LETTER

The purpose of the response letter is to advance your case (i.e., provide a justification) for accepting your manuscript by summarizing the revisions you have made based on the comments of the reviewers and the Associate Editor. The response letter is typically addressed to the Associate Editor or adjudicator who communicated to you the outcome of the review process. Response letters are often one to four typed pages in length, depending upon the number of issues that must be addressed.

You should understand that what seems to be a final decision about the fate of your manuscript by the Associate Editor is not really an ironclad decision that cannot be changed (Gustavii, 2003). If you have been asked to modify your manuscript, there are two possible outcomes. First, you may be asked to modify it, and then it will be accepted for publication. This is a good outcome for you, and you should do what you need to do to make sure the Associate Editor will be pleased with and will accept your modifications, and therefore, your manuscript. Avoid agitating the Associate Editor by picking a fight over a trivial detail or two. Many times it is better to accept the fact that you are not getting your way this time, and just move on. Second, if you have been asked to modify your manuscript, then it will come under "further consideration," and you should do your best to ensure you work with the comments of the reviewers and the Associate Editor to get your manuscript to a point of (i) acceptance as revised, (ii) acceptance with certain additional modifications, or (iii) need for another review either by the original reviewers or an entirely new set of reviewers. In all cases, you should complete the task of revision with politeness and respect for the reviewers and the Associate Editor and thank them for their time and effort (Gustavii, 2003). Persistence pays off. Gustavii (2003) has an excellent anecdotal story about *Nature* rejecting Hans Krebs's original submission covering the Krebs cycle, only to have it published elsewhere. Subsequently, Hans Krebs was awarded the Nobel Prize in 1952 for his work on the Krebs Cycle.

WHO WILL READ YOUR RESPONSE LETTER?

Although you will usually send the response letter only to the Associate Editor, she or he may not be the only person to read and evaluate it. Editors sometimes send your response letter and the revision of your manuscript to one or more of the original reviewers, especially if one of the reviewers had many concerns about the content of your manuscript and suggested extensive revisions. Editors also may send the response letter and the revised manuscript to at least one new reviewer. It is important that you recognize several people other than the Associate Editor may see your revised manuscript, the response letter, or both documents. Thus, your response letter should contain nothing that an original or new reviewer would consider insulting or in any way unprofessional.

WHAT ARE THE CHARACTERISTICS OF A GOOD RESPONSE LETTER?

The response letter should handle all of the business associated with the criticisms from all reviewers and the Associate Editor, but at the same time, it needs to contain several characteristics that will help in getting the manuscript through this resubmission process. We have identified six characteristics we believe should be reflected within the response letter.

Thoroughness

Thoroughness is a real key. You must deal with each and every reviewer comment in some way. A useful technique that limits thoroughness from making your letter excessively long is to write an encompassing statement concerning editorial and other relatively minor comments from the reviewers. You might say, "My coauthors and I have made all editorial and grammatical corrections suggested by the reviewers, except those delineated below." This statement dispenses with all those issues, but it also tells the Associate Editor you were attentive to them and made the suggested changes.

Well Written and Editorially Flawless

The degree to which your letter is written well and is editorially flawless reflects on your general attentiveness to detail as you write. A sloppy response letter sends a message to those reading it that your paper has not been treated carefully, with attention to detail, and is likely to remain problematic.

Respectfulness

Respectfulness toward all involved is a must. Do not agitate your Associate Editor. If you happen to indicate that reviewer #2 must be an idiot, based on his or her review, then you are telling the Associate Editor that she or he does not know how to select reviewers. Be polite and professional no matter how angry, insulted, or upset you were when you read and acted upon your reviews. *Often, the best thing you can do is to wait a day or two to regain your composure before you work on the response letter and the revision of the manuscript. This is a time to cultivate good will.*

Convincing

You must convince the Associate Editor, and anyone else who may review your response letter and revised manuscript, that you have resolved appropriately all controversies that arose during the review process. Citing literature that resolves an issue is sometimes a good technique. The Associate Editor and any reviewers who read the revised manuscript must be convinced that you have addressed every one of the criticisms appropriately.

Assertive, but Humble

Remain assertive. If you are not willing to argue on behalf of your paper, nobody will. Remind the Associate Editor of the major contributions your article will make to the field of study, yet remain humble in the process. No experiment is perfect, and there are always shortcomings and unanswered questions. Your paper does not resolve all the issues, but what it has resolved is done well and deserves to be published. Convey that you recognize the limitations contained in your manuscript, yet do not diminish the value of what you have accomplished.

Open-mindedness

Remain open for further questions, requirements, and criticisms. Relay in your letter that you have been reflective and appreciative of the work of the reviewers, and that you remain open to further criticism, if warranted. Remember that the goal of the response letter is to keep open the path to publication. Do not say, "We were willing to make some of these changes, but this is as far as we will go. Take the revision as it is or forget it." Instead, state that you are grateful for the opportunity to make revisions, are thankful for the efforts of the editorial staff, and remain open to additional suggestions. Again, this is a time to cultivate good will.

A METHOD FOR CONSTRUCTING THE RESPONSE LETTER

There probably are as many ways to construct a response letter as there are authors. However, we have found the method we have suggested to inexperienced writers works quite well until he or she can develop a method that works best for themselves.

The first thing you should do after you have read the comment sheets, evaluation forms, copies of the manuscript, and letter sent back by the Associate Editor is to put them aside for a few days. It is not uncommon to feel either angry or hurt by the comments of all who have reviewed your manuscript. If you try and respond to the comments and suggestions of the reviewers too soon, often the corresponding author develops an acrid keyboard and says things that they should not say and may regret having said some time later. Alternately, you may feel discouraged and develop a feeling that the manuscript cannot be salvaged. It is simply best to cool down for a few days before developing your revised manuscript and response letter. We would suggest that this "cool-down" period is mandatory. However, the length of time of this cool-down period must be evaluated and adjusted by each individual as they gain more experience writing manuscripts. If you feel for any reason you need to begin to develop a response immediately, then focus on those areas of the review that are not problematic and wait until last to deal with the ones that made you angry or were problematic.

REVISING THE MANUSCRIPT

After your mandatory cool-down period, you may begin to work on your revision of the manuscript and the response letter. Start by finding a large desk or table where you can spread everything out. If you do not already have a complete extra copy of your originally submitted manuscript on which you can make changes, then get one before you start trying to do changes. Place this copy directly in front of you as you sit at the work surface. Have with you a pen or thin-line marker of a color you can see readily. Most people find red is most visible for them, but any color highly visible to the author is fine. Also have with you a light-colored highlighter (that you can see through when you mark over something). Also have with you a writing tablet or a laptop computer where you can record all significant changes you have made in the manuscript so that at a later time you will be able to relay that information to the Associate Editor in the response letter.

If the reviewers for your journal write on the manuscript (most do not, but some may), then place all of the reviewers' copies and the Associate Editor's copy in front you on the work surface. Go through each line of each of the copies from the reviewers and the Associate Editor one line at a time so that, in the case where you get conflicting suggestions, you can resolve the conflict before you move on to the next item. As you address each item on each of the reviewers' copies, mark that item by striking through it with the highlighter. This light-colored strike will later tell you (and the Associate Editor who will look at your manuscript) that you have addressed each of those comments and suggestions one by one.

After you have completed all of your changes from the review copies of the manuscripts, it is time to move to the Associate Editor's letter. This letter will normally contain the broader aspects of what he or she and the reviewers would like you to change in the manuscript. There may, however, be some specific suggestions, and they also may be done at this time. Again, as you address each item and complete it, strike the item with a highlighter to indicate you have finished your work with that item. On your tablet or laptop computer next to you, record in very detailed form all changes you have made and justify all changes you feel you cannot make. Your detailed account should include reference to each page, paragraph, and line of the manuscript where you have made a change that was significant. These will become part of your response letter later.

Now it is time to move to the specific comments of each reviewer. In many cases, the Associate Editor may also have completed one of these comment sheets or written these out for the author to address. As it was with the copies of the manuscript, place each comment sheet from a reviewer next to your working copy of the manuscript, and work your way down the sheets line by line of the manuscript and line by line of their comment sheets. We suggest you do it this way so that you may take all reviewers' comments about a certain thing you said and deal with it at one time. It is not uncommon for three reviewers to offer three different ways to fix something they consider wrong in the manuscript. In this manner you can choose the one you believe is most correct for your situation, and then in your response letter justify why you chose one way to address the problem, and why you did not choose one of the other ways. As before, you should strike a mark with the highlighter through each item as you finish it so that the Associate Editor can see you have addressed each item from each reviewer when you resubmit your manuscript and its associated items. Likewise, you should record each item you addressed on your tablet or your laptop computer so that you can incorporate that information into your response letter later.

Now that you have made all the changes you were directed to make and felt you could, and justified those you did not make, it is time to synthesize your response letter. Next, we detail two structures for composing the response letter. These are two common ways to structure the response letter, but that does not mean you cannot develop your own method. However, if you are having problems getting started, you might want to try one of these structures until you gain enough experience to develop your own method. At the end of this chapter, we present a sample response letter that uses Option 2 below, and you should feel free to use it as a model to help you learn how to structure your response letter. Now you will use all the detailed information you accumulated on your tablet or laptop computer while you were addressing each item designated by each reviewer. If you have accumulated and kept that information organized properly throughout your revision of your manuscript, it should be much easier for you to incorporate those items into the response letter.

TWO OPTIONS FOR THE STRUCTURE OF THE RESPONSE LETTER

Other than addressing every concern of the reviewers and the Associate Editor, we know of no standard structure or content required for the response letter. We know of two structural approaches for preparing the response letter, and either one is fine to use. Within each structure, it is important to specify the location in the manuscript of every item of concern you have addressed. You must indicate whether the locations you cite are from the original draft or your revision, and page numbers, paragraph numbers, and/or line numbers should be cited for each concern you have addressed. Let us now look at these two structures.

- *Option 1.* Write a brief (one page or less) cover letter to the Associate Editor on letterhead. In this cover-letter portion of the response letter, you should identify the manuscript by the manuscript number/designation given by the journal upon your original submission. Here you should also reiterate that you are the corresponding author, and you should give the title and complete list of authors on the manuscript. You also might mention here your gratitude for the excellent review, some overarching themes of the review and of how you have responded to them, and your openness for further comment and subsequent revision. After this cover letter, discuss your responses to each of the concerns, and these pages should be separate sheets of paper without letterhead. You might title the first of these separate sheets "Detailed Responses to Reviewers." You should then list all of the concerns you have addressed, based on the comments of reviewer #1, then #2, and so on, under subheadings that identify each reviewer and his or her concern. It is within each of these subheaded sections that you should cite page numbers, paragraph numbers, and/or line numbers in an organized, easy-to-follow format.

- *Option 2*. In this option, simply merge the two separate parts described under Option 1 into one document, a multipaged letter to the editor with the first page on letterhead paper. It is quite common that the combined response letter under this option will be several pages long.

FINISHING DETAILS

After you have finished synthesizing your response letter, you should let it cool down for a period of time (maybe a day or two) before you return to do a final proofing of the letter to ensure you have addressed everything the Associate Editor requested. This also gives you time to catch any less-than-appropriate comments you may have made during your revision of the manuscript or the letter. During this time, you should make all the changes you recorded on your working copy of the manuscript that was used as you addressed the suggestions and comments of the reviewers. We would suggest you make a duplicate (do a save as) of your original manuscript submission and make the corrections on the new copy to be submitted. In this way, you can refer to your original submission if there ever is a need to clarify something the Associate Editor questioned. Let this new edition of your manuscript cool for a few days and return to proofing your response letter and printing it for submission. Finally, proof your revised manuscript to ensure you have correctly made all your changes, and then print the number of copies you have been directed to make by the Associate Editor.

SENDING THE RESPONSE LETTER AND REVISED MANUSCRIPT

Whether by traditional mail or electronic mail, it is best to act on the concerns of the reviewers, make the revisions, and submit the response letter and revised manuscript without unnecessary delay. However, do not push yourself to act so rapidly that it leads to needless errors or oversights, and be sure that you "cool down" before responding to a review that stung you. You should act as soon as you have adequate time to do a good job professionally. A rapid, high quality response sends a positive message to the editor, and a response that is returned to the Associate Editor at the time limit or after the time limit sends a message you probably do not want to send. Be sure to include everything the Associate Editor requested. If you must return the original reviewed copies of your manuscript, which is sometimes the case, then be sure to retain photocopies in case some minor revisions need to be done, and the Associate Editor would like to take care of them via telephone or e-mail.

FINAL ACCEPTANCE OF THE MANUSCRIPT

After the Associate Editor reads your revised manuscript and cover letter, and looks over the comment sheets and other items sent to you for addressing a problem, he or she will make a decision on whether the manuscript will be accepted or more work needs to be done on the manuscript. Certainly, you would hope that your revisions are deemed satisfactory and the Associate Editor makes a recommendation to accept the manuscript. If more work needs to be done, you will soon hear from the Associate Editor, and you will start another round of revisions. If your revision was acceptable to the Associate Editor, then she or he will send it to the person who makes the final decision on acceptance of the manuscript. The Science Editor, Technical Editor, Editor-in-chief, or other person in authority now reviews the manuscript thoroughly. If she or he finds it is acceptable, then they notify the corresponding author that the manuscript has been accepted for publication. The coauthor(s) can now refer to the manuscript as "accepted for publication." If the Science Editor, Technical Editor, Editor-in-chief, or other person in authority wants further revisions, they return the manuscript to the corresponding author for incorporation of those items into the manuscript. The corresponding author and coauthor(s) may not agree with the opinion of the Science Editor, Technical Editor, Editor-in-chief, or other person in authority, and they can request that those items not be changed. However, this must be done judiciously and carefully, because this person really has the final say-so on the fate of the manuscript. Usually, it is best to incorporate the suggestions and get on with the publishing process. It is usually not wise for an author to dig their heels into the ground and do battle with the Science Editor, Technical Editor, Editor-in-chief or other person in authority; rarely will they win, and this battle often creates much ill will between the parties. The author(s) should dig their heels in and fight only for items and issues they feel must not be changed. If revision is needed, the revisions are made and the manuscript is then returned to the Science Editor, Technical Editor, Editor-in-chief, or other person in authority, and hopefully, the revised manuscript now will be accepted for publication.

COPYEDITING

The Science Editor, Technical Editor, Editor-in-chief, or other person in authority now forwards the accepted manuscript to the copy editor, or equivalent, of the journal. If the copy editor suggests any changes, you should generally make them, because these changes only have to do with the format of the journal. Generally, these changes will have little effect on the scientific content or the written communication of the manuscript, but they may, depending upon the journal. Any and all final changes are made at this point, and usually, the final copy of the manuscript is submitted to the copy editor on a disk or via e-mail. The copy editor now finishes all of his or her work, and the manuscript is converted into a page proof known as a galley proof. The galley proof then is sent to the coauthor(s) for their final reading and final corrections. And this is exactly what it is—a final opportunity for the coauthor(s) to find and correct any errors in the manuscript that now has been typeset for publication. Chapter 26 will present information related to the galley proof. Changes other than correction of typographical errors are normally not allowed, because the Science Editor, Technical Editor, Editor-in-chief, or other person in authority does not see the article again once it has been approved for publication. At this point, the article now can be called "in press."

Example of a Response Letter (Shows Option 2 described previously)

December 31, 2002

Douglas A. Wilcox, Ph.D., PWS
United States Department of the Interior
U.S. Geological Survey
Great Lakes Science Center
1451 Green Road
Ann Arbor, MI 48105

Dear Dr. Wilcox:

My co-author and I appreciated the thoughtful review of *Wetlands* manuscript 02-48, and we have prepared the enclosed revision. We were pleased with the reviewers' positive response to the quality of the manuscript and with their ideas for how to improve the manuscript. We have considered each comment and have responded with what we believe are appropriate changes to the manuscript. I first will summarize our thoughts regarding a few key themes of the review. I then will highlight specific changes that are found in the revised manuscript.

Responses to Themes of the Review

Ecological Framework for the Research. Reviewer #1 stated that the paper does little to "advance our general understanding of wetland ecology," and the Associate Editor mentioned the lack of a "conceptual ecological framework." Our stated objective was to define how *A. maritima* responds to root-zone salt, a rather narrow but well defined goal. We believe we met our goal. We also believe the previous absence of a broader framework does not diminish the value of our work and should not preclude publication in *Wetlands*. That said, we agree framing the studies more broadly is both appropriate and helpful for interpreting the results. The work, particularly the multi-species comparison, was designed based on ecological concepts, but because of our desire to be focused and succinct, we had not emphasized this in the original submission. You will see that we have broadened the scope of the manuscript by describing the ecological framework within which our work was based. Changes in this regard are in both the Introduction and Discussion.

Description of Treatment Initiation for the First Experiment. Both reviewers questioned our description of the gradual increase in salinity and raised the possibility of moving the information to a table or figure. We believe the information in this degree of detail would be necessary for someone else to repeat the procedures, and we could not think of a way to present the information in a table or figure without it requiring considerable space. Thus, we elected to retain the text, which appears on page 6 of the revision.

Data Presentation. We agree with Reviewer #1 and the Associate Editor that the information in the original Figure 2 could be presented more efficiently in the text. The means and SE values are now in the Results on page 12. Reviewer #1 questioned the need for Table 1, but we consider it the clearest way to present the information. On page 12 of the original submission, the Associate Editor had suggested adding a visual image of typical symptoms. We liked this idea and created the new Figure 1 in response. Based on my e-mail communications with you, I understand that we will pay the fee for producing this figure in color. Reviewer #2 asked for a reference for what we referred to as "a preceding trial" and "a preliminary study" under the "Treatments" section within the Materials and Methods related to the first experiment. We did not add data, nor did we describe these procedures further because the work has not been published and was done as preliminary tests to verify the appropriate salinity range and to verify that use of ocean water and NaCl led to similar symptoms.

Statistical Presentation. We have added n, SE, F, df, and P in response to comments by the Associate Editor and both reviewers. The appropriate values were added at various locations throughout the text and in Table 1. Adding SE bars about the means in the figures was considered. For some figures, especially Figure 4, we feared the bars would detract from the clarity of the data presentation due to the number of lines and means, many of which overlap. Therefore, the figures now reflect what we consider a reasonable compromise. We have added n and the range of SE values. This approach retains clarity while showing more statistical information. The most critical statistic, the LSD value, which itself represents the degree of variation in the data set, remains the focus. If it is deemed essential to add bars for SE to each data point, we certainly will do so, but we question whether the benefit will justify the effect on visual clarity. While adding individual values would not be particularly detrimental for some figures, we thought it best to treat the statistical presentation uniformly in all figures and thus opted for our proposed compromise.

Additional, Specific Changes to the Manuscript

- Added references related to theoretical framework for the experiments and to potential future research. (Associate Editor and Reviewer #1)
- Described location of Delmarva Peninsula in Abstract. (Reviewer #1)
- Changed "trail" to "trial" in Materials and Methods. (Reviewer #1)
- Included percentage-reduction values to Results text in the experiment with five species. (Reviewer #1)
- Expanded consideration of conclusions in Discussion. (Reviewer #1)
- In Abstract and in last paragraph of Introduction, clarified that inundation was with salt water. (Associate Editor)
- Explained meaning of half-sibling in Materials and Methods. (Associate Editor)
- Changed terms to describe dimensions of square containers in first paragraph of Materials and Methods. (Associate Editor and Reviewer #2)
- In Materials and Methods, improved explanation of how blocking was done during the first experiment. (Associate Editor)
- Added degree of chlorination of tap water, which was quite low. (Associate Editor)
- Regarding the Materials and Methods for the first experiment, clarified that plants were irrigated regularly before salinity treatments began. (Associate Editor)
- Inserted "in" where it was missing on page 5 of the original version. (Reviewer #2)
- Changed syntax as noted on pages 8 and 9 of the original version. (Reviewer #2)
- Replaced "flooding" with "inundation" as requested on page 9 of the original version. (Associate Editor)

- Replaced awkward sentence (as marked on page 11 of the original version) with new sentence (now on page 12 of revision) to describe results that had been in the previous Figure 2 in the text. (Reviewer #2)
- On page 12 of the revision, clarified that day-7 and day-21 data for photosynthesis were averaged for each plant, thus preserving the correct number of replicates in the ANOVA. (Associate Editor)
- In response to comment on page 15 of the original submission, added information on ecology and habitat of *Cornus amomum* (and other species) to Introduction. (Associate Editor)
- Added "of" to the sentence where it was missing on page 15 of the original submission. (Associate Editor)
- Inserted "variation in" to sentence in last paragraph of the Discussion, as requested. (Associate Editor)
- Clarified the status of site 4 in study along Broadkill River (elected to retain coverage of site 4 based on Associate Editor's comment). Specifically, replaced zeros for site 4 in Table 2 with "—" as requested, and added explanation for missing numbers to title of table. (Associate Editor and Reviewer #1)

As I trust is apparent, my co-author and I have taken this review very seriously, and I hope you agree that the changes we have made to the manuscript are appropriate. Both of the reviewers and the Associate Editor identified many ways to improve the paper. We are most grateful for their time and attention.

If you have any questions or concerns about this revised version of manuscript 02-48, please do not hesitate to contact me.

Sincerely,
William R. Graves
Professor

REFERENCE

Gustavii, B. 2003. *How to write and illustrate a scientific paper*. Cambridge University Press, New York, NY.

CHAPTER 26

PROOFS, PROOFREADING, AND THE FINAL STEPS IN THE PUBLICATION PROCESS

The good observer is one who sees what he (she) is not looking for and who does not see what he (she) is looking for if it is not there.

—F. C. Gilbreth

DEFINITIONS AND NEED

You and your coauthors have traversed the steps in the publication process (see Chapter 2), and your manuscript has been accepted for publication. It is *almost* time to celebrate. The publisher has received your latest version, most probably as an electronic file you provided on a computer disk, CD, or via electronic mail. The publisher has copy-edited and formatted the manuscript to conform to the style of the journal. Within a few weeks to a few months before publication, you will receive a proof of your manuscript. Normally, the editor of the journal will alert you that the proofs will be coming soon, and they need your immediate attention. A proof is defined as "a copy (as of typeset text) made for examination or correction" (Mish, 2004). Proofreading differs from copy editing by the stage at which it will be done, its purpose, and the placement of the marks (Rude, 1991). In essence, copy editing is done early; it moves the manuscript from a keyboarded document to a typeset document, and it prepares the text for printing. Conversely, proofreading is done late and verifies that the text has been printed according to the specifications of both the journal and the author(s). Your final job on this path to publication is to proofread this proof document for mistakes and to provide the editor and printer with a final set of corrections before the document is sent to press.

Depending upon the journal, you may receive for final proofreading a galley proof (also known more simply as a galley), a page proof, or both. A galley proof is defined as "a proof of typeset matter especially in a single column before being made into pages" (Mish, 2004). Many times, galley proofs are printed on long, shiny paper, and they are used to check the accuracy of typesetting (Rude, 1991). A page proof is defined as a document of typeset text that consists of the corrected galley proof made into pages of the journal as they would appear after all articles have been proofread, returned to the publisher/printer, and made ready to be printed (Rude, 1991). The page proof is used to check the accuracy of the corrections made on the galley proof and the logic of the page breaks (Rude, 1991). Because most journals now have the final submission done electronically, most publishers and printers skip proofreading of the galley proof by the author(s) and go directly to page proofs, which are the only proofs proofread by the author(s). Some journals may have the corresponding author, or all coauthors, proofread and initial the galley proof, and then after pages are made, again proofread the page proof. In still other cases, the authors are responsible for creating the pages exactly as they are to be printed (i.e., camera-ready or phototypeset), and the journal simply reproduces those pages. In this last case, the author(s) have no chance to do a true page proof, but all authors should have done the page-proofreading step before the final copy was submitted. Usually, journals that use camera-ready page makeup will have detailed instructions about how the author(s) should proceed. In almost all instances, the editor will send the copy-edited manuscript along with the proof document so that the author(s) may see how the editor, copy editor, and printer (either as a compositor of typeset print or a software program that synthesizes the page) have changed your manuscript from what you submitted to them in your final form.

In the various forms of desktop publishing used today, most documents are proofread twice. The first time, the author(s) do the proofreading immediately after all aspects of word processing the document into manuscript form. The second proofreading is done after the pages have been designed. At each of these two stages, the process has been put in place to confirm (i) that previous errors have been corrected, (ii) that the addition of tables and figures to the text has been done correctly, and (iii) that the page breaks have been placed correctly (Rude, 1991). The process of proofreading and marking the galley and/or page proof is described later in this chapter. No changes other than the correction of errors should be made at the stage of proofreading. Proofreading is not revising, reviewing, and editing. It is correction of errors only.

Normally, this is your last chance to correct errors before the piece is published. You should take advantage of this opportunity because your next time to make corrections is after publication in an erratum page. Your credibility as a scientist may depend upon how well you have completed these acts associated with finishing the publication process, and as is so often the case in athletic events, only the final outcome counts. The erratum page is not a desirable place for a scientist to make corrections, because it implies some level of sloppiness in the entire process of conducting and reporting the research. You may discover errors you have made, and it is surprising how often that happens at this late stage in the publication process. It is also possible that the journal has deliberately (to meet in-house journal format criteria) or inadvertently altered something in your manuscript, and you want it changed back to the original way you said it. Now is the time to take out that fine-toothed comb and review your manuscript one last time with great care. Errors that are not caught at the proofreading stage shift the focus and alertness of the reviewer or reader from the scientific content to the mechanics of the presentation, and this is not good. Thus, take the time and effort needed to learn how to proofread. It will pay great dividends in the long run.

THE PROOFREADING PROCESS

The proof document, whether it is a galley proof or a page proof, should be compared word for word, phrase for phrase, sentence for sentence, and paragraph for paragraph with the accepted, final version of your keyboarded manuscript. In particular, you should check for letters, lines, spaces, punctuation, and words that might be missing from the original text or that may have been added inadvertently. Depending upon the journal, there may also be changes made by the copy editor. The proofreader needs to understand that there may be many corrections needed in this phase of the process.

Verify that your tables and figures are exactly as you intended. Your data are the heart of your article, and it is essential those data are reported correctly. Also, pay particularly close attention to the title, byline, and abstract, as those are read more frequently than any other parts of your article. Literature citations within the text and the resources listed in the references section should also be checked with special care because these parts of manuscripts are especially prone to errors.

SOME PRECAUTIONS

Proofreading is not easy, and should not be taken lightly. It is hard to proofread really well, but it can be learned if one gives forth the effort needed to learn the necessary skills and techniques. Here are some suggestions and caveats to consider, and some problems that may surface while you are proofreading your manuscript.

- Be sure you read the proof at least twice. You will probably not catch all errors in a single proofreading.
- Be sure you make marginal notations of all corrections. Printers, publishers, and editors often overlook corrections if they are made only within the line of text (Gustavii, 2003). Proper marking should contain both a marginal mark and an in-line mark.
- The most frequent changes made during proofreading are additions, deletions, or transpositions of capitalization, typeface, word order, punctuation, and number style (Matthews et al., 2000).
- One of the most common mistakes by writers proofreading their own material is that mentally they fill in missing letters and words. This also happens in reverse—one's eyes and brain omit letters and words that are there, but should not be (Davis, 2005). Although every author should proofread her or his own material, people not directly involved with the production of the manuscript should be used as proofreaders to avoid these tendencies (McMillan, 2006).
- A spelling error that does not change the shape of the printed word will be very hard to detect, especially when one is fatigued (Rude, 1991). An excellent example of this is the misspelling of the word evaluation as evalvation. The beginnings and ends of these words are identical, as are the elevation due to the "l" and "t," and this makes it very hard to find that a "v" was inserted rather than the correct "u."
- In a similar manner, it has been found that human eyes "zero in" on the first and last letters of a word, and the human brain can sort the incorrectly spelled word into the correct word. Again, fatigue will increase the frequency of this response. An example in which the correct first and last letters for the appropriate words have been used to bracket the jumbled central letters follows in the next paragraph. This example was received by one of the authors in an e-mail message that has an unknown origin. In the end, the human eye and brain can transpose letters positioned incorrectly into correct words, and the reader may not know it is happening. Before looking at this passage, allow your eyes and brain to rest for a moment, and then attempt to read it.

    Anicrdcog to rsrceaeh cnutedocd at a uinervisty in Elangnd, as lnog as the fsrit and lsat lteetr of a wrod are crecort, the hmuan eye can urmnlcsabe rdmalony pclaed ltreets taht flal bteewen tsehe crcerot frist and lsat letrets of the wrod. The rset of the wrod can be a taotl mses, but yuor biran rdaes, itperners, and iertsns the ltertes itno the crocert oedrr. Tihs pvoerd taht hmnaus do not raed eervy lteetr by isletf, but tehy raed the wrod as an erinte etnity.

- If you are having trouble with proofreading your manuscript or a proof, try reading aloud to yourself, or better yet, read to someone else or have them read it to you while you check it for errors (Davis, 2005).

- For figures, and especially photomicrographs, be sure you and the printer have the correct orientation on the piece. There is not a problem when the orientation is easy to determine, such as with a graph. With many photomicrographs of tissues, cells, or intracellular components, any one of the four sides of the photomicrograph could be the top. If the photomicrograph is physically separate from the manuscript, "top" should be written on the top of the back of the photomicrographic paper (Davis, 2005).

- Make use of computer spell-checking and grammar-checking functions provided with software packages. Errors are hard enough to find, but it is not wise to skip use of these two functions. However, these functions should not be the only proofreading of the piece because these functions cannot determine differences between words that have different meanings tied to different spellings, such as to, too, and two, and write, right, and rite.

- Only proofread for short periods of time. The intensity with which proofreading must be done precludes marathon sessions. Leave yourself and other proofreaders adequate time to get the job done without creating problems with fatigue and monotony.

- Keep in mind that the role of proofreading the manuscript is to find and correct errors made by the compositor, copy editor, or editor. The editor has the last say on whether a suggestion you make will be executed, and sometimes the editor may be stubborn about format or how the journal wants something to be done, and they will not approve your suggestion.

- How do these errors occur and where do they hide (Bolsky, 1988)? Mistakes tend to cluster, so if you find one mistake, heighten your awareness because additional mistakes may be lurking nearby. Many mistakes are found at the beginning of something such as a page, section, paragraph, or sentence. There may have been a change in font size, type, or style. Often, titles and headings do not match and flow properly, and this is a good sign there are errors nearby. Numbers in tables and figures can be deleted, inserted, or transposed very readily, and only a skillful proofreading technique will catch those errors. Many errors are associated with misplaced periods, commas, and decimal points. Finally, many errors hide within sets of items such as those that are held within brackets, parentheses, quotation marks, and dashes.

- Many manuscripts are changed from one computer platform to another and back again, and maybe back again, by the time the manuscript is ready to be published. Changing platforms, and also operating systems and versions of software, may induce errors. Proofreaders need to look diligently for additions, but especially deletions, as the manuscript is sent back and forth between the writer(s), editors, reviewers, and printers.

PROOFREADING MARKS

Understanding the four categories of proofreader's marks will help any one not experienced with using them (Rude, 1991). The first category is the general usage marks, which are associated with insertion, deletion, transposition, and so on, of letters, words, and sentences. Category two contains the punctuation marks such as periods, commas, semicolons, and so on. Category three deals with typography and includes operations that make letters, words, sentences, and so on, boldface, roman, italics, capitals, small capitals, lower case, etc. The last category contains those marks that concern spacing issues. These include marking a new paragraph, running-in two paragraphs into one, justifying right or left, centering, and so on.

There are several standardized systems used throughout the world, and the editors, publishers, and printers associated with the journal where your manuscript will be published will determine which system they will use. The most common systems are the American system (ANSI-NISO, 1991) and the British system (BSI, 1976; Butcher, 1992), as shown in the CSE Manual (CSE, 2006). There is also a continental European system, and several countries have their own standardized systems (Gustavii, 2003). For all intents and purposes, you will never use about 90 or 95% of the proofreading marks within these systems; only the printer, publisher, or the copy editor will use them. However, you need to be aware they exist and be able to refer to the systems when the need arises. We suggest that the editing and proofreader's marks you learned and used in Chapter 13 will handle almost all of your needs when editing and proofreading your manuscript, galley proof, or page proof. You really do not need to know and use an exhaustive list of proofreader's marks, but you should know the key proofreading marks and how to use them legibly and unequivocally (Gustavii, 2003). If you need access to the entire gamut of proofreader's marks, please see the CSE Manual (2006).

GUIDELINES FOR THE PROCESS OF PROOFREADING

The galley proof or page proof should be read and corrected by all authors, and one or more of the authors must sign off on the proof verifying they have read it, corrected it, and are responsible for its final content. In most instances, the required signature is that of the corresponding author, although some journals now require all coauthors to sign off that they have proofread and accepted the content of the proof. In addition, some journals also require all authors to sign a form that they agree with and share responsibility for the entire content of the published piece.

All changes you wish to make before publication are made as unequivocal, handwritten comments on the galley proof or page proof pages. The corrected proof pages should then be returned to the editorial office by traditional mail, some form of express mail, or FAX. If the journal's entire publication process is electronic, then your changes should be made on the galley proof or page proof pages via use of editing software or in an associated e-mail message returned with the electronic copy of the proof.

Two sets of marks are used for correcting errors during this final proofreading step. The first set of marks is called the marginal marks, and as the name implies, they are made in the margin of the proof. It is extremely important that a marginal mark be made for every item that needs correction. In almost all cases, the editor or compositor will scan the proof pages for marks in the margin, and they can often overlook in-text marks. In the case where the proofreader only makes an in-line mark, there is a reduced probability the correction will be made. In addition, when the compositor or editor has made a mistake repeatedly, each one of these mistakes must be met with a marginal mark. The marginal mark is almost always a statement of an operation to be done while correcting the error. For instance, insert, delete, or transpose a letter or word is a command to do an operation. In most instances, editors and compositors also suggest that this mark be circled when it designates an operation or provides instruction about what needs to be done.

The second set of marks is made in-line and tells the person making the corrections exactly where to make those corrections. Because these marks are usually made within single-spaced lines, often only a caret or deletion sign is made due to space limitations.

All markings should be done in such a manner that they are unequivocal and in accordance with the system the journal uses, so there is no question about what operation should be done. Many journals send an accompanying handout that delineates how the journal wants the proofreading to be done, what system of proofreader's marks are to be used, and what they should look like. If this is the case, you should follow these instructions to the letter. Any marks the author places on the proof should be done in colored ink that will draw the reader's attention. The brighter and hotter the color, the better. Be alert for queries and inquiries by the editor or the compositor found in the margin near where attention is needed. These must be addressed before you return the proof. Normally, if the editor sends you two copies of the galley proofs or page proofs, it is expected that you will make the corrections on one copy, copy those corrections onto the other copy, and then return one copy while retaining the other one. If more than one mistake must be corrected in a given line in the proof, the author should move from left to right in the printed line, make one marginal mark for each needed correction, and separate the corrections from each other with a slash mark (also called a virgule or solidus). When several lines of text must be added or changed, it is best to type the missing information on a separate piece of paper and then tape or staple it (do not use a paper clip) to the page proof in the appropriate place.

SPEED IS OF THE ESSENCE

Most journals will alert the corresponding author that the galley proof or page proof is coming, that it will be accompanied by the copy-edited manuscript, when it should arrive, and that the corresponding author must proofread as soon as possible, usually within a few days. This allows for the corresponding author to arrange with all the authors a time in which they will proofread the document. It also allows the corresponding author to arrange time for his or her own proofreading multiple times.

In every case we know, the author(s) should handle and return the galley proofs or page proofs with alacrity (yesterday, if possible!). We cannot overemphasize how important it is to deal with galley proofs or page proofs immediately. Now is not the time to execute changes you should have made before submission of the original manuscript or after the review process, but forgot to make them. You should only correct errors, and it is ethically appropriate that only essential changes are made. This is not the time to elect to change something simply for the sake of style or an evolving personal preference. *Substantial changes might force the journal to have the paper reviewed again and reconsidered for acceptance. In addition, you will probably be charged a fee for changes the journal considers more than minor or unwarranted.* One instance where this policy may be overridden is when a substantial, very important, piece of information must be added to the article. In this case, most journals will allow a "Note added in proof." This information is added at the end of the text of the article, and it should contain only the information needed to convey to the reader the important, new discovery that was not made previously.

There are good reasons why journals require authors to return proofs rapidly. Authors who abuse the privilege of a final proofreading can cause the issue of the journal to be delayed by days or weeks. At this point, all articles for a given issue have been slotted into all their pages so that the presses can run to print the issue. When the author(s) delay the process, it delays the printing of the journal. In addition, if an author decides to argue that several changes absolutely must be done, then more text may be added or deleted from a proof that has already been allotted pages, potentially changing the pagination and sequential formatting of the other articles in the issue.

SIGN ON THE DOTTED LINE

Most journals require the author(s) who reviewed and corrected the galley proof or page proof to place their initials in a specified place on the proof, provide their signature(s) at a specified location on the proof, or sign a separate form that must be returned with the proof. *By initialing or providing your signature, as required, you are agreeing that you, not the journal, are now responsible for all errors that exist in the printed article.* Save yourself from the embarrassment of having your name on a paper that contains mistakes or is listed in the errata of an issue the journal published subsequently. Check your galley proof and/or page proofs with extreme care!

COPYRIGHT

In addition to signing off on the galley proof or the page proof, either the corresponding author or all coauthors must sign the form(s) granting the copyright of the information in the manuscript from the author(s) to the publisher. The manuscript cannot be published until all necessary parties have signed this form and returned it to the publisher of the journal. Many journals include with the galley proof or page proof a copyright transfer agreement for you to sign and return with the galley or page proof. *You should understand that your signature is required on this form, and by virtue of your signature, you transfer ownership of the copyright from you and your coauthors to the journal.* You have no choice in this regard as the journal will not publish something that has another person or entity as the copyright holder.

PAGE CHARGES

Many journals charge the author(s) a fee to publish the article. Fees are usually assessed on the basis of a certain charge per fully or partially printed journal page. Often, charges range from $50 to $300 per printed page. An invoice or other form delineating the fee structure for the article may be included with the galley proof or page proof. Scientists affiliated with an institution use grant funds or some institutional funds to pay these page charges. In our experience, it is rare for a student, faculty member, or staff member of an institution to pay charges with personal funds. Rightly or wrongly, some authors decide on where to publish at least partly based on page fees.

The scenario described in the preceding paragraph is often associated with journals that are published by a professional society that has a relatively low cost for journal subscriptions and that relies on page charges for income. Journals not published by a professional society may have no or minimal page charges. Income to sustain such journals comes from subscription charges, and sometimes reprint charges.

REPRINTS/OFFPRINTS

Journals send to authors forms associated with reprint or offprint charges and the number of reprints or offprints ordered. If reprints or offprints are sought, author(s) must order them and arrange with their college, department, university, experiment station, or purchasing department for the payment.

Reprints and offprints are not the same. A reprint is simply a re-printing of the article that can be done at any time up to some time limit imposed by the journal (often about one year). An offprint is a copy of the article that can only be obtained (ordered) at the time the article is being printed. When the run of the presses has been completed, one cannot return to obtain an offprint at a later time. When a color photograph is part of the article, whether it is on the cover of the issue or embedded within it, an offprint normally ensues because the printer cannot usually go back to colored items and reprint them. These extra copies of the article, whether they are a reprint or an offprint, will be identical to the original article within the journal. Some journals provide a small number of reprints or offprints (10 to 50) free of charge, and then charge for additional copies, if the author(s) want them. You will probably receive a reprint order form with your galley proof or page proof, or it may come to you separately.

Before photocopy machines were generally available, reprints and offprints were a vitally important form of communication between scientists. With the present-day availability of photocopy machines, and more recently, the use of electronic transfers of information such as entire manuscripts and photographs in pdf files, the need for reprints and offprints has been reduced substantially. Most authors now receive reprint and offprint requests only from scientists in countries where photocopy machines and computers with Internet and e-mail access are limited. Thus, reprint and offprint requests are relatively infrequent, and for the most part, reprints and offprints are used almost solely for personal reasons such as the development of a dossier for promotion and tenure or a nomination for various achievement awards. Given these changes, an order of 50 to 100 reprints or offprints is generally more than enough for the coauthors to share.

REFERENCES

[ANSI-NISO] AMERICAN NATIONAL STANDARDS INSTITUTE, NATIONAL INFORMATION STANDARDS ORGANIZATION. 1991. *Proof corrections: American National Standard proof correction, ANSI/NISO Z39.22-1989.* NISO Press, Bethesda, MD. [Available from NISO Press Fulfillment, P.O. Box 338, Oxon Hill, MD 20750-0338, USA. FAX: 301-567-9553.]

BOLSKY, M.I. 1988. *Better scientific and technical writing.* Prentice Hall, Englewood Cliffs, NJ.

[BSI] BRITISH STANDARDS INSTITUTION. 1976. *British Standard copy preparation and proof correction, BS 3261. Part 2. Specifications for typographic requirements, marks for copy preparation and proof correction, proofing procedure.* British Standards Institution, London, UK. [Available from: American National Standards Institute, 1430 Broadway, New York, NY 10018.]

BUTCHER, J. 1992. *Copy-editing: The Cambridge handbook for editors, authors and publishers.* Cambridge University Press, Cambridge, UK.

[CSE] COUNCIL OF SCIENCE EDITORS, STYLE MANUAL COMMITTEE. 2006. *Scientific style and format: the CSE manual for authors, editors, and publishers.* Seventh edition. The Council, Reston, VA.

DAVIS, M. 2005. *Scientific papers and presentations.* Second edition. Academic Press, San Diego, CA.

GUSTAVII, B. 2003. *How to write and illustrate a scientific paper.* Cambridge University Press, New York, NY.

MATTHEWS, J.R., BOWEN, J.M. and MATTHEWS, R.W. 2000. *Successful scientific writing. A step-by-step guide for the biological and medical sciences.* Second edition. Cambridge University Press, New York, NY.

McMILLAN, V.E. 2006. *Writing papers in the biological sciences.* Fourth edition. Bedford/St. Martin's, Boston, MA.

MISH, F.C. 2004. *Merriam-Webster's collegiate dictionary.* Eleventh edition. Merriam-Webster, Inc., Springfield, MA.

RUDE, C.D. 1991. *Technical editing.* Wadsworth Publishing Co., Belmont, CA.

EXERCISE 26.1   Page Proof Exercise

The remaining pages of this chapter contain a hard copy of a rather brief manuscript that has been accepted for publication (used with permission). The manuscript is followed by a page proof of the paper. Try your hand at identifying all the errors you can find in the page proof by comparing it with the manuscript provided. Make marks directly on the page proof, as both in-line and marginal marks, as you would if you were returning the page proof to the journal that is publishing and printing the article. The next nine pages contain the accepted, final copy of the manuscript, and that document is followed by the three-page page proof. Because we wanted to illustrate many types of problems associated with proofreading, this exercise contains many more errors than we expect you would find in a typical galley or page proof. In general, we might expect anywhere from one to about ten mistakes that would need to be corrected in a typical galley or page proof. Take pride in your proofreading skill if you find at least 60 errors. If you don't find that many during your initial review, take a break and then try again with as much attention to detail as you can muster.

Kosteletzkya virginica Can be Rooted from Leafy or Leafless Stem Cuttings[1]

Cynthia L. Haynes and William R. Graves[2]

Department of Horticulture

Iowa State University, Ames, IA 50011-1100

Abstract

Protocols for producing Virginia mallow [*Kosteletzkya virginica* (L.) K. Presl. ex A. Gray] are needed to allow growers to meet the emerging demand for this herbaceous perennial. Virginia mallow has been propagated from seeds and by division, but the potential for using stem cuttings has not been evaluated. Two experiments were conducted to determine how indolebutyric acid (IBA) treatment affects rooting percentage and the number and length of primary roots on stem cuttings taken from different positions on stock plants. Rooting percentage was similar (mean = 68%) among single-leaf cuttings from nonterminal, distal positions on stock plants and leafless cuttings from basal positions. Averaged over these nonterminal cutting types, IBA treatment more than doubled rooting percentage, root count, and the length of the longest root. Terminal cuttings with leaves rooted more successfully (83%) than subtending cuttings with leaves during a second experiment, and IBA effects were less pronounced than during the first experiment. Results demonstrate that stem cuttings with or without leaves can be used to propagate Virginia mallow efficiently. Application of IBA is not necessary but enhances rooting and appears most beneficial for leafless cuttings.

Index words: herbaceous perennials, IBA, halophytes, Virginia mallow, Malvaceae.

[1]Received for publication May 11, 2004; in revised form July 2, 2004. Journal paper of the Iowa Agriculture and Home Economics Experiment Station, Ames, Iowa, Project No. 3603, and

supported by Hatch Act and State of Iowa funds. We gratefully acknowledge donation of seed by J.L. Gallagher, assistance of M.A. Kroggel, and data analysis of J.A. Schrader.
[2]Assistant Professor and Professor, respectively.

Significance to the Nursery Industry

This report presents the first evidence that *Kosteletzkya virginica* (Virginia mallow) can be propagated successfully by stem cuttings. The species typically loses its lower leaves and develops highly branched, open canopies. Our experiments show that stem cuttings with or without leaves can be rooted. The capacity for leafless stem cuttings to root greatly increases the number of potential propagules from a stock plant. Application of talc-based IBA at 8 g/kg is recommended to enhance rooting but is not necessary for root initiation, especially if cuttings have leaves. These results will facilitate production of this promising herbaceous perennial.

Introduction

Consumers are seeking additional taxa of herbaceous perennials, and many gardeners have a preference for regionally indigenous species. Virginia mallow is a North American perennial that is capturing the interest of gardeners (1). Although native to brackish or nearly fresh-water marshes and shores from Louisiana to Florida and north to Virginia (2), Virginia mallow also grows well in typical garden soils in USDA hardiness zones 5 to 10 (1). Plants grow to approximately 1 m (2-4 ft) in height and produce solitary pink flowers in late summer and autumn. Virginia mallow is uncommon in commerce, but a few specialty nurseries sell plants or seeds. It has appeared for several years in the Plant Delights (Raleigh, NC) mail-order catalog. In addition, Virginia mallow has been named the wildflower of the year by organizations in North Carolina and Virginia (see www.vnps.org/year.html).

Demand for Virginia mallow is likely to increase as the species is promoted, and methods for propagation need to be defined. Propagation by seed is possible, but high germination percentages are achieved only if seeds are scarified or are stored several years (8). Use of vegetative propagation methods to perpetuate desirable genotypes of Virginia mallow would avoid potential problems associated with variability among seedlings. Division can be used (1), but few plants can be obtained this way. Because Virginia mallow typically has an open growth habit with highly branched shoot systems, propagation by stem cuttings could generate many plants. The feasibility of propagating Virginia mallow by stem cuttings has not been evaluated, but stem cuttings are used for other genera of the Malvaceae (7). Therefore, our objective was to determine how IBA affects rooting percentage and the number and length of primary roots on stem cuttings taken from different positions on stock plants. We considered it important to compare root formation on cuttings obtained from terminal and basal positions on stock plants because leaves typically abscise from the numerous basal stems on mature plants of Virginia mallow. If roots can be induced on cuttings prepared from leafless, basal stems, the number of propagules possible from stock plants would be much greater than if success is possible only with leafy cuttings from near terminal positions on stock plants.

Materials and Methods

Potted stock plants propagated from seeds indigenous to Sussex County, DE were grown in a greenhouse in Ames, IA under natural photoperiod and irradiance. Stem cuttings from these plants were taken for the first experiment on April 26, 2002. All cuttings were stem sections (length = 8 cm [3 in]) that had two or three nodes. No shoot tips (terminal cuttings) were used. The cuttings were of two types based on their origin on stock plants and on the presence of leaves. Half of the cuttings (64) were from the upper portion of the shoot and had a single leaf

(distal cuttings). The remaining 64 cuttings were from basal areas of the shoots (basal cuttings)

where leaves had abscised naturally. Talc-based IBA at 8 g/kg (8000 ppm; Rhizopon AA #3

[Phytotronics, Earth City, MO]) was applied to the basipetal 2 cm (0.8 in) of half the stems of

both types, while no IBA was applied to the other cuttings. Each cutting was inserted singly into

a plastic container that was filled with coarse perlite (Strong-Lite, Seneca, IL). The perlite had

been flushed previously with tap water, and a dibble was used to create the holes into which

stems were inserted. Each container had drainage holes, a top diameter of 6 cm (2.4 in), a

volume of 160 cm^3 (9.8 in^3), and a depth of 8 cm (3.1 in). Each container with one cutting was

considered an experimental unit. The 32 replicates of each of the four treatment combinations (a

two-by-two factorial of cutting type and IBA treatment) were arranged in a completely

randomized design on a bench in a glass-glazed greenhouse. Tap-water mist from overhead

nozzles was delivered for 20 sec every 15 min during daylight hours. Data were collected on

June 7, six weeks after treatments were initiated. Cuttings with at least one primary root of any

length were considered rooted. Among rooted cuttings, roots were counted, and length of the

longest primary root was measured. The diameter of the stem at the base of all cuttings was

measured with a caliper to characterize a physical difference between the two cutting types.

Ninety-six stem cuttings for a second experiment were prepared from the same group of

stock plants on June 5, 2002. Half of these cuttings had an actively growing apex (terminal

cuttings). The other 48 cuttings were from immediately below the terminal cuttings (subtending

cuttings). Cuttings of both types were 8 cm (3 in) long and had two or three leaves. Half of the

cuttings of both types were treated with IBA as during the first experiment. Cuttings were

inserted singly into perlite in the same kind of containers used for the first experiment. Each container with one cutting was considered an experimental unit. The 24 replicates of each of the four treatments (a two-by-two factorial of cutting type and IBA treatment) were arranged in a completely randomized design on the same greenhouse mist bench used for the first experiment. Data were collected on July 17, six weeks after treatments were initiated. Rooting was quantified as during the first experiment.

A data logger with thermocouples and a LI190SB quantum sensor (LI-COR, Lincoln, NE) recorded air temperature and photosynthetically active radiation (*PAR*) every 15 min during both experiments. Mean daily extremes were 20/31C (68/88F). No supplemental irradiance was used, and mean *PAR* during the middle 5 hr of photoperiods was 199 $\mu mol \cdot s^{-1} \cdot m^{-2}$. Data from both experiments were subjected to analysis of variance. Data expressed as percentages were transformed to the arc sine of the square root before analysis, but nontransformed data are reported. No interactions between cutting type and IBA treatment were found, so main-effect means were determined and separated when appropriate using Fisher's least significant difference test at $P < 0.05$.

Results and Discussion

Nonterminal distal and basal cuttings rooted at similar percentages during the first experiment (Table 1). Cuttings of both types formed an average of five to six roots. We did not measure photosynthesis, but the greater length of the longest root of distal cuttings compared to basal cuttings (Table 1) may be attributed to the leaf on each distal cutting that provided carbohydrates to support growth of initiated roots (4). Averaged over cutting types, IBA more than doubled rooting percentage, root count, and the length of the longest root (Table 1).

In Experiment 2 rooting percentage was 27% higher for terminal cuttings than for subtending cuttings all of which had leaves (Table 2). Use of terminal cuttings also resulted in the most roots and in the greatest length of the longest root (Table 2). IBA had no effect on

rooting percentage or maximal root length during the second experiment, but IBA increased root

count by 134% (Table 2). There was 70% rooting overall, but the new shoots that had initiated

on some of the rooted cuttings appeared weak when data were collected. Such cuttings

frequently appeared to have been among the first to root, so we suspect that removing them

earlier from the mist bench would have prevented this problem. Commercial propagators are

advised to monitor root initiation of Virginia mallow carefully and use intermittent mist for the

minimum time necessary to induce a root system capable of sustaining plants in the production

environment.

We conclude that stem cuttings from throughout the shoot system of Virginia mallow can

be used to propagate this species. Both types of nonterminal stems used in our first experiment

rooted at > 60% (Table 1) despite lack of leaves on the basal cuttings. Because of the tendency

for Virginia mallow to abscise lower leaves, the capacity for leafless stem sections to root,

greatly increases the number of potential propagules from a stock plant. The highest rooting

percentage was achieved with terminal cuttings in Experiment 2 (Table 2). This is consistent

with previous reports that the position on a stock plant of stem tissue used as a cutting influences

rooting success (3, 5, 6), due in part to auxin from the terminal meristem or bud (4).

Interestingly, IBA affected rooting percentage only during the first experiment, when cuttings

with one or no leaves were used. While this suggests that propagators could be successful

without IBA if leafy cuttings are available, we recommend the use of talc-based IBA at 8 g/kg to

enhance the number of roots per cutting (Tables 1 and 2). Subsequent research should be

conducted to determine whether stems with smaller or larger diameters that those of our cuttings

(Table 1) can be rooted, and to define the optimal formulation and concentration of IBA for use on cuttings of Virginia mallow that vary in leaf surface area.

Literature Cited

1. Armitage, A. 1997. Herbaceous Perennial Plants: A Treatise on Their Identification, Culture, and Garden Attributes. 2nd ed. Stipes Publishing, Champaign, IL.

2. Fernald, M.L. 1950. Gray's Manual of Botany. 8th ed. American Book Co., New York.

3. Graves, W.R. 2002. IBA, juvenility, and position on ortets influence propagation of Carolina buckthorn from softwood cuttings. J. Environ. Hort. 20:57-61.

4. Hartmann, H.T., D.E. Kester, F.T Davies, Jr., and R.L. Geneve. 2002. Hartmann and Kester's Plant Propagation: Principles and Practices. 7th ed. Prentice Hall, Upper Saddle River, NJ.

5. Loreti, F. and H.T. Hartmann. 1964. Propagation of olive trees by rooting leafy cuttings under mist. Proc. Amer. Soc. Hort. Sci. 85:257-264.

6. O'Rourke, F.L. 1944. Wood type and original position on shoot with reference to rooting in hardwood cuttings of blueberry. Proc. Amer. Soc. Hort. Sci. 45:195-197.

7. Perry, L. 1998. Herbaceous Perennials Production: A Guide From Propagation to Marketing. Northeast Regional Agric. Eng. Serv. 93, Ithaca, NY.

8. Poljakoff-Mayber, A., G.F. Somers, E. Werker, and J.L. Gallagher. 1992. Seeds of *Kosteletzkya virginica* (Malvaceae): Their structure, germination, and salt tolerance. I. Seed structure and germination. Amer. J. Bot. 79:249-256.

Table 1. Rooting percentage, the root count and longest primary root per cutting among cuttings that rooted, and the diameter of the stem at its base for stem cuttings of *Kosteletzkya virginica* (Virginia mallow). Two types of cuttings were compared. Basal cuttings were leafless and originated from the lower portion of the shoot systems of stock plants. Distal cuttings had one leaf and were from nonterminal positions in the upper portion of shoot systems. Treatment combinations were a two-by-two factorial of type of cutting and indolebutyric acid (IBA) applied. There were no interactions between the two main effects. Values for rooting percentage and diameter of stem are means of 32 single-cutting replications. Values for root count and for the longest root within cutting types are means of the 20 (basipetal) and 23 (distal) cuttings that rooted. Values for root count and for the longest root within the two IBA treatments are means representing the 13 (IBA at 0 g/kg) and 30 (IBA at 8 g/kg) cuttings that rooted. Fisher's least significant difference (LSD) at $P \leq 0.05$ is shown unless the means did not differ significantly (NS).

Treatment	Rooting (%)	Root count	Longest root (mm)	Diameter of stem (mm)
Type of subtending cutting				
Basipetal, no leaves	63	5.1	25	6.7
Distal, one leaf	72	5.7	56	5.1
LSD	NS	NS	19	0.6
IBA applied (g/kg)				
0	41	2.5	24	6.1
8	94	6.7	49	5.7
LSD	20	3.3	20	NS

Table 2. Rooting percentage, the root count and longest single primary root per cutting among cuttings that rooted, and the diameter of the stem at its base for cuttings of *Kosteletzkya virginica* (Virginia mallow). Cuttings either included a stem apex (terminal cuttings) or were from stem sections subtending the apices. Treatments were applied as a two-by-two factorial of type of cutting and indolebutyric acid (IBA) applied. There were no interactions between the two main effects. Values for rooting percentage are means of 48 single-cutting replications. Values for root count and for the longest primary root within type of cutting are means representing the 40 (terminal) and 27 (subtending) cuttings that rooted. Values for root count and for the longest primary root within IBA treatments are means of the 33 (IBA at 0 g/kg) and 34 (IBA at 8 g/kg) cuttings that rooted. Fisher's least significant difference at $P < 0.05$ (LSD) is presented unless the means did not differ significantly (NS).

Treatment	Rooting (%)	Root count	Longest root (mm)
Type of cutting			
Terminal	83	6.0	127
Subtending	56	3.3	84
LSD	18	1.5	29
IBA applied (g/kg)			
0	69	2.9	119
8	71	6.8	100
LSD	NS	1.5	NS

Research Reports

Kostaletzkya virginiana Can be Rooted from Leafy or Leafless Stem Cuttings[1]

Cynthia L. Haynes and William R. Graves[2]

Department of Horticulture

Iowa State Unversity, Ames, IA 50011-1100

Abstract

Protocols for producing Virginia mallow [*Kosteletzkya virginica* (L.) K. Presl. ex A. Gray] are needed to allow growers to meet the emerging demand for this herbaceous perennial. Virginia mallow has been propagated from seeds and by division, but the potential for using stem cuttings has not been evaluated. Two experiments were conducted to determine how indolebutyric acid (IBA) treatment affects rooting percentage and the number and length of primary roots on stem cuttings taken from different positions on stock plants. Rooting percentage was similar (mean = 68%) among single-leaf cuttings from nonterminal, distal positions on stock plants and leafless cuttings from bottom positions. Averaged over these nonterminal cutting types, IBA treatment more than doubled rooting percentage, root count, and the length of the longest root. Terminal cuttings with leaves rooted more successfully (83) then subtending cuttings with leaves during a second experiment, and IBA affects were less pronounced than during the first experiment. Results demonstrate that stem cuttings with or without leaves can be used propagate virginia mallow efficiently. Application of IBA is not necessary but enhances rooting and appears most beneficial for leafless cuttings.

Index words: herbaceous perennials, IBA, halophytes, Virginia mallow, malvaceae.

Significance to the Nursery Industry

This report presents the first evidence that *Kosteletzkya virginica* (Virginia mallow) can be propagated successfully by stem cuttings. The species typically loses its lower leaves and develops highly branched, open canopies. Our experiments show that stem cuttings with or without leaves can be rooted. The capacity for leafless stem cuttings to root greatly increases the number of potential propagules from a stock plant. Application of talc-based IBA at 8 g/kg is recommended to enhance rooting but is not necessary for root initiation, especially if cuttings have leaves. These results will facilitate production of this promising herbaceous perennial.

[1]Received for publication May 11, 2004; in revised form July 2, 2004. Journal paper of the Iowa Agriculture and Home Economics Experiment Station, Ames, Iowa, Project No. 3603, and supported by Hatch Act and State of Iowa funds. We gratefully acknowledge donation of seed by J.L. Gallagher, assistance of M.A. Kroggel, and data analysis of J.A. Schrader.

[2]Professor, respectively.

Introduction

Consumers are seeking additional taxa of herbaceous perennials, and many gardeners have a preference for regionally indigenous species. Virginia mallow is a North American perennial that is capturing the interest of gardeners (1). Although native to brackish or nearly fresh-water marshes and shores from Lousiana to Florida and north to Virginia (2), Virginia mallow also grows well in typical garden soils in USDA hardiness zones 5 to 10 (1). Plants grow to approximately 1 m (2–4 ft) in height and produce solitary pink flowers in late summer and autumn. Virginia mallow is uncommon in commerce, but a few specialty nurseries sell plants or seeds. It has appeared for several years in the Plant Delights (Raleigh, N.C.) mail-order catalog. In addition, Virginia mallow has been named the wildflower of the year by organizations in North Carolina and Virginia (see www.vnps.org//year.html).

Demand for Virginia mallow is likely to increase as the species is promoted, and methods for propagation need to be defined. Propagation by seed is possible, but high

The *Journal of Environmental Horticulture* (ISSN 0738-2898) is published quarterly in March, June, September, and December by the Horticultural Research Institute, 1000 Vermont Avenue, NW, Suite 300, Washington, DC 20005. Subscription rate is $65.00 per year for scientists, educators and ANLA members; $95.00 per year for libraries and all others; add $25.00 for international (including Canada and Mexico) orders. Periodical postage paid at Washington, DC, and at additional mailing offices. POSTMASTER: Send address changes to Journal of Environmental Horticulture, 1000 Vermont Avenue, NW, Suite 300, Washington, DC 20005.

J. Environ. Hort. 22(4):173–175. December 2004

173

germination percentages are achieved only if seeds are scarified or are stored several years (8). Use of vegetative propagation methods to perpetuate desirable genotypes of Virginia mallow would avoid potential problems associated with variability among seedlings. Division can be used (1), but few plants can be obtained this way. Because Virginia mallow typically has an open growth habit with highly branched shoot systems, propagation by stem cuttings could generate many plants. The feasibility of propagating Virginia mallow by stem cuttings has not been evaluated, but stem cuttings are used for other genera of the Malvaceae (Perry, 1998). Therefore, our objective was to determine how IBA affects rooting percentage and the number and length of primary roots on stem cuttings taken from different positions on stock plants. We considered it important to compare root formation on cuttings obtained from terminal and basal positions on stock plants because leaves typically abscise from the numerous basal stems on mature plants of Virginia mallow. If roots can be induced on cuttings prepared from leafless, basal stems, the number of propagules possible from stock plants would be much greater than if success is possible only with leafy cuttings from near terminal positions on stock plants.

Material and methods

Potted stock plants propagated from seeds indigenous to Sussex County, Delaware, were grown in a greenhouse in Ames, IA, under natural photoperiod and irradiance. Stem cuttings from these plants were taken for the first experiment on April 26, 2002. All cuttings were stem sections (length = 8 cm [3 in]) that had two or three nodes. No shoot tips (terminal cuttings) were used. The cuttings were of two types based on there origin on stock plants and on the presence of leaves. Half of the cuttings (64) were from the upper portion of the shoot and had a single leaf (distal cuttings). The remaining 64 cuttings were from basal areas of the shoots (basal cuttings) where leaves had abscised naturally. Talc-based IBA at 8 g/kg (8000 ppm; Rhizopon AA #3 [Phytotronics, Earth City, Mo.]) was applied to the basipetal 2 cm (0.8 in) of half the stems of both types, while no IBA was applied to the other cuttings. Each cutting was inserted singly into a plastic container that was filled with coarse perlite (Strong-Lite, Seneca, IL). The perlite had been flushed previously with tap water, and a dibble was used to create the holes into which stems were inserted. Each container had drainage holes, a top diameter of 6 cm (2.4 in), a volume of 160 cm² (9.8 in³), and a depth of 8 cm (3.1 in). Each container with one cutting was considered an experimental unit. The 32 replicates of each of the four treatment combinations (a two-by-two factorial of cutting type and IBA treatment) were arranged in a completely randomized design on a bench in a glass-glazed greenhouse. Tap-water mist from overhead nozzles was delivered for 20 seconds every 15 min during daylight hours. Data was collected on June 7, six weeks after treatments were initiated. Cuttings with at least one primary root of any length were considered rooted. Among rooted cuttings, roots were counted, and length of the longest primary root was measured. The

diameter of the stem at the base of all cuttings was measured with a caliper to characterize a physical difference between the two cutting types. Ninety-six stem cuttings for a second experiment were prepared from the same group of stock plants on June 5, 2002. Half of these cuttings had an actively growing apex (terminal cuttings). The other 48 cuttings were from immediately below the terminal cuttings (subtending cuttings). Cuttings of both types were 8 cm (3 in) long and had two or three leaves. Half of the cuttings of both types were treated with IAA as during the first experiment. Cuttings were inserted singly into perlite in the same kind of containers used for the first experiment.. Each container with one cutting was considered an experimental unit. The 24 replicates of each of the four treatments (a two-by-two factorial of cutting type and IBA treatment) were arranged in a completely randomized design on the same greenhouse mist bench used for the first experiment. Data were collected on July 17, six weeks after treatments were initiated. Rooting was quantified as during the first experiment.

A data logger with thermocouples and a LI190SB quantum sensor (LI-COR, Lincoln, NB) recorded air temperature and photosynthetically active radiation (*PAR*) every 15 min during both experiments. Mean daily extremes were 20/31C (68/88F). No supplemental irradiance was used, and mean *PAR* during the middle 5 hr of photoperiods was 199 mmol·s⁻¹·m⁻². Data from both experiments were subjected to analysis of variance. Data expressed as percentages were transformed to the arc sine of the square root before analysis, but nontransformed data are reported. No interactions between cutting type and IBA treatment were found, so main-effect means were determined and separated when appropriate using Fisher's least significant difference test at $P < 0.05$.

Results & Discussion

Terminal distal and basal cuttings rooted at similar percentages during the first experiment (Table 1). Cuttings of both types formed an average of five to 6 roots. We did not measure photosynthesis, but the greater length of the longest root of distal cuttings compared to basal cuttings (Table 1) may be attributed to the leaf on each distal cutting that provided carbohydrates to support growth of initiated roots (4). Averaged over cutting types, IBA more than doubled rooting percentage, root count, and the length of the longest root (Table 2).

In Experiment 2 rooting percentage was 27% higher for terminal cuttings that for subtending cuttings, all of which had leaves (Table 2). Use of terminal cuttings also resulted in the most roots and in the greatest length of the longest root (Table 1). IBA had no effect on rooting percentage or maximal root length during the second experiment, but IBA increased root count by 134% (Table 2). There was 70% rooting overall, but the new shoots that had initiated on some of the rooted cuttings appeared weak when data were collected. Such cuttings frequently appeared to have been among the first to root, so we suspect that removing them earlier from the mist bench would have prevented this problem. Commercial propagators are advised to monitor root initiation of Virginia mallow carefully and

Table 1. Rooting percentage, the root count and longest primary root per cutting among cuttings that rooted, and the diameter of the stem at its base for stem cuttings of *Kosteletzkia virginica* (Virginia mallow). Two types of cuttings were compared. Basal cuttings were leafless and originated from the lower portion of the shoot systems of stock plants. Distal cuttings had one leaf and were from nonterminal positions in the upper portion of shoot systems. Treatments combinations were a two-by-two factorial of type of cutting and indolebutyric acid (IBA) applied. There were no indications between the two main effects. Values for rooting percentage and diameter of stem are means of 32 single-cuttings replications. Values for root count and for the longest root within cutting types are means of the 20 (basipetal) and 23 (distal) cuttings that rooted, Values for root count and for the longest root within the two IBA treatments are means representing the 13 (IBA at 0 g/kg) and 30 (IBA at 8 g/kg) cuttings that rooted. Fisher's least significant difference (LSD) at $P \leq 0.005$ is shown unless the means did not differ significantly (NS).

Treatment	Rooting (%)	Root count	Longest root (mm)	Diameter of stem (cm)
Type of subtending cutting				
Basipetal, no leaves	63	5.1	25	6.7
Distal, one leaf	72	5.7	56	5.1
LSD	NS	NS	19	0.6
IBA applied (g/kg)				
0	44	5,2	24	6.
8	94	6.7	49	5.7
LSD	20	3.3	20	ns

use intermittent mist for the minimum time necessary to induce a root system capable of sustaining plants in the production environment.

We conclude that stem cuttings from throughout the shoot system of Virginia mallow can be used to propagate this species. Both types of nonterminal stems used in our first experiment rooted at > 60% (Table 1) despite lack of leaves on the basal cuttings. Because of the tendency for Virginia mallow to abscise lower leaves, the capacity for leafless stem sections to root, greatly increases the number of potential propagules from a stock plant. The highest rooting percentage was achieved with terminal

Table 2. Rooting percentage, the root count and longest single primary root per cutting among cuttings that rooted, and the diameter of the stem at it's base for cuttings of *Kosteletzkya virginica* (Virginia mallow). Cuttings either included a stem apex (terminal cuttings) or were from stem sections subtending the apices. Treatments were applied as a two-by-two factorial of type of cutting and indolebutyric acid (IBA) applied. There were no interactions between the two main effects. Values for rooting percentage are means of 48 single-cutting replications. Values for root count and for the longest primary root within type of cutting are means representing the 40 (terminal) and 27 (subtending) cuttings that rooted. Values for root count and for the longest primary root within IBA treatments are means of the 33 (IBA at 0 g/kg) and 34 (IBA at 8 g/kg) cuttings that rooted. Fisher's least significant difference at $P < 0.05$ (LSD) is presented unless the means did not differ significantly (NS).

Treatment	Rooting (%)	Root counts	Longest Root (mm)
Type of cutting			
Terminal	83	6.0	127
subtending	56	3.3	84
LSD	18	1.3	29
IBA applied (g/kg)			
0	69	2.9	119
8	71	6.8	100
LSD	NS	1.5	NS

cuttings in Experiment 2 (Table 2). This is consistent with previous reports that the position on a stock plant of stem tissue used as a cutting influences rooting success (3, 5, 6), due in part to auxin from the terminal meristem or bud (4). Interestingly, IBA affected rooting percentage only during the first experiment, when cuttings with one or no leaves were used. While this suggests that propagators could be successful without IBA if leafy cuttings are available, we recommend the use of **talc-based IBA** at 8 kg/g to enhance the number of roots per cutting (Tables 1 and 2). Subsequent research should be conducted to determine whether stems with smaller or larger diameters that those of our cuttings (Table 1) can be rooted, and to define the optimal formulation and concentration of IBA for use on cuttings of Virginia mallow that vary in leaf surface area.

Literature cited

1. Armitage, A. 1997. Herbaceous Perennial Plants: A Treatise on Their Identification, Culture, and garden attributes. 2nd ed. Stipes Publishing, Champaign, IL.

2. Fernald, M.L. 1950. Gray's Manual of Botany. 8th ed. American Book Co., New York.

3. Graves, W.R. 2002. IBA, juvenility, and position on ortets influence propagation of Carolina buckthorn from softwood cuttings. J. 7th ed. Prentice Hall, Upper Saddle River, NJ.

4. Hartmann, H.T., D.E. Kester, F.T Davies, Jr., and R.L. Geneve. 2002. Hartmann and Kester's Plant Propagation: Principles and Practices. Environ. Hort. 20:57–61.

4. Loreti, F. and H.T. Hartmann. 1964. Propagation of olive trees by rooting leafy cuttings under mist. Proc. Amer. Soc. Hort. 85:257–2264.

6. O'Rourke, F.L. 1944. Wood type and original position on shoot with reference to rooting in hardwood cuttings of blueberry. Proc. Amer. Soc. Hort. Sci. 45:195–197.

7. Perry, l. 1998. Herbaceous perennials Production: A Guide From Propagation to Marketing. Northeast Regional Agric. Eng. Serv. 93, Ithaca, NY.

8. Poljakoff-Mayber, A., G.F. Somers, E. Werker, and J.L. Gallagher. 1992. Seeds of Kosteletzkya virginica (Malvaceae): Their structure, germination, and salt tolerance. I. Seed structure and germination. Amer. Journal of Bot. 79:249–256.

J. Environ. Hort. 22(4):173–175. December 2004

175

INDEX

Printed in the USA
CPSIA information can be obtained
at www.ICGtesting.com
CBHW051018200424
7036CB00011B/8